Unless Recalled

Integrated Optical Circuits and Components

OPTICAL ENGINEERING

Series Editor

Brian J. Thompson

Provost
University of Rochester
Rochester, New York

Laser Engineering Editor:
Peter K. Cheo
United Technologies Research Center
Hartford, Connecticut

Laser Advances Editor:
Leon J. Radziemski
Head, Department of Physics
New Mexico State University
Las Cruces, New Mexico

Optical Materials Editor:
Solomon Musikant
Paoli, Pennsylvania

LASER HANDBOOKS—*Edited by Peter K. Cheo*

Handbook of Molecular Lasers

Other Volumes in Preparation

Handbook of Optical Fibers and Cables, *by Hiroshi Murata*

Integrated Optical Circuits and Components

Design and Applications

edited by

Lynn D. Hutcheson
APA Optics, Inc.
Blaine, Minnesota

MARCEL DEKKER, INC.   New York and Basel

Library of Congress Cataloging-in-Publication Data

Integrated optical circuits and components.

    (Optical engineering ; v. 13)
    Includes bibliographies and index.
    1. Integrated optics. I. Hutcheson, Lynn D.
II. Series: Optical engineering (Marcel Dekker, Inc.) ;
v. 12
TA1660.I542 1987 621.38'0414 87-9178
ISBN 8-8247-7575-9

MARCEL DEKKER, INC.
270 Madison Avenue, New York, New York 10016

Current printing (last digit):
10 9 8 7 6 5 4 3 2 1

PRINTED IN THE UNITED STATES OF AMERICA

About the Series

Optical science, engineering, and technology have grown rapidly in the last decade so that today optical engineering has emerged as an important discipline in its own right. This series is devoted to discussing topics in optical engineering at a level that will be useful to those working in the field or attempting to design systems that are based on optical techniques or that have significant optical subsystems. The philosophy is not to provide detailed monographs on narrow subject areas but to deal with the material at a level that makes it immediately useful to the practicing scientist and engineer. These are not research monographs, although we expect that workers in optical research will find them extremely valuable.

Volumes in this series cover those topics that have been a part of the rapid expansion of optical engineering. The developments that have led to this expansion include the laser and its many commercial and industrial applications, the new optical materials, gradient index optics, electro- and acousto-optics, fiber optics and communications, optical computing and pattern recognition, optical data reading, recording and storage, biomedical instrumentation, industrial robotics, integrated optics, infrared and ultraviolet systems, etc. Since the optical industry is currently one of the major growth industries this list will surely become even more extensive.

Brian J. Thompson
University of Rochester
Rochester, New York

Preface

The field of integrated optics has been in full force for more than a decade, with the number of government, industrial, university, and research laboratories involved in the field increasing rapidly. Integrated optics was first developed in tandem with fiber optics and still is in some instances. However, more recently the field has commanded the attention of researchers involved in almost all aspects of optics. One need only notice the many technical meetings and journals that deal with the subject, either in whole or in part, to realize the potential of integrated optics.

This volume is written for the applied scientist, practicing engineer, and manager who wants to gain insight into how and where integrated optics can be used in practical applications. This book deals with the practical aspects of designing, fabricating, and testing integrated optical circuits and devices. It is expected that this treatise will be helpful to the researcher and engineer trying to learn the finer points of fabricating integrated optical devices that people have spent many years mastering.

Integrated optics, as its name implies, concerns the incorporation of all the optical components needed to perform the necessary optical circuit function onto a single chip. Over the past several years, researchers have come to realize that near term guided-wave applications will be hybrid integrated optics, where only partial integration of optical components is attained. This is not alarming because significant progress has been made in butt coupling of laser diodes and detectors, which makes integrated optical systems more promising. A number of complex devices have been demonstrated

in optical signal processing, optical communication, and optical
sensing applications. This has encouraged the search for new
applications and the application of this technology to systems.

An area of research that shows promise for expanding the
capabilities and applications of integrated optics is the integration
of optical functions and electronics on a common substrate. This
has been an active area of research during the past few years
using III-V semiconductors, which have the potential for total
integration.

The present book represents the joint effort of eleven contributors. This cooperative effort has some distinct advantages. As
the field of integrated optics expands and encompasses broader
technologies and applications, it is almost impossible for one person
to have an in-depth knowledge of all aspects of the subject. Each
of these contributors is an expert and has both a comprehensive
knowledge of and extensive experience in the subject he has
written on. An extensive set of references supplements each
chapter.

The editor wishes to thank all of the contributors for providing
an interesting and productive experience in connection with the
preparation of this book. In particular, the editor appreciates the
suggestions and critical comments by Dr. D. G. Hall and Dr. J. J.
Burke on the introductory chapter.

Lynn D. Hutcheson

Contents

Contributors

Rodney C. Alferness Photonic Circuits Department, AT&T Bell Laboratories, Holmdel, New Jersey

James J. Burke Optical Sciences Center, University of Arizona, Tucson, Arizona

James K. Carney GaAs Electronics Design, Honeywell Physical Sciences Center, Honeywell, Bloomington, Minnesota

Bor-Uei Chen PCO, Inc., Chatsworth, California

Talal K. Findakly* TRW Electro Optics Research Center, TRW, Redondo Beach, California

Dennis G. Hall The Institute of Optics, University of Rochester, Rochester, New York

Lynn D. Hutcheson APA Optics, Inc., Blaine, Minnesota

Steven K. Korotky Photonic Circuits Research Department, AT&T Bell Laboratories, Holmdel, New Jersey

---

Current affiliation: United Technologies Research Center, United Technologies, East Hartford, Connecticut

Virginia M. Robbins Center for Compound Semiconductor Micro-electronics, University of Illinois at Urbana-Champaign, Urbana, Illinois

Colin T. Seaton Optical Sciences Center, University of Arizona, Tucson, Arizona

George I. Stegeman Optical Sciences Center, University of Arizona, Tucson, Arizona

Gregory E. Stillman Center for Compound Semiconductor Micro-electronics, University of Illinois at Urbana-Champaign, Urbana, Illinois

Integrated Optical Circuits and Components

1
Introduction

LYNN D. HUTCHESON   *APA Optics, Inc., Blaine, Minnesota*

1.1 EVOLUTION OF THE TECHNOLOGY

Since the invention of the transistor in 1947, the development of integrated electronic circuits has led to increased reliability with increased performance and reduced cost, size, weight, and power requirements. The same advantages and the enthusiasm they generate are the justification for pursuing a corresponding effort in integrated optical circuit technology, with the expectation of realizing similar benefits for signal processing, computing, sensors, and communications systems. For the first decade or so, the field of integrated optics (IO) grew very slowly, but since that time the number of governmental, industrial, university, and other research laboratories involved in this field has increased rapidly. The result has been enormous progress in materials, fabrication processes, component development, and device demonstration.

Bulk optical devices in general have tight alignment tolerances that render them vulnerable to such external effects as temperature and vibration, making them useful for laboratory experiments, but difficult and expensive to implement in a system. Like integrated electronics, the promise of integrated optics lies in the realization of devices and systems that would otherwise be too cumbersome or expensive to be utilized in bulk form. In addition to

TABLE 1.1 Material Comparisons for Optical Circuits

Property	$LiNbO_3$	GaAs/AlGaAs	Thin film on silica
Form	Single crystal	Epitaxially grown	Thin-film evaporation or sputtering
Refractive index	2.2	3.5	Depends on Materials (ZnO and Si_3N_4, n = 2)
Attenuation	0.5 dB/cm	1-2 dB/cm λ = 1.3 μm, 5 dB/cm λ = 0.85 μm	0.1 dB/cm
Birefringence	Very high	Low	High
Electro-optic coefficient	High r_{33} = 30 × 10^{-12} m/V	Medium r_{41} = 1.5 × 10^{-12} m/V	Medium r_{33} = 2 × 10^{-12} m/V (ZnO)
Piezoelectric	Good	No (but techniques exist for generating acoustic waves)	Material dependent— yes for ZnO
Electronic compatibility	No	Yes	Yes
Laser compatibility	No	Yes	No

guiding the light in a material having thicknesses approximate to a wavelength, integrated optics also encompasses the generation, manipulation, and detection of light waves, and the coupling of optical waves into and out of IO circuits.

Integrated optics in general does not rest on new concepts but requires innovative design and fabrication techniques along with advanced material technology. For instance, the availability of room-temperature continuous-wave (CW) laser diodes has revolutionized the entire field of optical communication while it provides an impetus for device and system concepts using guided-wave optics.

Optical waveguides, components, and devices have been fabricated in various materials, using a variety of techniques. For many IO devices, $LiNbO_3$ remains the prime candidate because of its excellent piezoelectric, electro-optical, and waveguiding properties. Just about every IO component, with the exception of emitters and detectors, has been demonstrated in $LiNbO_3$.

For higher levels of integration the development of integrated optics in semiconductor materials is developing very rapidly. Developing integrated optical components in semiconductors has the additional advantage of integration compatibility with electronics if silicon or gallium arsenide (GaAs) are the chosen materials. In particular, GaAs and its related compounds are particularly attractive because both lasers and detectors can be accommodated, which means that total integration is possible. Some researchers believe that for integrated optics to have a large impact in high-speed systems, the design, fabrication, and processing must be compatible with existing electronic processing. A comparison of the semiconductor and $LiNbO_3$ technology is shown in Table 1.1. The materials listed by no means provide an exhaustive list. See Chapter 3 for more detail regarding waveguide materials and fabrication techniques. There is no clear answer to which material will be the most important. The particular application for which the optical circuit is being considered will dictate which material is chosen. In the next section a brief discussion of integrated optic applications is presented.

1.2 APPLICATIONS OF INTEGRATED OPTICAL CIRCUITS

Research on integrated optical circuits has been increasing at a rapid rate in recent years. The principal applications of integrated optics are optical communication, optical sensing, and optical data processing, with the latter application including both analog and digital processing. A large variety of single components have been demonstrated experimentally, and more recently, the integration of multiple components on a single substrate.

One of the more complex optical integrated circuits is the radio-frequency (RF) spectrum analyzer, which has been demonstrated at a number of laboratories [1–6]. One implementation of the RF spectrum analyzer is shown in Fig. 1.1. This integrated optic device consists of a hybrid package that includes a butt-coupled laser diode, a LiNbO$_3$ substrate with a Ti diffused waveguide, two guided-wave lenses, a surface acoustic wave transducer array, and a butt-coupled linear photodiode array.

The optical waveguide is formed on the highly polished surface of the LiNbO$_3$ substrate by depositing a thin film of titanium metal and then diffusing this film into the substrate at a temperature of 1000°C in an oxygen atmosphere. The waveguide diffusion is performed in the presence of LiNbO$_3$ powder to compensate for the out-diffusion of Li$_2$O, so as not to produce a deeper-lying out-diffused waveguide capable of supporting additional modes. The laser diode radiation is butt coupled into the waveguide through the polished edge, expands to the desired width, and is collimated by the first guided-wave lens. A portion of this collimated beam is then deflected by Bragg interaction with a surface acoustic wave generated by an interdigital transducer array. A second lens focuses the deflected beam onto the polished output edge of the waveguide, to which the photodiode array is butt coupled.

For the device to achieve high resolution, a nearly diffraction-limited optical image quality is required for the guided-wave lens.

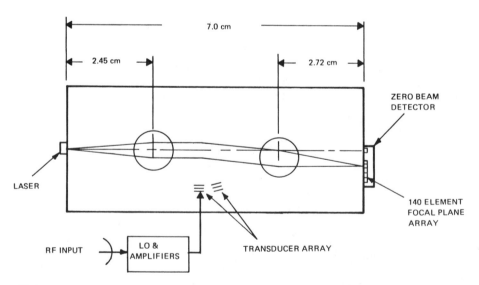

FIG. 1.1 Schematic of integrated optical RF spectrum analyzer (from Ref. 1).

A number of techniques for lens fabrication have been developed, three of which are the geodesic [7], Luneburg [8], and diffraction [9] lens techniques.

The RF signals to be analyzed are converted to the operating bandwidth of the surface acoustic wave (SAW) Bragg cell via a local oscillator and applied to the transducer. The deflection of the optical beam by the SAW is proportional to the frequency of the applied RF signal. The second guided-wave lens focuses the optical beam onto the output edge of the optical waveguide, and the beam is detected by the photodiode array. Since the deflection angle is proportional to the frequency of the RF signal, its frequency can be determined from the position of the focused beam on the detector array.

The first IO spectrum analyzer demonstrated used an end-fire-coupled HeNe laser as the optical source and a self-scanned photodiode array [10] consisting of 140 pixels with a 12-μm pitch with an access time of 2 μs. The next-nearest-neighbor crosstalk was down by 15 dB. The SAW transducers are two-element titled arrays designed to have a combined 400-MHz bandwidth centered at 600 MHz. Measurement of the transducer performance showed a 3-dB bandwidth of 400 MHz and a 5% deflection efficiency with an RF power of 60 mW. The total device dynamic range of the RF input was measured in excess of 20 dB. The device was tested with two 0.3-μs simultaneous pulses separated in frequency by 20 MHz, which were fully distinguishable with no sign of intermodulation between the signals. Other signal-processing applications of integrated optical circuits include correlation [11], analog-to-digital conversion [12–14], signal sampling [15,16], and optical logic functions [17–24].

1.3 ORGANIZATION OF THE BOOK

In developing and planning this book, an attempt was made to give the applied scientist, practicing engineer, and manager an insight into how and where integrated optics can be used in practical applications. This book evolves from a historical perspective to give the reader a feeling for the importance of integrated optics, the direction in which integrated optics is going, and some of its important applications. Subjects covered include all aspects of integrated optic circuit development, including materials, design considerations, fabrication techniques, component development, and monolithic integration and application requirements, as well as new and novel developments.

Chapter 2 is intended to give the reader an understanding of the terminology and concepts of optical waveguides and devices. Planar waveguide geometries are considered for both the step-index and graded-index waveguides. Two different formalisms have been

been developed over the years to describe light propagation in waveguides: geometrical ray optics and electromagnetic wave theory.
The latter approach is used in Chpater 2 and begins with the application of Maxwell's equations to determine the field characteristics
associated with a uniform dielectric slab bounded by semi-infinite
media.   Coupled-mode theory for both TE and TM modes is discussed
and the author points out that there is a basic error for TM polarization when there is a perturbation on the surface of the waveguide
such as a grating.   Some results are presented that illustrate the
size of the discrepancy when using the formulas for the coupling
coefficient.   In addition the reader will find a discussion of how
coupled-mode theory is applied to the design of waveguide diffraction
gratings and directional couplers.   The reader will be left with an
excellent understanding of the importance of coupled mode theory
and how it can be used in practical applications.

The fabrication of optical waveguides with a review of the fabrication techniques, materials, and state-of-the-art results for both
active and passive waveguides is covered in Chapter 3.   This chapter
is an excellent compendium of the multitude of waveguide materials
with results of their use and a discussion of advantages and disadvantages.   The fabrication techniques include spin and dip coating,
chemical vapor deposition, sputtering, evaporation, epitaxial growth,
ion exchange, diffusion, ion implantation, and others.   If one wants
to compare the results of various optical waveguides, this chapter
should provide the means.

A theoretical and experimental foundation for scattering and
bending losses in optical waveguides can be found in Chapter 4.
Scattering and attenuation in waveguides can often limit device performance.   It is important to be able to estimate the magnitude of
the various scattering effects and bending loss in terms of measure-
able properties of the waveguide.   In this chapter, scattering is considered from the basis of two common mechanisms: (1) attenuation
caused by surface-roughness-induced radiation from planar optical
waveguides, and (2) in-plane scattering caused by refractive index
fluctuations and surface roughness.   On the other hand, the optical
transmission characteristics of optical waveguides having directional
changes plays an important role in determining the density of optical
components on a single integrated optic chip.   Two broad categories
of loss mechanisms due to waveguide bends are discussed in this
chapter.   The first is a power loss due to curvature of the waveguide, and the second is the power scattered from the fundamental
mode when incident upon a junction between two nonidentical waveguides.   Techniques that have been employed to reduce the effects
of bending losses are also presented.

Research in integrated optic lasers and detectors has been quite
intense in recent years.   Chapter 5 shows the evolution of lasers

and detectors in the GaAs/AlGaAs and the InP/GaInAsP materials system. A selection of the more promising laser structures that have led to high-speed, long-lifetime, and low-current-threshold demonstrations are discussed. Several different photoreceiver designs are compared, as are the experimental and theoretical performance of the detectors in optical receiver configurations.

As the title of Chapter 6 indicates, the fundamentals, design considerations, and capabilities of titanium-diffused lithium niobate (Ti:LiNbO$_3$) technology are discussed. An overview of the key ingredients of what is needed to make LiNbO$_3$ integrated optics possible is presented, with an emphasis on issues pertaining to the attainment of large figures of merit in the areas of insertion loss, operating voltage, and speed. Included is an easy-to-follow sequence of steps for designing IO LiNbO$_3$ devices, which includes selecting all the critical waveguide parameters: refractive index profile, mode characteristics, optical coupling to and from the waveguide, electro-optic interaction, and device performance. High-speed waveguide modulators and switches and a comparison of the various implementations, crystal orientation, electrode geometry, and optical isolation from the electrodes are presented. This chapter provides a good insight into what is important for packaging these devices and for reproducibly manufacturing high-figure-of-merit components. The relationship between this technology and high-speed optical communications, and what it takes to achieve good system performance, are also discussed.

Lithium niobate promises to be important in long-distance optical communications; however, the development of optoelectronic GaAs/AlGaAs-material-based systems is critical to the future of high-speed computing systems, as discussed in Chapter 7. The integration of high-speed electronics with optoelectronic devices on a common substrate is being developed in a number of laboratories around the world. Interested parties include major computer manufacturers, defense contractors, telecommunications companies, and a number of universities. Japan announced that they are establishing a research and development company for integrated optoelectronic circuits (IOCs) to be staffed by 13 different companies. In this chapter, IOCs are examined from the standpoint of what it takes to fabricate the device and what performance can be expected. Special emphasis is placed on the use of IOCs in high-speed processors. The components that make up an IOC, including materials, electronics and optoelectronics, are presented. The parameters that are important to a system designer—bandwidth, electrical power, density, and bit error rate—are described. Other operating characteristics, such as temperature sensitivity, are discussed to provide an appreciation of what must be considered when the IOC is taken out of the laboratory and is designed into a system. The present status of IOCs is described and a few examples are given of the expected performance of IOCs and their impact on the system.

Using integrated optics for high-speed digital optical processing and computing is discussed in Chapter 8. Presently, processing systems using optics are hybrid in nature, with only some of the functions performed in the optics domain and the other functions using electronics. The ability to perform logic operations in the optics domain will enhance system flexibility because many electrical-to-optical and optical-to-electrical conversions can be eliminated. In this chapter, the integrated optical logic devices that have been demonstrated and logic devices that are capable of being implemented in an integrated optics format are discussed. The logic devices discussed include types for which input and output data are in a combination of electrical and optical formats. Logic devices require nonlinear functions, and several methods are available to introduce appropriate nonlinearities into an optical device. Various implementations of optical logic devices are presented, with the different types of nonlinearity being employed as the basis for categorizing the optical devices.

Chapter 9, discusses nonlinear interactions in optical waveguides. There are a number of characteristics of nonlinear interactions that lend themselves to the utilization of integrated optics waveguides. The strength of the interactions is proportional to the power, and in optical waveguides the guiding dimensions are limited to approximately a few optical wavelengths. This reduction in beam cross-sectional area, which can be maintained for centimeter distances, reduces the power required to produce large effects. The theoretical premise of achieving high efficiency with low optical power is discussed together with optimized material parameters. Experimental results and a comparison with theoretical predictions are presented. This is an active area of research which should soon evolve into practical applications of nonlinear integrated optics to picosecond serial signal processing.

REFERENCES

1. D. Mergerian, E. C. Malarkey, R. P. Pautienus, J. C. Bradley, G. E. Marx, L. D. Hutcheson, and A. L. Kellner, Operational integrated optical R. F. spectrum analyzer, *Appl. Opt.*, *19*:3033 (1980).

2. T. R. Ranganath, T. R. Joseph, and J. Y. Lee, "Integrated-Optic Spectrum Analyzer: A First Demonstration," Technical Digest of the International Electron Devices Meeting, p. 843 (Dec. 1980).

3. R. L. Davis and F. S. Hickernell, "An IO Spectrum Analyzer with Thin Film Lenses," Post Deadline Paper at the 3rd International Conference on Integrated Optics and Optical Fiber Communications, Paper WE6-1 (Apr. 1981).

4. L. Thylen and L. Stensland, Electro-optic approach to an integrated optics spectrum analyzer, *Appl. Opt.*, *20*:1825 (1981).

5. L. Thylen and L. Stensland, Lensless integrated optics spectrum analyzer, *IEEE J. Quantum Electron.*, *QE-18*:381 (1982).

6. T. Suhara, N. Nishihara, and J. Kayama, A folded-type integrated-optic spectrum analyzer using butt-coupled chirped grating lenses, *IEEE J. Quantum Electron. QE-18*:1057 (1982).

7. J. C. Bradley, L. D. Hutcheson, A. L. Kellner, E. C. Malarkey, D. Mergerian, and R. P. Pautienus, Geodesic lens performance characteristics, *Proc. SPIE*, *239*:84 (1980).

8. F. Zernike, Luneburg lens for optical waveguide use, *Opt. Commun. 12*:379 (1974).

9. W. S. C. Chang, S. Forouhar, J.-M. Delavaux, and R.-X. Lu, Fabrication and performance of diffraction lenses, *Proc. SPIE*, *306* (1981).

10. G. M. Borsuk, Photodetectors for acousto-optic signal processors, *Proc. IEEE*, *69*:100 (1981).

11. C. M. Verber, R. P. Kenan, and J. R. Bush, Design and performance of an integrated optical digital correlator, *IEEE J. Lightwave Technol. LT-1*:256 (1983).

12. F. J. Leonberger, C. E. Woodward, and D. L. Spears, Design and development of a high speed electrooptic A/D converter, *IEEE Trans. Circuits and Syst.*, *CAS-26*:1125 (1979).

13. S. Yamade, M. Minakata, and J. Noda, High-speed 2-bit analog-digital converter using LiNbO$_3$ waveguide modulators, *Electron. Lett.*, *17*:259 (1982).

14. C. L. Chang and C. S. Tsai, Electro-optic analog to digital converter using channel waveguide Fabry-Perot modulator array, *Appl. Phys. Lett.*, *43*:22 (1983).

15. H. A. Haus, "Picosecond Sampling in Optical Waveguides," Proceedings of the Topical Meeting on Integrated and Guided-Wave Optics, Paper WA-1 (1982).

16. M. Izutsu, H. Haga, and T. Sueta, Picosecond signal sampling and multiplication by using integrated tandem light modulators, *IEEE J. Lightwave Technol.*, *LT-1*:285 (1983).

17. H. F. Taylor, Guided wave electrooptic devices for logic and computation, *Appl. Opt.*, *17*:1493 (1978).

18. P. W. Smith, E. H. Turner, and P. J. Maloney, Electro-optic nonlinear Fabry-Perot devices, *IEEE J. Quantum Electron.*, *OE-14*:207 (1978).

19. P. W. Smith, I. P. Kaminow, P. J. Maloney, and L. W. Stulz, Integrated bistable optical devices, *Appl. Phys. Lett.*, *33*:24 (1978).

20. H. M. Gibbs, S. L. McCall, T. N. C. Venkatesan, A. C. Gossard, A. Passner, and W. Wiegmann, Optical bistability in semiconductors, *Appl. Phys. Lett.*, *35*:451 (1979).

21. E. Garmire, S. D. Allen, and J. Marburger, Bistable optical devices for integrated optics and fiber optics applications, *Opt. Eng.*, *18*:194 (1979).

22. J. A. Copeland, J. C. Campbell, A. G. Dentai, and S. E. Miller, Wavelength-multiplexed AND gate: a building block for monolithic optically coupled circuits, *Appl. Phys. Lett.* *39*:197 (1981).

23. R. Normandin and G. I. Stegeman, A picosecond transient digitizer based on nonlinear integrated optics, *Appl. Phys. Lett.*, *40*:759 (1982).

24. C. T. Seaton, X. Mai, G. I. Stegeman, and H. G. Winful, Nonlinear guided-wave applications, *Opt. Eng.*, *24*:593 (1985).

2

Theory of Waveguides and Devices

DENNIS G. HALL   *The Institute of Optics, University of Rochester, Rochester, New York*

2.1 INTRODUCTION

From the geometrical-optics point of view, dielectric optical waveguides confine light by the mechanism of total internal reflection.
Two types of waveguiding structures are important for integrated
optics: (1) the step-index waveguide, and (2) the graded-index
waveguide. Both consist of a region of elevated refractive index
bounded on at least two sides by media of lower refractive index.
For step-index waveguides, the index of refraction is constant
within the confinement region, while for graded-index waveguides,
the index of refraction is a function of the coordinates and thus
varies both inside and (possibly) outside of the confinement region.
Optical waveguides can also be categorized by the number of spatial
dimensions that provide confinement. Planar (or slab) waveguides
provide index boundaries along only one coordinate axis and thus
are one-dimensional waveguides. Optical fibers and channel (or
strip) waveguides provide index boundaries along two coordinate
axes and are thus two-dimensional waveguides. Both step-index
and graded-index planar waveguides are used as the basic substrates for the fabrication and development of multicomponent optical
circuits. Channel waveguides formed in or on a host crystal by
diffusion are the basic ingredients for a class of optical switches
based on an optical tunneling process.

There have been a number of textbook-level treatments of the theory of optical waveguides. Notable among these are the books by Kapany and Burke [1], Marcuse [2], and Adams [3], and Kogelnik's chapter in the text *Integrated Optics* edited by Tamir [4]. We have chosen to add yet another treatment to the literature in order to make the present text a self-contained unit that will be useful to those seeking an introduction to the field as well as experienced scientists and engineers. The notation for this chapter parallels rather closely that of Ref. 4, both to avoid introducing a new notation into the literature and because of the convenience of that notation. There are, however, certain differences between the present treatment and Kogelnik's and the reader should take some care in comparing the two.

2.2 STEP-INDEX PLANAR WAVEGUIDES

In this section we consider the simplest type of optical waveguide: a one-dimensional waveguide for which the refractive index of each material is a constant, except at the boundaries, where it changes value abruptly. Light propagation in such a structure can be discussed in terms of either rays or waves. The former will not be treated here and the reader is urged to consult ref. 3 and 4 for details. Our approach consists of a straightforward application of Maxwell's equations to determine the characteristics of the fields associated with a uniform dielectric slab bounded by semi-infinite media. A brief review of Maxwell's equations is included as a convenience for the reader.

The slab waveguide can be said to support both bound modes and radiation modes. For the former, light is trapped within the dielectric layer by total internal reflection (TIR) at both boundaries. Outside the physical boundaries of the waveguide, one finds an exponentially decaying (or evanescent) field that is characteristic of total internal reflection. For the latter, the TIR requirement is violated at one or both boundaries, leading to propagating (instead of decaying) fields outside the waveguide. Since it is the intent of this chapter to provide the background necessary to understand components and devices, we consider explicitly only the bound modes of the slab waveguide. The radiation modes are important in connection with coupling light into or out of a waveguide and in connection with radiation losses (attenuation) that arise from compositional or structural fluctuations in the waveguide. Both of these topics are addressed separately in Chapter 4.

2.2.1 Review of Maxwell's Equations

The theory of dielectric optical waveguides begins by applying Maxwell's equations to suitably defined structures. The four Maxwell equations are

$$\underline{\nabla} \cdot \underline{D} = 0 \tag{2.1}$$

$$\underline{\nabla} \cdot \underline{H} = 0 \tag{2.2}$$

$$\underline{\nabla} \times \underline{E} = -\frac{\partial B}{\partial t} \tag{2.3}$$

$$\underline{\nabla} \times \underline{H} = \frac{\partial \underline{D}}{\partial t} \tag{2.4}$$

where $\underline{E}$ is the electric-field vector, $\underline{D}$ the displacement vector, $\underline{H}$ the magnetic-field vector, and $\underline{B}$ the magnetic flux density. In isotropic, linear media, $\underline{D}$ and $\underline{E}$ are related by the permittivity ε according to

$$\underline{D} = \varepsilon\underline{E} \tag{2.5}$$

and ε is frequently expressed in terms of the electric susceptibility χ_e:

$$\varepsilon = \varepsilon_0(1 + \chi_e) \tag{2.6}$$

where $\varepsilon_0 = 8.85 \times 10^{-12}$ F/m is the permittivity of free space. In a similar way, $\underline{B}$ and $\underline{H}$ are related by the permeability μ according to

$$\underline{B} = \mu\underline{H} \tag{2.7}$$

with μ expressible in terms of the magnetic susceptibility χ_m:

$$\mu = \mu_0(1 + \chi_m) \tag{2.8}$$

where $\mu_0 = 4\pi \times 10^{-7}$ H/m is the permeability of free space. In this chapter we deal exclusively with nonmagnetic materials for which χ_m is very small compared with unity, or $\mu \simeq \mu_0$. The vectors in Eqs. (2.1)–(2.4) will, unless otherwise stated, be assumed to have a time dependence of the form

$$\underline{E} = \underline{E}_0(x,y,z)e^{-i\omega t} \tag{2.9}$$

where $\omega = 2\pi f$ and f is the optical frequency. For this choice, using Eqs. (2.5) and (2.7), the set of four Maxwell equations becomes

$$\underline{\nabla} \cdot \underline{D} = 0 \tag{2.10}$$

$$\underline{\nabla} \cdot \underline{H} = 0 \tag{2.11}$$

$$\underline{\nabla} \times \underline{E} = i\mu\omega\underline{H} \tag{2.12}$$

$$\underline{\nabla} \times \underline{H} = -i\varepsilon\omega\underline{E} \tag{2.13}$$

If μ and ε are constants, Eqs. (2.10)–(2.13) lead to the usual wave equations for $\underline{E}$ and $\underline{H}$:

$$\nabla^2\underline{E} + n^2 \left(\frac{\omega^2}{c^2}\right) \underline{E} = 0 \tag{2.14}$$

$$\nabla^2\underline{H} + n^2 \left(\frac{\omega^2}{c^2}\right) \underline{H} = 0 \tag{2.15}$$

where the index of refraction n is given by

$$\varepsilon = n^2\varepsilon_0 \tag{2.16}$$

and $\mu\varepsilon_0 = 1/c^2$, as usual. In analyzing propagation in a dielectric waveguide, the major tasks consist of finding solutions of Eqs. (2.14) and (2.15) for each medium and using the boundary conditions to match the solutions at the boundaries between these media. Each of Eqs. (2.10)–(2.13) provides a boundary condition, and the details of the derivation of such conditions can be found in any of the standard texts on electromagnetic theory. For uncharged dielectrics, the boundary conditions are: (1) the component of $\underline{E}$ tangent to the boundary is continuous across the boundary; (2) the component of $\underline{H}$ tangent to the boundary is continuous across the boundary; (3) the component of $\underline{D}$ normal to the boundary is continuous across the boundary; and (4) the component of $\underline{B}$ normal to the boundary is continuous across the boundary. If spatial variations in ε are included, the wave equations in Eqs. (2.17) and (2.18) are modified by the inclusion of an additional term involving $\underline{\nabla}\varepsilon$, but a discussion of that class of problems is not included here.

The time-averaged Poynting vector $\underline{S}$ specifies the (average) power per unit cross-sectional area carried by a light wave. For fields of the form given in Eq. (2.9), one finds that S can be written as

$$\underline{S} = \frac{1}{2} \, \text{Re}(\underline{E} \times \underline{H}^*) \qquad (2.17)$$

where Re designates the real part of the quantity and the asterisk indicates the complex conjugate. One calculates the average power carried by an electromagnetic wave by integrating Eq. (2.17) over the area of a surface perpendicular to the direction of propagation. For a simple plane wave of the form

$$\underline{E} = \underline{E}_0 e^{i(\underline{k} \cdot \underline{r} - \omega t)} \qquad (2.18)$$

with $\underline{E}_0$ constant and $\underline{r} = x\hat{x} + y\hat{y} + z\hat{z}$ ($\hat{x}$, $\hat{y}$, and $\hat{z}$ are unit vectors), one can show that $\underline{S}$ has the simple form

$$\underline{S} = \frac{n}{2} \sqrt{\frac{\varepsilon_0}{\mu_0}} \, |\underline{E}_0|^2 \hat{k} \qquad (2.19)$$

where $\hat{k}$ is a unit vector in the direction of propagation. The quantity $(\mu_0/\varepsilon_0)^{1/2}$ is often called the impedance of free-space and has the approximate value 377 Ω.

2.2.2 Bound Modes of the Dielectric Slab Waveguide: TE-polarization

The geometry for this discussion is shown in Fig. 2.1. A layer of refractive index n_f and thickness h is bounded by semi-infinite substrate and cover media of refractive indices n_s and n_c, respectively. The waveguide boundaries are located at $x = 0$ and $x = h$ and we seek solutions of the wave equations in Eqs. (2.14) and (2.15) that correspond to waves propagating in the positive z direction. For

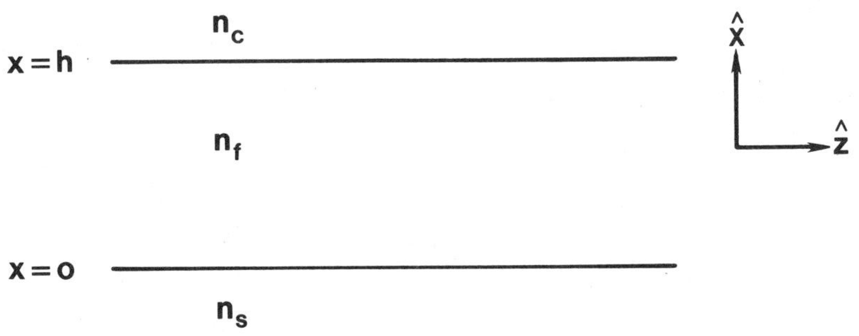

FIG. 2.1 Side view of the basic slab waveguide geometry.

bound modes we require that the fields decay exponentially with distance from the interface into the substrate and cover media. Suitable fields are

$$\underline{E} = \hat{y}E_c \exp[-\gamma_c(x - h)] \exp[i(\beta z - \omega t)] \qquad x \geqslant h$$

$$(2.20)$$

$$\underline{E} = \hat{y}[A \exp(ik_f x) + B \exp(-ik_f x)] \exp[i(\beta z - \omega t)] \quad 0 \leqslant x \leqslant h$$

$$(2.21)$$

$$\underline{E} = \hat{y}E_s \exp(\gamma_s x) \exp[i(\beta z - \omega t)] \qquad x \leqslant 0$$

$$(2.22)$$

Each field has the form $\underline{E} = E_y\hat{y}$ and consists of a single component along the y direction. The constants k_f and β can be viewed as the x and z components, respectively, of the propagation vector $\underline{k}$ in the slab region, and so the electric field $\underline{E}$ is completely transverse (TE). Inserting Eqs. (2.20)–(2.22) into the wave equation gives the following expressions:

$$\gamma_c = (\beta^2 - n_c^2 k_o^2)^{1/2} \qquad (2.23)$$

$$\gamma_s = (\beta^2 - n_s^2 k_o^2)^{1/2} \qquad (2.24)$$

$$k_f = (n_f^2 k_o^2 - \beta^2)^{1/2} \qquad (2.25)$$

where $k_o = w/c$. Since γ_c and γ_s must be positive for proper behavior of the fields at infinity, β must be greater than or equal to the larger of $n_c k_o$ or $n_s k_o$. It is common to define the effective index of refraction N according to

$$\beta = Nk_o \qquad (2.26)$$

and to assume that $n_s \geqslant n_c$, so that $N \geqslant n_s$. Further, since β can be viewed as only one component of $k = n_f k_o$, we find that

$$n_s \leqslant N \leqslant n_f \qquad (2.27)$$

The boundary conditions on the tangential components of $\underline{E}$ and $\underline{H}$ at both $x = h$ and $x = 0$ yield the following four equations:

$$E_c = A \exp(ik_f h) + B \exp(-ik_f h) \qquad (2.28)$$

$$\gamma_c E_c = -ik_f[A \exp(ik_f h) - B \exp(-ik_f h)] \qquad (2.29)$$

$$E_s = A + B \qquad (2.30)$$

$$\gamma_s E_s = ik_f(A - B) \qquad (2.31)$$

where H_z, the tangential component of $\underline{H}$, is obtained from Eq. (2.12) with $\underline{E} = \hat{y}E_y$:

$$H_z = (i\mu\omega)^{-1} \frac{\partial E_y}{\partial x} \qquad (2.32)$$

Equations (2.28)–(2.31) have a nontrivial solution if

$$2k_f h - 2 \tan^{-1} \frac{\gamma_s}{k_f} - 2 \tan^{-1} \frac{\gamma_c}{k_f} = m(2\pi) \qquad (2.33)$$

where m is an integer. Equation (2.33) is known as the waveguide dispersion relation. It determines the allowed values of β for given n_f, n_c, n_s, h, and the optical frequency ω. The presence of the integer m in the dispersion relation means that the allowed values of β form a discrete set, not a continuum. In terms of ray optics, this means that only certain angles of propagation are allowed for the ray in its zigzag path between the two waveguide boundaries. The second and third terms on the left-hand side of Eq. (2.33) are the phase shifts associated with total internal reflection from the film/substrate and film/cover interfaces. In a ray-optics treatment, the dispersion relation is derived by requiring that the round-trip change in phase (along x) must be an integer multiple of 2π.

Equations (2.30) and (2.31) can be combined to give

$$B = A \exp(i2\phi_s) \qquad (2.34)$$

where $\tan \phi_s = \gamma_s/k_f$. Equations (2.20)–(2.22) can be put in complete agreement with Kogelnik's notation [4] by setting $E_f = 2A \times \exp(i\phi_s)$ so that

$$E = \hat{y}E_c \exp[-\gamma_c(x - h)] \exp[i(\beta z - \omega t)] \qquad x \geqslant h \qquad (2.35)$$

$$\underline{E} = \hat{y}E_f \cos(k_f x - \phi_s) \exp[i(\beta z - \omega t)] \qquad 0 \leqslant x \leqslant h \qquad (2.36)$$

$$\underline{E} = \hat{y}E_s \exp(\gamma_s x) \exp[i(\beta z - \omega t)] \qquad x \leqslant 0 \qquad (2.37)$$

By defining ϕ_c in analogy with ϕ_s as $\tan \phi_c = \gamma_c / k_f$, the dispersion relation becomes

$$2k_f h - 2\phi_s - 2\phi_c = m(2\pi) \tag{2.38}$$

An alternative and often-quoted form of this equation [2] is

$$\tan k_f h = \frac{k_f(\gamma_c + \gamma_s)}{k_f^2 - \gamma_c \gamma_s} \tag{2.39}$$

The dispersion relation determines the values of the propagation constant β that are permitted for a given waveguide structure. It is a transcendental equation in β, so numerical methods are required to obtain solutions. Kogelnik and Ramaswamy [5] have shown that one can introduce a set of normalized variables V, b, and a and generate universal curves from Eq. (2.33). This technique is particularly useful and illustrative for the TE modes. In terms of the normalized frequency,

$$V = 2\pi \frac{h}{\lambda_0} \sqrt{n_f^2 - n_s^2} \tag{2.40}$$

where λ_0 is the wavelength in vacuum, the normalized guide index

$$b = \frac{N^2 - n_s^2}{n_f^2 - n_s^2} \tag{2.41}$$

and the asymmetry parameter

$$a = \frac{n_s^2 - n_c^2}{n_f^2 - n_s^2} \tag{2.42}$$

the dispersion relation takes on the form

$$V \sqrt{1 - b} - \tan^{-1} \sqrt{\frac{b}{1 - b}} - \tan^{-1} \sqrt{\frac{b + a}{1 - b}} = m\pi \tag{2.43}$$

Figure 2.2 shows plots of b versus V for two values of the mode integer m and two values of the asymmetry parameter a. Each value of V and m leads to a family of curves, with each curve

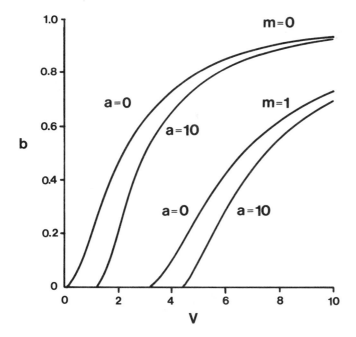

FIG. 2.2 Plots of the normalized guide index b as a function of the normalized frequency V for the two lowest-order TE modes. Two values of the asymmetry parameter a are shown for each value of the mode integer m.

associated with a specific value of a. Each value of m corresponds to a mode of the waveguide and defines a specific shape for the electric field distribution.

The condition $N = n_s$ defines the cutoff condition for the wave-guide mode. In essence, the cutoff condition gives the minimum ratio h/λ_0 required to support a mode of order m. For h/λ_0 less than this minimum, the total internal reflection requirement is not fulfilled and the mode is not bound. From Eq. (2.41), cutoff occurs for b = 0, for which Eq. (2.43) becomes

$$V_m - \tan^{-1} \sqrt{a} = m\pi \tag{2.44}$$

where the subscript m has been added to indicate the cutoff value of V for the mode of order m. A symmetric waveguide is one for which $n_s = n_c$ or a = 0. In that case, the smallest ratio h/λ_0 for a given mode is

$$\left(\frac{h}{\lambda_0}\right)_m = \frac{m}{2\sqrt{n_f^2 - n_c^2}} \tag{2.45}$$

It is significant to note that the m = 0 mode has a cutoff given by $(h/\lambda_0)_0 = 0$. This means that for a given wavelength λ_0, there is no minimum thickness required to support the m = 0 mode of a symmetric slab waveguide. For a ≠ 0, the minimum thickness or maximum wavelength is obtained from Eq. (2.44). The number of modes supported by a given structure at a given wavelength also emerges from Eq. (2.43) and the condition b = 0 ($N = n_s$). Since the lowest-order mode is labeled m = 0 and not m = 1, the total number of modes is just M = m + 1 or

$$M = 1 + \frac{1}{\pi}\left(\frac{2\pi h}{\lambda_0}\sqrt{n_f^2 - n_s^2} - \tan^{-1}\sqrt{a}\right) \tag{2.46}$$

Figure 2.2 shows that for a given structure and a given m, the effective index N increases with V, starting with $N = n_s$ at cutoff and approaching $N = n_f$ as h/λ_0 becomes large. For a given waveguide and a given wavelength, V is constant and N decreases with increasing m. The operation of many waveguide components and devices depends strongly on the value of N, a fact that suggests that integrated optics is restricted to single-mode waveguide technology. We will return to this point later in this chapter.

Because the constants γ_c, γ_s, and k_f that appear in Eqs. (2.35)–(2.37) depend on β, it follows that each value of the mode integer m specifies a unique shape for the electric-field distribution. Figure 2.3 shows plots of the electric field for the m = 0, 1, 2 modes of a slab waveguide for the parameters given in the figure caption. It is clear that m labels the number of zeros contained between x = 0 and x = h, and that m + 1 labels the number of extrema in the same interval. By applying the boundary conditions on the tangential components of $\underline{E}$ and $\underline{H}$ to the fields given in Eqs. (2.35)–(2.37), one can show that the ratio of the values of the fields on the interfaces, E_c/E_s, is given by

$$\frac{E_c}{E_s} = \left(\frac{n_f^2 - n_s^2}{n_f^2 - n_c^2}\right)^{1/2} \tag{2.47}$$

which is independent of film thickness, wavelength, and mode integer. For a symmetric waveguide, $n_s = n_c$ and $E_c/E_s = 1$, as demanded by symmetry. For $n_s > n_c$, one finds that $E_c/E_s < 1$. As

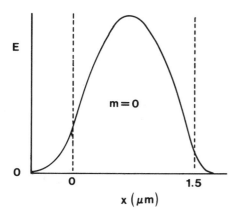

(a)

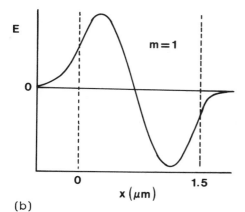

(b)

(c)

FIG. 2.3 Electric field profiles for TE modes with (a) m = 0, (b)
m = 1, and (c) m = 2. The parameters for this example are λ =
0.6328 μm, n_c = 1.0, n_f = 1.6, n_s = 1.47, and h = 1.5 μm.

Kogelnik [4] points out, the boundary conditions provide a useful set of relationships between E_c, E_s, and E_f:

$$E_c^2 = E_f^2 \frac{n_f^2 - N^2}{n_f^2 - n_c^2} \tag{2.48}$$

$$E_s^2 = E_f^2 \frac{n_f^2 - N^2}{n_f^2 - n_s^2} \tag{2.49}$$

The power per unit width (along y) carried by a guided wave is obtained by integrating Eq. (2.17) over the x coordinate:

$$P = \int_{-\infty}^{\infty} \underline{S} \cdot \hat{z}\, dx \tag{2.50}$$

where $\hat{z}$ is a unit vector in the z direction. For TE modes, $\underline{E}$ is strictly in the $\hat{y}$-direction and one requires the x component of $\underline{H}$ to evaluate Eq. (2.50). Using

$$H_x = -(i\mu\omega)^{-1} \frac{\partial E_y}{\partial z} \tag{2.51}$$

which follows from Eq. (2.12) with $\underline{E} = \hat{y}\, E_y$, the fields given in Eqs. (2.35)–(2.37) and a fair amount of algebraic manipulation lead to

$$P = \frac{N}{4} \sqrt{\frac{\varepsilon_0}{\mu_0}}\, E_f^2\, h_{eff} \tag{2.52}$$

where

$$h_{eff} = h + \frac{1}{\gamma_c} + \frac{1}{\gamma_s} \tag{2.53}$$

is called the effective thickness of the planar waveguide. It should be noted that this equation for P differs from that in Ref. [4] by a factor of 1/4 due to differences in the form of $\underline{S}$ used in each case. This is essentially a difference in normalization and leads to no contradictions as long as either approach is used consistently.

It is often important to consider what fraction of the energy carried by a waveguide mode is actually contained within the physical boundaries of the waveguide. The ratio of the power contained within the waveguide boundaries to the total power is called the confinement factor Γ and is defined as

$$\Gamma = \frac{\displaystyle\int_0^h \underline{S} \cdot \hat{z} \, dx}{(N/4)\sqrt{\mu_0/\varepsilon_0} \, E_f^2 \, h_{eff}} \qquad (2.54)$$

where we have used Eq. (2.52). After a bit of algebra, Γ can be expressed in terms of the variables V, b, and a as

$$\Gamma = \frac{h}{h_{eff}} \left[1 + \frac{\sqrt{b}}{V} + \frac{\sqrt{b+a}}{V(1+a)} \right] \qquad (2.54a)$$

If the waveguide mode's energy were spread uniformly over h_{eff}, Γ would be given by the first term in the above. The second and third terms in Eq. (2.54a) appear because of the mode shape. It is instructive to consider the behavior of Γ in the limit of a symmetric waveguide, a = 0, for which it can be shown that

$$\Gamma = \frac{V\sqrt{b} + 2b}{V\sqrt{b} + 2} \qquad (2.55)$$

Since $0 \leqslant b \leqslant 1$, it is clear that Γ varies from zero at cutoff (b = 0) to unity for large h/λ_0. This is physically reasonable since we expect the waveguide mode to be well contained within the waveguide boundaries when it is far beyond cutoff. To offer a specific numerical example, consider the case of a gallium aluminum arsenide ($Ga_{1-x}Al_xAs$) layer of thickness h = 2 µm bounded on both top and bottom by a different composition of the same materials ($Ga_{1-y}Al_yAs$). The presence of aluminum in GaAs decreases its refractive index, so one requires y > x to make an acceptable waveguide. Assuming a wavelength $\lambda_0 = 0.84$ µm, x = 0.05, and y = 0.3, one finds that $n_f = 3.56$ and $n_c = 3.38$. As a result, V = 1.67, and for the TE_0 mode b ≅ 0.375, for which $\Gamma = 0.59$. That is, only a little over half of the modal energy is carried within the higher-index layer.

2.2.3 Bound Modes of the Dielectric Slab Waveguide: TM Polarization

The TM modes of the slab waveguide are characterized by a magnetic field vector $\underline{H} = H_y\hat{y}$ that points in the y direction. The analysis proceeds in the same manner as in the TE case, and this section simply provides a statement of the principal formulas. For the geometry of Fig. 2.1, the fields are [4]

$$\underline{H} = \hat{y}H_c e^{-\gamma_c(x-h)} e^{i(\beta z - \omega t)} \qquad x > h \qquad\qquad (2.56)$$

$$\underline{H} = \hat{y}H_f \cos(k_f x - \phi_s) e^{i(\beta z - \omega t)} \qquad 0 \leqslant x \leqslant h \qquad (2.57)$$

$$\underline{H} = \hat{y}H_s e^{\gamma_s x} e^{i(\beta z - \omega t)} \qquad x < 0 \qquad\qquad (2.58)$$

By applying the boundary conditions on the tangential components of $\underline{E}$ and $\underline{H}$, one can derive the familiar form of the dispersion relation

$$2k_f h - 2\phi_c - 2\phi_s = m(2\pi) \qquad\qquad (2.59)$$

While the form of Eq. (2.59) is the same as that of Eq. (2.38), some care is required because the phases ϕ_c and ϕ_s are different for the TE and TM cases. For the TM polarization, the phases are given by

$$\tan \phi_s = \frac{(n_f^2/n_s^2)\gamma_s}{k_f} \qquad\qquad (2.60)$$

$$\tan \phi_c = \frac{(n_f^2/n_c^2)\gamma_c}{k_f} \qquad\qquad (2.61)$$

where k_f, γ_s, and γ_c are as defined in Eqs. (2.23)–(2.25).

The power (per unit width) carried by a TM mode is obtained from Eq. (2.50) and can be expressed as

$$P = \frac{(N/4)\sqrt{\varepsilon_0/\mu_0}\; H_f^2\, h_{eff}}{n_f^2} \qquad\qquad (2.62)$$

where the effective thickness is given by [4]

$$h_{eff} = h + (\gamma_c q_c)^{-1} + (\gamma_s q_s)^{-1} \qquad\qquad (2.63)$$

and

$$q_c = \frac{N^2}{n_c^2} + \frac{N^2}{n_f^2} - 1 \qquad (2.64)$$

$$q_s = \frac{N^2}{n_s^2} + \frac{N^2}{n_f^2} - 1 \qquad (2.65)$$

Once again, the factor of 1/4 in Eq. (2.62) is absent in Kogelnik's treatment [4].

It is considerably more complicated to express the dispersion relation for TM modes in terms of the parameters V, b, and a than it was for TE modes. This topic has been discussed by Kogelnik and Ramaswamy [5], and a useful simplification occurs when $n_s/n_f \overset{\sim}{=}$ 1. In that case, with the asymmetry parameter redefined as

$$a = \left(\frac{n_f}{n_c}\right)^4 \frac{n_s^2 - n_c^2}{n_f^2 - n_s^2} \qquad (2.66)$$

the curves in Fig. 2.2 can be used for TM modes as well as TE modes. In this limit it is easy to see from Eq. (2.44) that since $n_f/n_c > 1$, the cutoff value of h/λ_0 is larger (for a given mode) for TM than for TE. A single-mode waveguide, then, is one for which the ratio h/λ_0 is chosen so that the TM_0 mode is not bound In practice, it is often difficult to do this because the TE and TM cutoff thicknesses can be quite close together.

2.3 GRADED-INDEX PLANAR WAVEGUIDES

Graded-index waveguides currently play a prominent role in integrated optics. Much of the present interest can be attributed to the widespread use of titanium-indiffused lithium niobate (Ti:LiNbO$_3$) waveguides for various types of switching and modulation applications. These waveguides are formed by diffusing titanium into a polished LiNbO$_3$ crystal. The diffusion process creates a higher-index region in the crystal in which the refractive index is largest at the surface and diminishes with increasing depth into the crystal until it reaches the normal value for LiNbO$_3$. Specific details about the fabrication process can be found in any of Refs. 6–8. Unfortunately, the presence of a graded refractive index introduces a

new level of complexity into the problem of obtaining the modes of
the waveguide. The wave equation can only be solved to yield analytic expressions for the fields in a few simple cases, and one is
forced to rely on numerical techniques. This section describes two
very common "working approximations" that appear in the literature:
(1) the WKB approximation, and (2) the modeling of the refractive
index variation as a simple exponential. Each result provides a certain level of insight into propagation in graded-index waveguides.

2.3.1 The Wave Equations

Figure 2.4 shows the basic geometry and the behavior of the refractive index for a graded-index planar waveguide. It is convenient
for this discussion to let x = 0 denote the cover/waveguide interface and let x increase with depth into the waveguide (just the opposite of Fig. 2.1). As in Sec. 2.2, the basic problem consists of
obtaining an appropriate solution of the wave equation. The derivation of the familiar wave equations that appear in Eqs. (2.14) and
(2.15) assumes that ε does not depend on the coordinates x, y,
and z. If one repeats the derivation without making that assumption, one finds that the wave equation for $\underline{E}$ is given by

$$\nabla^2 \underline{E} + n^2(x,y,z) \frac{\omega^2}{c^2} \underline{E} = -\underline{\nabla} \frac{\underline{E} \cdot \underline{\nabla} \varepsilon}{\varepsilon} \tag{2.67}$$

It is common to argue that $\underline{\nabla}\varepsilon/\varepsilon$ is small and can therefore be neglected, thus returning the right-hand side of Eq. (2.67) to zero.
For TE modes, the right-hand side will vanish if we can assume that
$\varepsilon = \varepsilon(x,z)$ but does not depend on y. Since, for TE modes, $\underline{E} =$
$\hat{y}E_y$, we find that $\underline{E} \cdot \underline{\nabla} \varepsilon = 0$ if ε is independent of y. This is usually
a very reasonable assumption for planar waveguides, so Eq. (2.14)

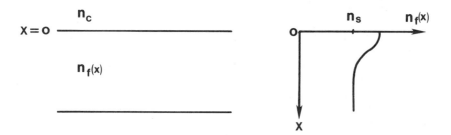

FIG. 2.4 Graded-index slab waveguide geometry.

can be used to solve for the TE modes of the graded-index slab waveguide.

The situation is less simple for TM modes for which $\underline{H} = \hat{y}H_y$. Retaining terms in $\underline{\nabla}\varepsilon$, we find that the wave equation for the magnetic field is

$$\nabla^2\underline{H} + n^2(x,y,z)\frac{\omega^2}{c^2}\underline{H} = -i\omega\underline{E} \times \underline{\nabla}\varepsilon \qquad (2.68)$$

Using Eq. (2.13) to eliminate $\underline{E}$ gives

$$\nabla^2\underline{H} + n^2(x,y,z)\frac{\omega^2}{c^2}\underline{H} = (\underline{\nabla} \times \underline{H}) \times \frac{\underline{\nabla}\varepsilon}{\varepsilon} \qquad (2.69)$$

This equation does not simplify significantly by assuming $\varepsilon \neq \varepsilon(y)$. The usual approach is to return to the argument that $\underline{\nabla}\varepsilon/\varepsilon$ is presumed small and thus solve Eq. (2.15). An example of the use of this approximation appears in the 1973 paper by Conwell [9].

2.3.2 The WKB Approximation

The WKB approximation is a familiar one in quantum mechanics, although it was first introduced by Lord Rayleigh in 1912 in an investigation of wave propagation (see, e.g., Ref. 10, in particular, footnote 2, p. 216). As is well known, the time-independent Schrödinger equation has the same form as the time-independent wave equations (2.14) and (2.15), so the WKB approximation has found widespread use in classical problems as well as in quantum mechanical problems. The use of the WKB (Wentzel, Kramers, and Brillouin) method in connection with multimode optical fibers is discussed (with references) in the review article by Olshansky [11]. Its use in connection with structures of importance in integrated optics is discussed (with references) in the review article by Tien [12]. The treatment presented here is not a formal one, but serves only to motivate the basic results.

For TE modes, the usual assumption that $E_y = E(x) \exp[i(\beta z - \omega t)]$ leads to the differential equation

$$\left\{\frac{d^2}{dx^2} + \left[n_f^2(x)\frac{\omega^2}{c^2} - \beta^2\right]\right\}E(x) = 0 \qquad (2.70)$$

where $n_f(x)$ is constant for $x < 0$, and depends on x (only) for $x \geqslant 0$. When $n_f(x)$ is constant for all x, the solution to Eq. (2.70) is of the form $\exp(\pm ikx)$ with $k^2 = n_f^2(\omega^2/c^2) - \beta^2$. Now that n_f depends on x, a reasonable guess about the form of $E(x)$ is

$$E(x) = \exp[iu(x)] \tag{2.71}$$

which transforms Eq. (2.70) into

$$\left(\frac{du}{dx}\right)^2 = k^2(x) + i \frac{d^2u}{dx^2} \tag{2.72}$$

where

$$k(x) = n_f^2(x) \frac{\omega^2}{c^2} - \beta^2 \tag{2.73}$$

When $k(x)$ is a constant, the solution of Eq. (2.72) is trivial and is given by

$$u_0(x) = \pm \int k(x) \, dx + C \tag{2.74}$$

where C is a constant of integration and the subscript zero desig-nates the zeroth-order solution of Eq. (2.72). In the WKB approxi-mation, we use u_0 to evaluate the righthand side of Eq. (2.72) and then obtain the first-order solution u_1 by integration:

$$u_1(x) = \pm \int k(x) \, dx + \frac{i}{2} \ln[k(x)] + C \tag{2.75}$$

where it has been assumed that

$$\frac{dk(x)/dx}{k^2(x)} \ll 1$$

Referring to Eq. (2.71), we obtain

$$E(x) = \frac{C_1}{\sqrt{k(x)}} \exp\left[\pm i \int k(x) \, dx\right] \tag{2.76}$$

as the basic WKB solution of Eq. (2.70), where C_1 is a constant.

The significant feature in Eq. (2.76) is the argument of the ex-ponential. The quantity $k(x)$ is the x component of the propagation constant of the "ray" one envisions propagating in the graded-index slab; it is the analog of the parameter k_f that appears in Eq. (2.25). As the ray encounters the refractive index gradient, $k(x)$ takes on a different value at each x, and the integral in Eq. (2.76) keeps track of the total phase accumulated along the x component of the ray's path. Returning to Eq. (2.38), we see that the first term $2k_f h$ gives the total phase accumulated along the x component of a ray's round trip through a waveguide of thickness h. The integral in Eq. (2.76) is the generalization of the step-index result. With this background in mind, we find that the WKB dispersion relation is

$$2 \int_0^{x_t} k(x)\ dx - 2\phi_s - 2\phi_c = m(2\pi) \qquad (2.77)$$

The limits of integration define the excursion limits for the ray in the waveguide; they are the turning points of the motion. The upper limit is defined by the condition

$$k(x_t) = 0 \qquad (2.78)$$

In 1974, Tien et al. [13] used Eqs. (2.77) and (2.78) to calculate the effective index of refraction for a simple (exponential) model for $n_f(x)$. The calculation proceeds by choosing a specific form for $n_f(x)$, solving Eq. (2.78) for x_t, and then integrating Eq. (2.77) for a given choice of the mode integer m. The exponential distribution is one for which the wave equation can be solved directly, and Tien et al. found very good agreement between the WKB and exact values of the effective index N. Many other authors, notably Hocker and Burns [14] and Noda et al. [15], have used the WKB approximation with reasonable success to analyze the results of measured values of N. In the absence of detailed exact solutions, the WKB method provides a tractable entry point into the problem of propagation in a graded-index waveguide.

2.3.3 The Exponential Model

The WKB method outlined in Sec. 2.3.2 gives a useful approximate description of the dispersion relation for guided waves in a graded-index waveguide. The WKB solution for $E(x)$ is of limited usefulness, however, since it is badly behaved at the turning points. There are several ways to model $n_f(x)$ that lead to analytic solutions of the wave equation for the modal fields. Three such distributions (parabolic, $sech^2$, and exponential profiles) were discussed briefly by Kogelnik in Ref. 4. In this section we consider only the exponential profile, as it is widely used to model propagation in planar graded-index waveguides.

The first use of the exponential model in an analysis of propagation in graded-index planar waveguides was given by Conwell [9]. The calculation proceeds by considering the geometry shown in Fig. 2.4. The refractive index distribution is given by

$$n_f^2(x) = n_c^2 \qquad\qquad\qquad x < 0 \qquad\qquad (2.79)$$

$$n_f^2(x) = n_s^2 + \Delta\varepsilon \, \exp\left(\frac{-x}{d}\right) \quad x \geqslant 0 \qquad\qquad (2.80)$$

where, as usual, n_c is the refractive index of the cover medium, n_s the refractive index of the host medium, $n_f^2(x) = n_s^2 + \Delta\varepsilon$ gives the value of the refractive index at the surface ($x = 0$), and $1/d$ gives the falloff rate of the exponential. Limiting our attention to the TE polarization for which $\underline{E} = \hat{y}E_y$, we follow the usual procedure and find solutions to Eq. (2.14) for $x < 0$ and for $x \geqslant 0$. For the latter, we use Eq. (2.80) and write $E_y = F(x)\exp(i\beta z)$ so that

$$\left\{ \frac{d^2}{dx^2} + \left[\frac{n_f^2(x)\omega^2}{c^2} - \beta^2 \right] \right\} F(x) = 0 \qquad\qquad (2.81)$$

A change of variables in which we write $s = \exp(-x/d)$ transforms this equation into

$$\left\{ s^2 \frac{d^2}{ds^2} + s \frac{d}{ds} + \left[\frac{d^2 \Delta\varepsilon\omega^2}{c^2} s + \left(\frac{n_s^2\omega^2}{c^2} - \beta^2 \right) d^2 \right] \right\} F(s) = 0$$

$$(2.82)$$

which is a form of Bessel's equation. By requiring that $F(s)$ be well behaved as $s \to 0$ ($x \to \infty$), we find that

$$F(x) = C_1 J_{2p_0 d}\left(2d \, \frac{\omega}{c} \, \sqrt{\Delta\varepsilon} \, e^{-x/d} \right) \qquad\qquad (2.83)$$

where

$$p_0 = \sqrt{ \beta^2 - \frac{n_s^2\omega^2}{c^2} } \qquad\qquad (2.84)$$

J is a Bessel function of the first kind, and C_1 is a constant.

For $x < 0$, the solution of the wave equation is the familiar decaying exponential. The complete electric field profile, then, is

$$E_y = \begin{cases} C \exp(p_2 x) \exp[i(\beta z - \omega t)] & x \leqslant 0 \\ C_1 J_{2p_0 d}(2d \frac{\omega}{c} \sqrt{\Delta \epsilon}\, e^{-x/d}) \exp[i(\beta z - \omega t) & x \geqslant 0 \end{cases}$$

(2.85)

where C is a constant and

$$p_2 = \sqrt{\beta^2 - \frac{n_c^2 \omega^2}{c^2}}$$

(2.86)

The application of the boundary conditions on the tangential components of $\underline{E}$ and $\underline{H}$ yields the dispersion relation

$$p_2 = -\frac{v}{4d} \frac{J_{2p_0 d-1}(v) - J_{2p_0 d+1}(v)}{J_{2p_0 d}(v)}$$

(2.87)

where $v = 4\pi \sqrt{\Delta \epsilon}\, d/\lambda_0$. It is significant to note that while the exponential model leads to a closed-form solution for both the modal fields and the dispersion relation, the results are nonetheless complicated. Upon examining Eq. (2.87) one finds that the propagation constant appears not only in p_2 but also in the (fractional) order number of the Bessel functions. It is not particularly difficult, however, to solve Eq. (2.87) by numerical techniques and subsequently to construct the electric field profiles according to Eq. (2.85). Conwell's original paper includes an expression for the dispersion relation for TM modes, and a subsequent paper [16] treats the case of an anisotropic planar waveguide and an exponential refractive index profile.

2.4 APPLICATIONS OF COUPLED-MODE THEORY

Sections 2.2 and 2.3 emphasized the fundamental properties of the modes of planar optical waveguides. Device considerations dictate an examination of the way in which these guided modes interact with each other. The most widely used theoretical tool for studying such interactions is known as coupled-mode theory, various aspects of which have been discussed by Marcuse [2], Kogelnik [4], and Yariv [17]. In a typical application, the waveguide structure of Fig. 2.1

is changed (or perturbed) in some way (a surface corrugation, an
applied electric field that leads to a change in the local refractive
index, etc.), resulting in an exchange of energy among the allowed
modes of propagation. As Marcuse [2] points out, there are two
formulations of coupled-mode theory, one based on the modes of the
"ideal" waveguide, the other based on "local" normal modes. The
ideal-mode version of coupled-mode theory is by far the more widely
used of the two and is therefore the one that is described in this
section. It deserves mention, however, that although this formulation is quite successful for TE modes, it can lead to serious errors
for TM modes, particularly when the quantity $n_f^2 - n_c^2$ is large. This
point has been discussed by Streifer et al. [18], Stegeman et al.
[19], and Gruhlke and Hall [20], and is considered in Sec. 2.4.4.
Because of this problem, the detailed discussion below will concentrate on the TE polarization. More sophisticated treatments [19,20]
are required for the TM polarization.

2.4.1 Coupled-Mode Theory (TE)

The waveguide structure shown in Fig. 2.1 can be described in
terms of an x-dependent dielectric constant constant $\varepsilon(x)$ defined as

$$\varepsilon(x) = \begin{cases} n_c^2 \varepsilon_0 & x > h \\ n_f^2 \varepsilon_0 & 0 \leqslant x \leqslant h \\ n_s^2 \varepsilon_0 & x < 0 \end{cases} \tag{2.88}$$

The modes of this structure emerge from a solution of the wave
equation

$$\nabla^2 \underline{E} + \mu \varepsilon(x) \omega^2 \underline{E} = 0 \tag{2.89}$$

as discussed in Secs. 2.2 and 2.3. If the basic structure of Eq.
(2.88) is changed in some small way, the effect of the perturbation
can be introduced through the quantity $\Delta \varepsilon$, which may depend on
coordinates, so that Eq. (2.89) becomes

$$\nabla^2 \underline{E} + \mu (\varepsilon + \Delta \varepsilon) \omega^2 \underline{E} = 0 \tag{2.90}$$

It is common to rearrange this new wave equation, define the polarization vector $\underline{P}$ as

$$\underline{P} = (\Delta\varepsilon)\underline{E} \tag{2.91}$$

and write

$$\nabla^2\underline{E} + \mu\varepsilon\omega^2\underline{E} = -\mu\omega^2\underline{P} \tag{2.92}$$

which is the starting point for the derivation of the couple-mode equations.

Coupled-mode leads to relatively simple formulas only for the case in which only two modes are of any significance. Fortunately, a number of important applications can be treated this way. The term "mode" that appears here is being used in a very general sense. For example, in the case of a waveguide diffraction grating, a portion of the incident energy is reflected while the remainder is transmitted through the grating. In spite of the fact that both the reflected and transmitted waves might carry energy in the same TE_0 mode of the waveguide, one commonly refers to them as two separate modes: the reflected mode and the transmitted mode. In the ideal-mode version of coupled-mode theory, the solution of Eq. (2.92) is expressed in terms of the solutions of Eq. (2.89) in the following way:

$$\underline{E} = a_1(y,z)\underline{E}_1(x) + a_2(y,z)\underline{E}_2(x) \tag{2.93}$$

where $a_1(y,z)$ and $a_2(y,z)$ are amplitudes to be determined, and $\underline{E}_1(x)\exp(\pm i\underline{\beta}_1 \cdot \underline{R})$ and $\underline{E}_2(x)\exp(\pm i\underline{\beta}_2 \cdot \underline{R})$ are solutions of Eq. (2.89), where $\underline{R} = y\hat{y} + z\hat{z}$ is a vector parallel to the plane of the waveguide. It is well known [2,4] that the components of $\underline{E}_1(x)$ and $\underline{E}_2(x)$ that are transverse to the direction of propagation satisfy an orthogonality relation. One can use this orthogonal set of functions as a basis for an expansion of $\underline{E}$, and it is in this spirit that one writes Eq. (2.93). It is assumed that only two modes have appreciable amplitudes, however, so the sum over all modes collapses to only two terms. For TE modes, the field $\underline{E}(x)$ is completely transverse to $\underline{\beta}$, so the expansion in Eq. (2.93) can be written in terms of the entire vector $\underline{E}(x)$. TM modes must be treated differently, however, since there are problems with the TM formalism, as will be discussed in Sec. 2.4.4. The remainder of this section, together with Secs. 2.4.2 and 2.4.3 deal exclusively with TE modes.

The notation $\underline{E}(x)$ is a shorthand notation for the functions in Eqs. (2.35)–(2.37):

$$\underline{E}(x) = \begin{cases} E_c \exp[-\gamma_c(x - h)] & x \geqslant h \tag{2.94} \\[2mm] E_f \cos(k_f x - \phi_s) & 0 \leqslant x \leqslant h \tag{2.95} \\[2mm] E_s \exp(\gamma_s x) & x \leqslant 0 \tag{2.96} \end{cases}$$

For TE modes, the orthogonality condition can be put in the form

$$\frac{1}{2} \int_{-\infty}^{\infty} \underline{E}_1(x) \cdot \underline{E}_2^* \, dx = \frac{\mu\omega}{\beta_1} \delta_{12} \tag{2.97}$$

where δ_{12} is the Kronecker delta and the factor of $1/2$ originates from Eq. (2.17). $\underline{E}_1(x)$ satisfies the equation

$$\left[\frac{\partial^2}{\partial x^2} + \mu\varepsilon(x)\omega^2\right] \underline{E}_1(x) = \beta_1^2 \underline{E}_1(x) \tag{2.98}$$

and similarly for $\underline{E}_2(x)$. After one substitutes Eq. (2.93) into Eq. (2.92), and uses Eq. (2.98), one obtains

$$\left(\frac{\partial^2}{\partial y^2} + \frac{\partial^2}{\partial z^2} + \beta_1^2\right) a_1(y,z)\underline{E}_1(x)$$

$$+ \left(\frac{\partial^2}{\partial y^2} + \frac{\partial^2}{\partial z^2} + \beta_2^2\right) a_2(y,z)\underline{E}_2(x) = -\mu\omega^2 \underline{P} \tag{2.99}$$

After multiplying Eq. (2.99) by $\underline{E}_1^*$ and integrating the result over x, the orthogonality relation in Eq. (2.97) leads to

$$\left(\frac{\partial^2}{\partial y^2} + \frac{\partial^2}{\partial z^2} + \beta_1^2\right) a_1(y,z) = -\frac{\beta_1\omega}{2} \int_{-\infty}^{\infty} \underline{P} \cdot \underline{E}_1^* \, dx \tag{2.100}$$

In a similar way, the analogous equation for $a_2(y,z)$ is

$$\left(\frac{\partial^2}{\partial y^2} + \frac{\partial^2}{\partial z^2} + \beta_2^2\right) a_2(y,z) = -\frac{\beta_2\omega}{2} \int_{-\infty}^{\infty} \underline{P} \cdot \underline{E}_2^* \, dx \tag{2.101}$$

Equations (2.100) and (2.101) form a pair of coupled second-order differential equations. They are coupled because $\underline{P} = (\Delta\varepsilon)\underline{E}$ contains both $a_1(y,z)$ and $a_2(y,z)$. The equations can be simplified to a pair of coupled first-order differential equations by introducing the slowly varying envelope approximation, as shown in the sections that follow.

2.4.2 Waveguide Diffraction Gratings (TE)

The waveguide diffraction grating is perhaps the most versatile of the integrated optical components demonstrated to date. Gratings have been used as input-output couplers [21-23], beam splitters [24,25], reflection filters [26,27], polarization converters [28], and dispersive elements for wavelength-division demultiplexing [29-31]. There are two principal types of waveguide gratings: surface corrugations and index gratings. The former can be fabricated by recording an interference pattern in a layer of photoresist deposited on top of the waveguide. The photoresist pattern is then used as a mask for ion milling or reactive ion etching. The latter can be prepared by electron-beam writing in a suitable material. Nishihara et al., for example, used a scanning electron microscope to write lines of elevated refractive index in As_2S_3 planar waveguides [32]. The discussion that follows will concentrate on surface corrugations.

Nearly all of the theoretical treatments of the interaction between guided modes and waveguide gratings consider the case in which the guided modes propagate in a direction perpendicular to the grating lines. This activity has been motivated by the use of such a geometry in distributed feedback semiconductor lasers [33]. The few analyses of the case of nonnormal incidence to the grating lines that have appeared in the literature have used a variety of methods and notations, and the results do not always agree with each other [19, 34-38]. The paragraphs that follow develop the solution to the nonnormal incidence problem within the framework of coupled-mode theory. The treatment considers only the case in which the incident and diffracted waves are bound modes of the waveguide; radiation modes are not considered.

Figure 2.5 shows the basic geometry for the analysis. The grating "lines" have a period Λ, are oriented parallel to the $\hat{y}$ axis, and are taken to be of essentially infinite extent along $\hat{y}$. β_1 and β_2 are the propagation vectors of the incident and reflected waves, respectively. The lines shown in Fig. 2.5 merely represent a grating that is actually a surface corrugation. Figure 2.6 shows a side view

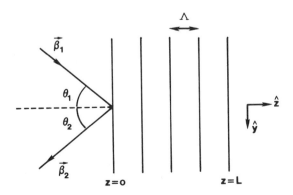

FIG. 2.5 Waveguide grating geometry.

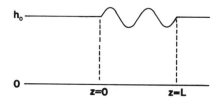

FIG. 2.6 Side view of corrugated waveguide.

of the waveguide structure. The upper surface is corrugated between $z = 0$ and $z = L$, where the waveguide thickness h is taken
to be of the form

$$h = h_o + (\Delta h) \cos K_o z \qquad (2.102)$$

with h_O the mean thickness, Δh the strength of the corrugation, and
$K_O = 2\pi/\Lambda$ the grating constant.

The amplitudes $a_1(y,z)$ and $a_2(y,z)$ are related through Eqs.
(2.100) and (2.101). These amplitudes, however, are rapidly varying functions of y and z, and it is convenient to write them in terms
of the slowly varying amplitudes $A_1(z)$ and $A_2(z)$:

$$a_1(y,z) = A_1(z) \exp[i(\beta_{1y} y + \beta_{1z} z)] \qquad (2.103)$$

$$a_2(y,z) = A_2(z) \exp[i(\beta_{2y} y - \beta_{2z} z)] \qquad (2.104)$$

A_1 and A_2 can be considered to be functions of z only since both
the grating rulings and the waveguide are assumed to be essentially
infinite in extent along $\hat{y}$. After inserting Eqs. (2.103) and (2.104)
into Eqs. (2.100) and (2.101), the coupled-mode equations take on
the form

$$\left(2i\beta_{1z} \frac{dA_1}{dz} + \frac{d^2 A_1}{dz^2}\right) \exp[i(\beta_{1y} y + \beta_{1z} z)] = -\frac{\beta_1 \omega}{2} \int_{-\infty}^{\infty} \underline{P} \cdot \underline{E}_1^* \, dx$$

$$(2.105)$$

$$\left(2i\beta_{2z} \frac{dA_2}{dz} + \frac{d^2 A_2}{dz^2}\right) \exp[i(\beta_{2y} y - \beta_{2z} z)] = \frac{\beta_2 \omega}{2} \int_{-\infty}^{\infty} \underline{P} \cdot \underline{E}_2^* \, dx$$

$$(2.106)$$

Since the amplitudes A_1 and A_2 vary slowly with z, we make the often used slowly varying envelope approximation and neglect the second derivatives in Eqs. (2.105) and (2.106):

$$\frac{dA_1}{dz} = i \frac{\omega}{4} \frac{\beta_1}{\beta_{1z}} \int_{-\infty}^{\infty} \underline{P} \cdot \underline{E}_1^* \, dx \, \exp[-i(\beta_{1y}y + \beta_{1z}z)] \qquad (2.107)$$

$$\frac{dA_2}{dz} = -i \frac{\omega}{4} \frac{\beta_2}{\beta_{2z}} \int_{-\infty}^{\infty} \underline{P} \cdot \underline{E}_2^* \, dx \, \exp[-i(\beta_{2y}y - \beta_{2z}z)] \qquad (2.108)$$

Equations (2.107) and (2.108) reduce to those obtained for normal incidence by Marcuse [2] and Kogelnik [4], who use more formal methods that do not require the use of the slowly varying envelope approximation. In fact, Sipe and Stegeman [39] have criticized the use of the slowly varying envelope approximation. They claim that Eqs. (2.107) and (2.108) are exact but have been derived incorrectly. We have chosen the present route for this discussion because of its straightforward nature, but it is clear that the results do not depend on the explicit use of the slowly varying envelope approximation. The interested reader should consult Ref. 39 for more details concerning the subtleties of the derivation.

A careful study of the derivation of Eqs. (2.107) and (2.108) reveals that the results hold only if $\underline{E}_1$ and $\underline{E}_2$ satisfy the orthogonality relation in Eq. (2.97). This means that $\underline{E}_1$ and $\underline{E}_2$ must correspond to different spatial modes of the waveguide. The most common applications in integrated optics require that both the incident and reflected waves correspond to the *same* spatial mode of the waveguide. That Eqs. (2.107) and (2.108) do not apply to this case is merely an artifact of the simplified derivation used in this treatment. More sophisticated methods show that Eqs. (2.107) and (2.108) are also valid when $\underline{E}_1$ and $\underline{E}_2$ represent the same spatial mode. For the remainder of this chapter we assume that Eqs. (2.107) and (2.108) are completely general.

The right-hand sides of Eqs. (2.107) and (2.108) can be simplified considerably by recognizing that the grating acts as an efficient reflecting element only if the Bragg condition is at least approximately satisfied. In this case, the first-order Bragg condition requires that

$$\beta_{1y} = \beta_{2y} \qquad (2.109)$$

$$\beta_{1z} + \beta_{2z} = 2\delta + K_o \qquad (2.110)$$

where δ is a small parameter. If $\delta = 0$, Eqs. (2.109) and (2.110) state the exact Bragg condition. Inserting δ in Eq. (2.110) allows us to examine how the grating reflectivity decreases as either the grating spacing or the incident wavelength is detuned from the Bragg condition. The components of $\underline{\beta}_1$ and $\underline{\beta}_2$ can be written in terms of the angles θ_1 and θ_2 as follows: $\beta_{1y} = \beta_1 \sin \theta_1$, $\beta_{2y} = \beta_2 \sin \theta_2$, $\beta_{1z} = \beta_1 \cos \theta_1$, and $\beta_{2z} = \beta_2 \cos \theta_2$. After using Eqs. (2.91), (2.93), (2.103), and (2.104) to form $\underline{P}$, and invoking the Bragg condition, Eqs. (2.107) and (2.108) take on the form

$$\frac{dA_1}{dz} = iK_{11}A_1(z) + iK_{21}A_2(z) \, \exp[-i(2\delta + K_o)z] \qquad (2.111)$$

$$\frac{dA_2}{dz} = -K_{12}A_1(z) \, \exp[i(2\delta + K_o)z] - iK_{22}A_2(z) \qquad (2.112)$$

where

$$K_{11} = \frac{\omega}{4} \frac{\int_{-\infty}^{\infty} (\Delta\epsilon)\underline{E}_1 \cdot \underline{E}_1^* \, dx}{\cos \theta_1} \qquad (2.113)$$

$$K_{22} = \frac{\omega}{4} \frac{\int_{-\infty}^{\infty} (\Delta\epsilon)\underline{E}_2 \cdot \underline{E}_2^* \, dx}{\cos \theta_2} \qquad (2.114)$$

$$K_{12} = \frac{\omega}{4} \frac{\int_{-\infty}^{\infty} (\Delta\epsilon)\underline{E}_1 \cdot \underline{E}_2^* \, dx}{\cos \theta_2} \qquad (2.115)$$

$$K_{21} = \frac{\omega}{4} \frac{\int_{-\infty}^{\infty} (\Delta\epsilon)\underline{E}_2 \cdot \underline{E}_1^* \, dx}{\cos \theta_1} \qquad (2.116)$$

These four coupling coefficients can be evaluated in a straightforward manner as shown below.

It is the goal of the ideal-mode version of coupled-mode theory to express the properties of the perturbed waveguide in terms of those of the unperturbed structure. For the grating geometry of Fig. 2.6, the unperturbed waveguide is taken to be a planar waveguide of thickness h_o and dielectric constant $\epsilon(x)$ as given in Eq. (2.88) with h replaced by h_o. The perturbation is given by Eq. (2.102). The modal fields designated $\underline{E}_1$ and $\underline{E}_2$ are those of the unperturbed waveguide and their properties were discussed earlier. To evaluate the integrals in the coupling coefficients, we assume that the waveguide supports only a single TE_0 spatial mode, so that $\beta_1 = \beta_2$, for which the Bragg condition demands that $\theta_1 = \theta_2$. In a region in which the actual waveguide surface bends outward ($h > h_0$) from that of the unperturbed waveguide, $\Delta \varepsilon$ is given by

$$\Delta \varepsilon = \begin{cases} \varepsilon_0 (n_f^2 - n_c^2) & h_0 \leqslant x \leqslant h \\ 0 & \text{otherwise} \end{cases} \qquad (2.117)$$

In a region for which the actual surface bends inward ($h < h_0$), $\Delta \varepsilon$ is given by

$$\Delta \varepsilon = \begin{cases} \varepsilon_0 (n_c^2 - n_f^2) & h \leqslant x \leqslant h_0 \\ 0 & \text{otherwise} \end{cases} \qquad (2.118)$$

If we take Δh to be small compared to h_0, the fields $\underline{E}_1$ and $\underline{E}_2$ do not depart significantly from their values at $x = h_0$, so we can evaluate them at $x = h_0$ and remove them from each integral. With these simplifications and assumptions, the coupling coefficients become

$$K_{11} = K_{22} = \frac{\omega}{4} \frac{E_c^2 \varepsilon_0 (n_f^2 - n_c^2)(\Delta h) \cos K_0 z}{\cos \theta_1} \qquad (2.119)$$

$$K_{12} = K_{21} = \frac{\omega}{4} \frac{E_c^2 \varepsilon_0 (n_f^2 - n_c^2)(\Delta h) \cos(K_0 z) \cos 2\theta_1}{\cos \theta_1} \qquad (2.120)$$

where we have used the facts that $E_c = E_1(h_0)$ and that the sign difference between Eqs. (2.117) and (2.118) is compensated for by the order of the limits of integration.

It is clear that the coupling coefficients are periodic functions of z, with period Λ. If we were to average Eq. (2.112), for example, over the grating period Λ and make use of the fact that $A_2(z)$ does not vary appreciably over that length, we would find that the second term on the right-hand side is nearly zero. However, because the first term contains a factor of $\exp(iK_0 z)$, it does not necessarily average to zero. We will return to this point later, but it is important here to realize that the right-hand sides of Eqs. (2.111) and (2.112) must have nearly zero phase in order to drive a significant reflected wave. Any term on the right-hand side of either equation that does not have essentially zero phase can be neglected. This is the so-called synchronous approximation, and it leads to the neglect of the terms containing K_{11} and K_{22}. The two remaining terms involve $\cos K_0 z$, which is itself a sum of two exponentials. The synchronous approximation again demands that we retain only that part of cos $K_o z$ that leaves the phase of the right-hand sides of Eqs. (2.111) and (2.112) nearly zero. The resulting coupled-mode equations are

$$\frac{dA_1}{dz} = i\kappa A_2(z)e^{-i2\delta z} \qquad\qquad (2.121)$$

$$\frac{dA_2}{dz} = -i\kappa A_1(z)e^{i2\delta z} \qquad\qquad (2.122)$$

where

$$\kappa = \frac{\omega}{8}\frac{E_c^2\varepsilon_0(n_f^2 - n_c^2)(\Delta h)\cos 2\theta_1}{\cos \theta_1} \qquad\qquad (2.123)$$

The orthonormality condition in Eq. (2.97) normalizes the power per unit width carried by each mode to unity (1 W/m). This in turn fixes the value of E_f through Eq. (2.52) and E_c through Eq. (2.48) such that

$$E_c^2 = \frac{4\sqrt{\mu_0/\varepsilon_0}\,(Nh_{eff})^{-1}(n_f^2 - N^2)}{n_f^2 - n_c^2} \qquad\qquad (2.124)$$

After some algebra, one can show that the coupling coefficient κ is given by

$$\kappa = \frac{\pi}{\lambda}\frac{\Delta h}{h_{eff}}\frac{n_f^2 - N^2}{N}\frac{\cos 2\theta_1}{\cos \theta_1} \qquad\qquad (2.125)$$

which reduces to that given by Kogelnik [4] for normal incidence ($\theta_1 = 0$).

With the coupling coefficient κ expressed solely in terms of experimental parameters, the only task that remains is to solve Eqs. (2.121) and (2.122) for the amplitudes $A_1(z)$ and $A_2(z)$. This is a very straightforward task that has been discussed elsewhere [4] and will only be outlined here. First, we replace A_1 and A_2 by R_1 and S_1 through the substitutions $A_1 = R_1 \exp(-i\,\delta z)$ and $A_2 = S_1 \exp(i\,\delta z)$. Second, we write $R_1 = F_1 \sinh(\alpha z) + F_2 \cosh(\alpha z)$ and $S_1 = G_1 \cosh(\alpha z) + G_2 \sinh(\alpha z)$ and require $R_1(z = 0) = 1$ and $S_1(z = L) = 0$, for a grating of length L. Quite a bit of algebra leads to the results

$$A_1(z) = \frac{\alpha \cosh[\alpha(L - z)] - i\delta \sinh[\alpha(L - z)]}{\alpha \cosh \alpha L - i\delta \sinh \alpha L} e^{-i\delta z} \qquad (2.126)$$

$$A_2(z) = \frac{i\kappa \sinh[\alpha(L - z)]}{\alpha \cosh \alpha L - i\delta \sinh \alpha L} e^{i\delta z} \qquad (2.127)$$

where $\alpha^2 = \kappa^2 - \delta^2$. The reflectivity $R = |A_2(z = 0)/A_1(z = 0)|^2$ is given by

$$R = \frac{\sinh^2(\kappa L \sqrt{1 - \delta^2/\kappa^2})}{\cosh^2(\kappa L \sqrt{1 - \delta^2/\kappa^2}) - \delta^2/\kappa^2} \qquad (2.128)$$

For perfect Bragg matching ($\delta = 0$), Eq. (2.128) specializes to

$$R = \tanh^2 \kappa L \qquad (2.129)$$

The coupling coefficient κ is a function of all the waveguide parameters, the corrugation depth Δh, and the angle of incidence θ_1. For a given wavelength λ_0 and a given set of materials, there is an optimum value for the film thickness h_0 that maximizes κ and hence R. Figure 2.7 shows a plot of K' versus h_0, where

$$\kappa = \frac{K'(\Delta h) \cos 2\theta_1}{\cos \theta_1} \qquad (2.130)$$

defines K' for the parameters $n_c = 1.0$, $n_f = 1.56$, $n_s = 1.47$, and $\lambda_0 = 0.83$ μm. For this example, the coupling coefficient exhibits a distinct maximum near $h_0 = 0.4$ μm. Figure 2.8 shows the behavior of the reflectivity R as a function of h_0 for the parameters used in Fig. 2.7 along with $\delta = 0$, $\Delta h = 250$ Å, $\theta_1 = 30°$, and $L = 422$ μm. For sufficiently large values of κL, $R = \tanh^2 \kappa L$ is relatively flat, so the maximum in Fig. 2.8 is much broader than that in Fig. 2.7. The variation of R with detuning δ is given by Eq. (2.128) and is illustrated in Fig. 2.9 for $\kappa L = 3$ and $\kappa L = 1$. Both curves show a central lobe and sidelobes; only the first three sidelobes for $\kappa L = 3$ and one sidelobe for $\kappa L = 1$ appear in the figure. In general, an increase in the grating reflectivity at $\delta = 0$ is accompanied by a narrowing of the central lobe. For the parameters used in Fig. 2.8, the first zero in the curve labeled $\kappa L = 3$ corresponds to a 4-Å detuning of the optical wavelength from that which satisfies the Bragg condition ($\delta = 0$). It is this narrowband characteristic of the grating reflection that justifies the neglect of nonsynchronous terms on the right-hand sides of Eqs. (2.111) and (2.112). Such terms would correspond to large detuning from the

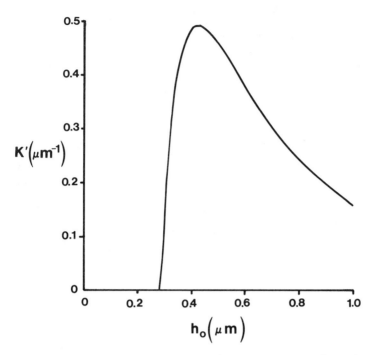

FIG. 2.7 Plot of K' versus h_o for parameters given in the text.

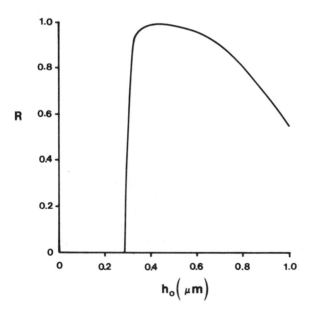

FIG. 2.8 Plot of the reflectivity R as a function of h_o for same parameters used in Fig. 2.7 and $\delta = 0$, $\theta_1 = 30°$, $\Delta h = 250$ Å, and $L = 422$ μm.

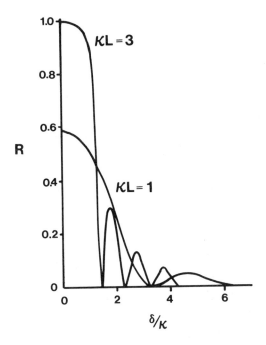

FIG. 2.9 Grating reflectivity R as a function of δ/κ for $\kappa L = 1$ and $\kappa L = 3$.

Bragg condition and would make a negligible contribution to the grating reflectivity.

2.4.3 Coupled Waveguides (TE)

Several devices of importance for applications of integrated optics make use of what is often called evanescent coupling between modes. When two waveguides are placed in close proximity, the exponential (or evanescent) tails of the waveguide modes overlap, allowing a type of optical tunneling to transfer energy between the two structures. The interaction is actually another manifestation of the familiar phenomenon of frustrated total reflection. Figure 2.10 shows the geometry for the simple example that this section considers. Two planar waveguides are oriented parallel to each other and separated by a distance s. The lower (upper) waveguide has thickness $h_1(h_2)$ and refractive index $n_1(n_2)$, and the surrounding medium has refractive index n_c. For sufficiently small values of s, light injected into the lower waveguide at z = 0 will oscillate back and forth

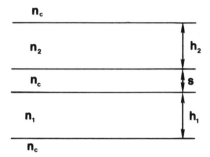

FIG. 2.10 Parallel slab waveguides separated by a distance s.

between the two waveguides, with complete transfer occurring for $n_1 = n_2$ and $h_1 = h_2$. The details of this periodic (in z) coupling between the two waveguides have been studied by a number of authors [40,41], and two review articles describe the history and performance characteristics of devices based on this coupling [42,43]. Our objective in this section is to analyze, using coupled-mode theory, the simple geometry in Fig. 2.10 and develop the standard formulas that describe what has come to be called the directional coupler switch.

If $A_1(z)$ represents the amplitude of the mode with propagation constant β_1 supported by the lower waveguide, and $A_2(z)$ represents the amplitude of the mode with propagation constant β_2 supported by the upper waveguide, the coupling of these two amplitudes is given by

$$\frac{dA_1}{dz} = i\kappa A_2 e^{-i2\delta z} \tag{2.131}$$

$$\frac{dA_2}{dz} = i\kappa A_1 e^{i2\delta z} \tag{2.132}$$

where δ is, again, a detuning parameter given by

$$2\delta = \beta_1 - \beta_2 \tag{2.133}$$

and κ is the coupling coefficient given by

$$\kappa = \frac{\omega}{4} \int_{-\infty}^{\infty} (\Delta\varepsilon)\, \underline{E}_1 \cdot \underline{E}_2^* \, dx \tag{2.134}$$

where the subscripts 1 and 2 identify the lower and upper wave-guides, respectively. Equations (2.131) and (2.132) are often called the codirectional coupled-mode equations since β_1 and β_2 are both directed along +z. They should be compared to Eqs. (2.121) and (2.122), the so-called contradirectional coupled-mode equations. Note that they differ in form only by a minus sign in the second equation of each pair.

The coupling coefficient can be evaluated in much the same manner as that in Sec. 2.4.2. From the point of view of waveguide 2, it is the presence of waveguide 1 that serves as the perturbation, so

$$\Delta\epsilon = \begin{cases} \epsilon_0(n_1^2 - n_c^2) & 0 \leqslant x \leqslant h_1 \\ 0 & \text{otherwise} \end{cases} \tag{2.135}$$

The coupling coefficient becomes

$$\kappa = (n_1^2 - n_c^2) \frac{\epsilon_0 \omega}{4} \int_0^{h_1} \underline{E}_1 \cdot \underline{E}_2^* \, dx \tag{2.136}$$

When $\delta = 0$, the two waveguides are identical, and as we will show shortly, complete energy transfer is possible. For identical wave-guides ($h_1 = h_2$, $n_1 = n_2$), we insert

$$\underline{E}_1 = \hat{y} E_f \cos(k_f x - \phi_c) \tag{2.137}$$

$$\underline{E}_2 = \hat{y} E_c \exp(x - h_1 - s) \tag{2.138}$$

into Eq. (2.136) and integrate to obtain

$$\kappa = \frac{n_1^2 - N^2}{N h_{\text{eff}}} \sin(k_f h_1) e^{-\gamma_c s} \tag{2.139}$$

where we have used Eq. (2.48) and the normalization condition (P = 1) in Eq. (2.52). The coupling coefficient displays an exponential dependence on s, as expected, and is a constant. The coupled-mode equations themselves can be solved by the same method as in Sec. 2.4.2 by requiring $A_1(0) = 1$ and $A_2(0) = 0$, which just means that all the power is initially in the lower waveguide. The solutions are

$$A_1 = \left[\cos\left(\sqrt{\kappa^2 + \delta^2}\, z\right) + \frac{i\delta}{\sqrt{\kappa^2 + \delta^2}} \sin\left(\sqrt{\kappa^2 + \delta^2}\, z\right) \right] e^{-i\delta z}$$

$$(2.140)$$

$$A_2 = \frac{i\kappa \sin\left(\sqrt{\kappa^2 + \delta^2}\, z\right)}{\sqrt{\kappa^2 + \delta^2}}\, e^{i\delta z} \qquad\qquad (2.141)$$

The power (per unit width) is proportional to $|A|^2$, so

$$\frac{P_1(z)}{P_1(0)} = \cos^2\left(\sqrt{\kappa^2 + \delta^2}\, z\right) + \frac{\delta^2}{\delta^2 + \kappa^2} \sin^2\left(\sqrt{\kappa^2 + \delta^2}\, z\right)$$

$$(2.142)$$

$$\frac{P_2(z)}{P_1(0)} = \frac{\kappa^2}{\kappa^2 + \delta^2} \sin^2\left(\sqrt{\kappa^2 + \delta^2}\, z\right) \qquad\qquad (2.143)$$

where $P_1(z)$ is the power per unit width in the lower waveguide, $P_2(z)$ is that in the upper waveguide, and $P_1(0)$ is that originally injected into the lower waveguide.

The nature of the solution is most clearly revealed for the case of identical waveguides, $\delta = 0$, for which

$$\frac{P_1(z)}{P_1(0)} = \cos^2 \kappa z \qquad\qquad (2.144)$$

$$\frac{P_2(z)}{P_1(0)} = \sin^2 \kappa z \qquad\qquad (2.145)$$

At $z = 0$, as required by our choice of initial conditions, all of the power is in the lower waveguide. For $z > 0$, the power oscillates back and forth (periodically) between the waveguides. In particular, complete transfer from waveguide 1 to waveguide 2 occurs in a distance

$$L = \frac{\pi}{2\kappa} \qquad\qquad (2.146)$$

known as the transfer length. As δ increases from zero, the maximum possible power that can be transferred decreases as $\kappa^2/(\kappa^2 + \delta^2)$ and the transfer length decreases. If a device is fabricated with a length L that has been chosen according to Eq. (2.146), and if δ can then be varied by means of a control signal, the choice

$$\sqrt{\kappa^2 + \delta^2} \, L = \pi \tag{2.147}$$

selects the value of δ for which $P_2(z)/P_1(0) = 0$. Since L satisfies Eq. (2.146), Eq. (2.147) requires that

$$\delta = \sqrt{3} \, \kappa \tag{2.148}$$

Switching δ between 0 and $\sqrt{3} \, \kappa$ changes the output of the lower waveguide between $P_1(0)$ and zero. If the waveguides are formed in an electro-optic material, δ can be varied by an applied voltage. This is the basic concept behind the directional-coupler switch.

It is very important to note that complete energy transfer from the lower to upper waveguide in Fig. 2.10 occurs only for $\delta = 0$. If the fabrication process introduces differences between the two, δ will exceed zero, and only a fraction of $P_1(0)$ will transfer. We can investigate how sensitive the process is to δ by estimating how large of a difference in the effective indices N_1 and N_2 is required to reduce the maximum coupled power from $P_1(0)$ to 90% of that value. Typical devices have a length L = 0.5 cm, which, from Eq. (2.146) means that $\kappa \cong 3 \times 10^{-4} \ \mu m^{-1}$. For $P_2(z)/P_1(0) = 0.9$, we find that $\delta = \kappa/3$. It follows from Eq. (2.133) that $2\delta = (\omega/c)(N_1 - N_2) \equiv (\omega/c)(\Delta N)$, where c is the speed of light in vacuum. Equating the two expressions for δ gives

$$\Delta N = \frac{2}{3} \left(\frac{\kappa}{k_o} \right) \tag{2.149}$$

for 90% energy transfer, where $k_o = \omega/c$. Using $\kappa = 3 \times 10^{-4} \ \mu m^{-1}$ and a wavelength $\lambda = 0.84 \ \mu m$, we find that $\Delta N \cong 2.7 \times 10^{-5}$, so the waveguides must be very similar for efficient energy transfer.

The example presented above is only an illustrative one. Actual devices are based on parallel "channel" waveguides formed in either $LiNbO_3$ or GaAs crystals. In the case of $LiNbO_3$, the channels are formed by diffusing two parallel strips of titanium into the surface of the crystal. The result consists of two parallel graded-index waveguides, each of which is a two-dimensional waveguide, the modes of which can be known only in an approximate way. Nevertheless, the major qualitative features of the simple example of this section are preserved. The advantage of using $LiNbO_3$ (or GaAs) as a host crystal rests in the relatively large electro-optic effect exhibited by that crystal. By depositing surface electrodes near the diffused channels, the fringing fields generated by a control voltage can be used to vary the refractive index in one or both channels. By using a segmented-electrode arrangement, it is actually possible to overcome the obstacle imposed by the requirement that the waveguides be identical. The result is a device in which both the straight-through and crossover states can be selected by the control signal. Such devices are said to use the "alternating $\Delta\beta$" technique [42,43]. Additional details about these switches appear in Chapter 8.

2.4.4 Coupled-Mode Theory (TM)

In Secs. 2.4.2 and 2.4.3 we examined the performance of waveguide diffraction gratings and coupled waveguides for the case of TE mode propagation. Both examples showed the importance of the coupling coefficient κ, which contains the essential details needed to design an integrated optical component. The ideal-mode version of coupled-mode theory contains a basic error for the case of the TM polarization when the perturbation $\Delta\varepsilon$ consists of a surface deformation such as a grating. This error can lead to incorrect results for the coupling coefficient κ. The existence of this difficulty was first pointed out by Streifer et al. [18] in connection with an investigation of the reflectivity of waveguide gratings. Stegeman et al. [19] subsequently showed that this problem arises because the basic ideal-mode expansion of the electric and magnetic fields violates the boundary conditions. It is not our intention in this section to describe the theoretical aspects of coupled-mode theory for TM modes; instead, we limit ourselves to a discussion of the various results that appear in the literature for the TM-mode coupling coefficient for waveguide gratings. We first present the formulas for κ given in the literature in a common notation and then present plots of both κ and the reflectivity R for a particular sample geometry as an illustration of the size of the disagreement among the several results.

Wagatsuma et al. [35], Van Roey and Lagasse [37], and Stegeman et al. [19] have each presented an analysis of the TM_0-TM_0 grating reflection problem for an arbitrary angle of incidence (see Fig. 2.5). According to Wagatsuma et al., the coupling coefficient κ_W is given by

$$\kappa_W = \frac{\pi}{\lambda} \frac{\Delta h}{h_{eff}} \frac{n_f^2 - N^2}{Nq_c \cos\theta_1} \left[\frac{N^2}{n_c^2} \left(1 + \frac{n_c^4}{n_f^4}\right) - \left(\frac{N^2}{n_c^2} - 1\right) 2\cos 2\theta_1 \right]$$

$$(2.150)$$

where

$$q_c = \frac{N^2}{n_f^2} + \frac{N^2}{n_c^2} - 1$$

$$(2.151)$$

and the subscript W that appears on κ refers to the first author's last name. Equation (2.150) should be compared to the analogous result for TE reflection given in Eq. (2.125). Whereas the TE result reduces to that given by Kogelnik [4] for $\theta_1 = 0$, Eq. (2.150) does not so reduce. Kogelnik's TM result κ_K is given by

$$\kappa_K = \frac{\pi}{\lambda} \frac{\Delta h}{h_{eff}} \frac{n_f^2 - N^2}{N} p_c \tag{2.152}$$

where

$$p_c = \frac{1}{2q_c} \left(\frac{n_f^2}{n_c^2} + \frac{n_c^2}{n_f^2} \right) \left(\frac{N^2}{n_f^2} - \frac{N^2}{n_c^2} + 1 \right) \tag{2.153}$$

The second result for arbitrary angle of incidence is that of Van Roey and Lagasse [37]:

$$\kappa_V = \kappa_W \cos \theta_1 \tag{2.154}$$

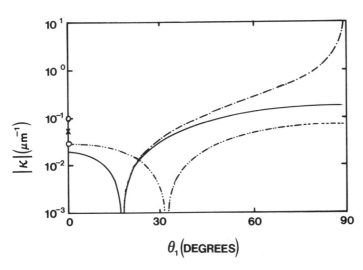

FIG. 2.11 Plots of $|\kappa|$ as a function of θ_1 for all theoretical descriptions of TM_0–TM_0 grating reflection. The results are those of Wagatsuma et al. (–·–·–) [35], Van Roey et al. (———) [37], Stegeman et al. (–··–··) [19], Marcuse (●) [2], Kogelnik (x) [4], and Streifer et al. (○) [18]. The parameters used to construct the plots are given in the text.

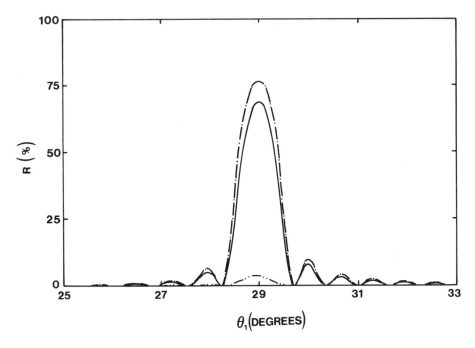

FIG. 2.12 Angular dependence of the grating reflectivity for the theories of Wagatsuma et al. ($-\cdot-\cdot-\cdot$), Van Roey et al. (———), and Stegeman et al. ($-\cdot\cdot-\cdot\cdot$) for the same parameters used for Fig. 2.11.

It is clear that $\kappa_V = \kappa_W$ for $\theta_1 = 0$. Stegeman et al. obtain the expression

$$\kappa_s = \frac{\pi}{\lambda} \frac{\Delta h}{h_{eff}} \frac{n_f^2 - N^2}{N} \left[\frac{N^2 - (n_f^2/n_c^2)(N^2 - n_c^2) \cos 2\theta_1}{(N^2/n_c^2)(n_f^2 + n_c^2) - n_f^2} \right] \quad (2.155)$$

In the limit $\theta_1 = 0$, Eq. (2.155) reduces to that derived by Marcuse using the so-called local normal-mode approach to coupled-mode theory [2]. A final result that appears in the literature is that given by Streifer for normal incidence ($\theta_1 = 0$):

$$\kappa_{ST} = \frac{\pi}{\lambda} \frac{\Delta h}{h_{eff}} \frac{n_f^2 - N^2}{N} \frac{n_f^2(n_c^2 + n_f^2)(1 - N^2) + n_c^4}{N^2(n_f^2 + n_c^2) - n_f^2 n_c^2(n_c^2 - n_f^2)} \quad (2.156)$$

All the formulas given above are similar in form. To gain some insight into the importance of the differences in the various results, Fig. 2.11 shows a plot of $|\kappa|$ as a function of the angle of incidence θ_1 for the following parameters: $n_c = 1.0$, $n_f = 1.85$, $n_s = 1.5$, $h_o = 0.475$ µm, $\Delta h = 0.05$ µm, Λ = grating period = 0.274 µm, L = 41 µm, and $\lambda = 0.83$ µm [44]. Curves are shown for each of the three theories that apply to the case of arbitrary angle of incidence. Results obtained from the three normal incidence theories are shown as symbols on the vertical axis. It is clear that there are major differences among the theories. Figure 2.12 shows a plot of the reflectivity R as a function of θ_1 for the three nonnormal incidence theories. The plots were constructed by using the same parameters as in Fig. 2.11 together with Eq. (2.128). Since the theories of Wagatsuma et al. and Van Roey and Lagasse differ only by a factor of $\cos \theta_1$, the resulting reflectivity curves are quite similar. However, both disagree significantly with the theory of Stegeman et al. At the present time there is no experimental basis for deciding which theory gives the best description of the grating-reflection process for TM modes.

REFERENCES

1. N. S. Kapany and J. J. Burke, *Optical Waveguides*, Academic Press, New York (1972).

2. D. Marcuse, *Theory of Dielectric Optical Waveguides*, Academic Press, New York (1974).

3. M. J. Adams, *An Introduction to Optical Waveguides*, Wiley, New York (1981).

4. H. Kogelnik, Theory of dielectric waveguides, Topics in Applied Physics, vol. 7, *Integrated Optics* (T. Tamir, ed.), Springer-Verlag, New York (1979).

5. H. Kogelnik and V. Ramaswamy, Scaling rules for thin-film optical waveguides, *Appl. Opt.*, *13*:1857 (1974).

6. R. V. Schmidt and I. P. Kaminow, Metal-diffused optical waveguides in LiNbO$_3$, *Appl. Phys. Lett.*, *25*:458 (1974).

7. L. W. Stulz, Titanium in-diffused LiNbO$_3$ optical waveguide fabrication, *Appl. Opt.*, *18*:2041 (1979).

8. C. T. Mueller, C. T. Sullivan, W. S. C. Chang, D. G. Hall, J. D. Zino, and R. R. Rice, An analysis of the coupling of an injection laser diode to a planar LiNbO$_3$ waveguide, *IEEE J. Quantum Electron.*, *QE-16*:363 (1980).

9. E. M. Conwell, Modes in optical waveguides formed by diffusion, *Appl. Phys. Lett.*, *23*:328 (1973).

10. A. Messiah, *Quantum Mechanics*, Vol. 1, North-Holland, Amsterdam (1964).

11. R. Olshansky, Propagation in glass optical waveguides, *Rev. Mod. Phys. 51*:341 (1979).

12. P. K. Tien, Integrated optics and new wave phenomena in optical waveguides, *Rev. Mod. Phys.*, *49*:361 (1977).

13. P. K. Tien, S. Riva-Sanseverino, R. J. Martin, A. A. Ballman, and H. Brown, Optical waveguide modes in single-crystalline $LiNbO_3$-$LiTaO_3$ solid-solution films, *Appl. Phys. Lett. 24*:503 (1974).

14. G. B. Hocker and W. K. Burns, Modes in diffused optical waveguides of arbitrary index profile, *IEEE J. Quantum Electron.*, *QE-11*:270 (1975).

15. J. Noda, M. Minakata, S. Saito, and N. Uchida, Precise determination of refractive index and thickness in the Ti-diffused $LiNbO_3$ waveguide, *J. Opt. Soc. Am.*, *68*:1690 (1979).

16. E. M. Conwell, Modes in anisotropic optical waveguides formed by diffusion, *IEEE J. Quantum Electron.*, *QE-10*:608 (1974).

17. A. Yariv, Coupled-mode theory for guided-wave optics, *IEEE J. Quantum Electron.*, *QE-9*:919 (1973).

18. W. Streifer, D. Scifres, and R. D. Burnham, TM-mode coupling coefficients in guided-wave distributed feedback lasers, *IEEE J. Quantum Electron.*, *QE-12*:74 (1976).

19. G. I. Stegeman, D. Sarid, J. J. Burke, and D. G. Hall, Scattering of guided waves by surface periodic gratings for arbitrary angles of incidence: perturbation field theory and implications to normal-mode analysis, *J. Opt. Soc. Am.*, *71*:1497 (1981).

20. R. W. Gruhlke and D. G. Hall, Comparison of two approaches to the waveguide scattering problem: TM polarization, *Appl. Opt.*, *23*:127 (1984).

21. M. L. Dakss, L. Kuhn, P. F. Heidrich, and B. A. Scott, Grating coupler for excitation of optical guided waves in thin films, *Appl. Phys. Lett.*, *16*:523 (1970).

22. C. G. Ghizoni, B. Chen, and C. L. Tang, Theory and experiments on grating couplers for thin-film waveguides, *IEEE J. Quantum Electron.*, *QE-12*:69 (1976).

23. F. Stone and S. Austin, A theoretical and experimental study of the effect of loss on grating couplers, *IEEE J. Quantum Electron.*, *QE-12*:727 (1976).

24. K. S. Pennington and L. Kuhn, Bragg diffraction beam splitters for the thin-film optical guided waves, *Opt. Commun.*, *3*:357 (1971).

25. Y. Handa, T. Suhara, N. Nishihara, and J. Koyama, Integrated grating circuit for guided-beam multiple division fabricated by electron-beam writing, *Opt. Lett.*, *5*:309 (1980).

26. D. C. Flanders, H. Kogelnik, R. V. Schmidt, and C. V. Shank, Grating filters for thin-film optical waveguides, *Appl. Phys. Lett.*, *24*:194 (1974).

27. R. V. Schmidt, D. C. Flanders, C. V. Shank, and R. D. Standley, Narrow-band grating filters for thin-film optical waveguides, *Appl. Phys. Lett.*, *25*: 651 (1974).

28. T. Fukuzawa and M. Nakamura, Mode coupling in thin-film chirped gratings, *Opt. Lett.*, *4*:343 (1979).

29. A. C. Livanos, A. Katzir, A. Yariv, and C. S. Hong, Chirped-grating demultiplexers in dielectric waveguides, *Appl. Phys. Lett.*, *30*:519 (1977).

30. A. Yi-Yan, C. D. Wilkinson, and P. J. R. Laybourne, Two-dimensional grating unit cell demultiplexer for thin-film optical waveguides, *IEEE J. Quantum Electron.*, *QE-16*:1089 (1980).

31. J. D. Spear-Zino, R. R. Rice, J. K. Powers, D. A. Bryan, and D. G. Hall, Multiwavelength monolithic integrated fiber optics terminal: an update, *Proc. SPIE*, *239*:293 (1980).

32. Y. Handa, T. Suhara, H. Nishihara, and J. Koyama, Scanning-electron-microscope-written gratings in chalcogenide films for optical integrated circuits, *Appl. Opt.*, *18*:248 (1979).

33. W. Streifer, D. R. Scifres, and R. D. Burnham, Coupling coefficients for distributed feedback single- and double-heterostructure diode lasers, *IEEE J. Quantum Electron.*, *QE-11*:867 (1975).

34. H. M. Stoll, Distributed Bragg deflector: a multifunctional integrated optical device, *Appl. Opt.*, *17*:2562 (1978).

35. K. Wagatsuma, H. Sakaki, and S. Saito, Mode conversion and optical filtering of obliquely incident waves in corrugated waveguide filters, *IEEE J. Quantum Electron.*, *QE-15*:632 (1979).

36. J. Marcou, N. Gremillet, and G. Thomin, Polarization conversion by Bragg deflection in isotropic planar integrated optics

waveguides: Part I, Theoretical study, *Opt. Commun.*, *32*:423 (1981).

37.   J. Van Roey and P. E. Lagasse, Coupled wave analysis of obliquely incident waves in thin film gratings, *Appl. Opt.*, *20*: 423 (1981).

38.   S. R. Seshadri, TE-TE mode coupling at oblique incidence in a periodic dielectric waveguide, *Appl. Phys.*, *25*:211 (1981).

39.   J. E. Sipe and G. I. Stegeman, Comparison of normal mode and total field analysis techniques in planar integrated optics, *J. Opt. Soc. Am.*, *69*:1676 (1979).

40.   L. V. Iogansen, Theory of resonant electromagnetic systems with total internal reflection, *Sov. Phys. Tech. Phys.*, *11*: 1529 (1967).

41.   E. A. J. Marcatili, Dielectric rectangular waveguide and directional coupler for integrated optics, *Bell Syst. Tech. J.*, *48*: 2071 (1969).

42.   H. Kogelnik and R. V. Schmidt, Switched directional couplers with alternating $\Delta\beta$, *IEEE J. Quantum Electron.*, *QE-12*:396 (1976).

43.   R. V. Schmidt and R. C. Alferness, Directional coupler switches, modulators, and filters using alternating $\Delta\beta$ techniques, *IEEE Trans. Circuits Syst.*, *CAS-26*:1099 (1979).

44.   L. A. Weller-Brophy and D. G. Hall, Waveguide diffraction gratings in Integrated Optics, *Proc. SPIE 578*:173 (1985).

3

Optical Waveguide Fabrication

BOR-UEI CHEN   *PCO, Inc., Chatsworth, California*

TALAL FINDAKLY*   *TRW Electroc Optics Research Center, TRW, Redondo Beach, California*

3.1 INTRODUCTION

Optical waveguides are structures that confine and steer optical waves in a region of higher index of refraction than that of their surrounding medium. One good example of an optical waveguide is the optical fiber, which is widely used for low-loss transmission of wideband optical information over long distances. The optical confinement in the fiber core is achieved by increasing the index of refraction in the core region relative to the cladding in a circularly symmetric cross-sectional configuration in most cases. On the other hand, integrated optical waveguides are usually asymmetric either in the planar or channel waveguide configurations. Most waveguides considered in this book are single-mode waveguides for both transverse electric (TE) and transverse magnetic (TM) polarizations. Figure 3.1 illustrates typical cross sections of various optical waveguide structures that have been investigated in integrated optics. Figure 3.1a–c depicts three different structures of planar waveguides, in which the optical energy is confined only in the depth direction. The thin film sandwiched between the substrate and superstrate has the highest index of refraction, thus providing the guiding region.

*Current affiliation: United Technologies Research Center, United Technologies, East Hartford, Connecticut

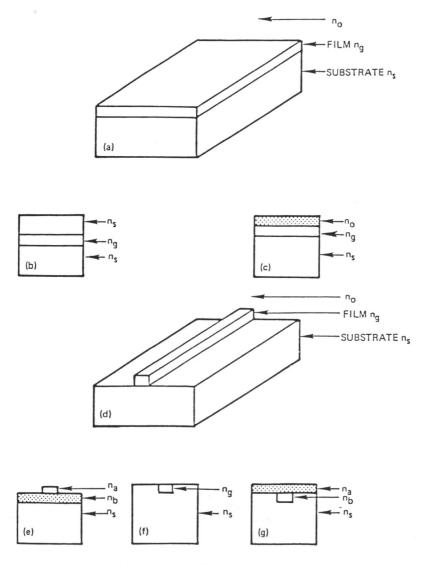

FIG. 3.1 (a), (b), and (c) illustrate various planar waveguide structures: (a) asymmetric air-guide-substrate, $n_g > n_s > n_o$; (b) symmetric optical waveguide, $n_g > n_s$; (c) composite waveguide structure, $n_g > n_s$, and $n_g > n_o$; (d), (e), (f), and (g) illustrate various channel waveguide structures: (d) and (e) are ridge waveguides with $n_g > n_s$ for (d), and n_a, $n_b < n_o$ for (e); (f) is an embedded channel waveguide, $n_g > n_s$; and (g) is an inverted ridge waveguide, n_a, $n_b > n_s$.

In the case of Fig. 3.1a, the superstrate is typically air with an index of refraction of 1. The modal properties of optical waveguides are determined by the index difference between the waveguide and its surroundings and waveguide dimensions. For single-mode operation, the thickness is typically on the order of one to a few operating wavelengths. The index of refraction of surface waveguides can be either homogeneous or graded with various profiles. The index profile is determined primarily by the waveguide fabrication process. Channel waveguides (Fig. 3.1d-g) provide two-dimensional confinement with cross sections of a few square micrometers. In the channel waveguide case, the guided modes are not exactly TE or TM modes. However, the longitudinal component of the electrical field (or magnetic field) is much smaller than the transverse component. One can therefore approximate the mode structure by a TE (or TM) polarization. The waveguide symmetry in both planar and channel waveguides is determined by the fabrication process and material selection as illustrated in Fig. 3.1.

3.2 WAVEGUIDE FABRICATION

Since the name "integrated optics" was coined in 1968, many waveguide fabrication techniques have been proposed and used to form various optical waveguides on different substrate materials. The usefulness of these waveguides is determined largely by their losses. In most practical applications, the propagation losses should be less than 1 dB/cm. In this section we summarize the fabrication processes reported to date, with emphasis on those yielding low-loss waveguides. The fabrication processes can be classified into three general categories: (1) thin-film deposition, (2) epitaxial growth, and (3) modification. A fourth category is provided to include other possible methods.

3.2.1 Thin-Film Deposition

Within this category, the optical waveguides are formed by the deposition of a layer of higher index of refraction on a lower index substrate. This deposition can be done by evaporation (thermal evaporation, electron-beam evaporation), radio-frequency sputtering, spin and dip coating, and chemical vapor deposition.

A. Evaporation

The two most common methods of thin-film deposition by evaporation are thermal and electron-beam evaporation. Evaporation techniques have been employed in the optical coating field for a long time. Various dielectric films, have been used for coatings, interference filters, etc. These thin-film coatings are more than adequate for most bulk optics applications. For integrated optics applications, however, the requirements on the absorption losses and scattering losses due to surface roughness are much more stringent. This is simply because the optical light is traveling within the thin-film waveguide rather than perpendicular to the thin film. As a result, the evaporation technique is used mainly for electrode fabrication, deposition of masks, or diffusion sources.

Amorphous ZnS films have been deposited on glass substrates at room temperature by electron-beam evaporation. The films had a loss of >5 dB/cm, attributed to the long tail of the absorption edge [1]. E-beam evaporation has also been used to fabricate low-loss glass waveguides on fused-quartz substrate [2]. A special glass, Planar CAS 10, was used as the evaporation source. Results of two runs were reported, the first with a deposition rate of 3.8 nm/s and a thickness of 2.05 μm, and the second with 2.6 nm/s and a 0.91-μm thickness. The latter supported only one mode, while the former supported two modes. The refractive index of the glass film was estimated to be 1.469 at λ = 632.8 nm. The measured waveguide losses were 1.2 ± 1 dB/cm at λ = 6.76.4 nm and 4.3 dB/cm at λ = 476.2 nm.

B. Radio-Frequency Sputtering

Radio-frequency (RF) sputtering is a popular technique of fabricating low-loss dielectric waveguides. The sputtering process is quite well understood [3] and precise control of various sputtering parameters is obtainable with commercial sputtering systems. Several sputtering modes can be obtained by making appropriate electrical connections to the J head (substrate table). These sputtering modes include RF sputtering, RF bias sputtering, RF sputtering with J head grounded, and RF sputter-etch. Depending on the sputtering gases used, reactive sputtering can modify the chemical composition of the deposited films.

RF sputtering was first used to deposit 7059 glass waveguide on ordinary glass slides [4]. Corning 7059 glass is a Pyrex type of glass with the following bulk composition: SiO_2 50.2%, BaO 25.1%, B_2O_3 13.0%, Al_2O_3 10.7% and As_2O_3 0.4%. The refractive index of bulk 7059 glass is about 1.53. Because of the change of composition incurred during the sputtering process the refractive index of sputtered film was determined to be 1.62 by the Brewster angle measurement. Films sputtered with 100% argon gas were brown and lossy because of oxygen deficiency. When 100% oxygen gas was used, the waveguide loss was reduced to less than 1 dB/cm. The use of oxygen as sputtering gas reduces the deposition rate. This is attributed to the formations of high concentration of negative ions of O or O_2, which act as electron traps [5]. The electron affinities of O^- and O_2^-, are 1.47 and 0.43 eV, respectively. These are the additional energy values required to support the discharge. In practice, excellent waveguides can be formed using a mixture of argon and oxygen, with oxygen concentration in the range 10 to 20%.

In the steady state of the sputtering, the ratio of the effective sputtering rates of differenc constituents must be equal to the ratio of their concentrations in the target. If this is not the case, then for the faster sputtering constituents, the number of atoms on the surface available for sputtering becomes less and less. Consequently, their sputtering rates are reduced gradually until equilibrium is reached. For lower-sputtering-rate constituents the situation is just the opposite. The sputtering rates are increased to reach the equilibrium value. On the deposition side, however, the ratio of deposition rates of different constituents does not follow this rule since the sticking coefficients (the probability of staying on the surface when the atom or molecule strikes the substrate) are different for different constituents. Even for a single-component target, the sticking coefficient may change with the substrate material, substrate temperature, and sputtering conditions. This effect explains the fact that sputtered 7059 glass films have different composition and refractive indices from that of the bulk material. It was observed that sputtered 7059 glass films have a different barium oxide content than the bulk material [6]. It was also observed that the refractive index of sputtered films varied as a function of the power level at which the deposition was done [7]. The film refractive index at λ = 632.8 nm could vary from 1.53 to 1.585 as the RF sputtering power density was changed from 0.5 W/cm^2 to 4.0 W/cm^2. It was also observed that the film refractive index depended on the substrate material used. Under similar sputtering conditions, the film index was 1.61 for fused-quartz substrate, 1.55 to 1.56 for soda-lime glass slides, and 1.55 for Nd-doped glass substrates [8]. The 7059 waveguide is one of the most successful materials fabricated by the RF reactive sputtering technique and has been used to fabricate various passive waveguide components. Barium silicate was also used to form thin-film glass waveguides by RF reactive sputtering [9]. The sputtering target was formed by hot pressing of a mixture of barium carbonate and silica. The sputtered films exhibited low waveguide loss, and the refractive index of the film was controlled by varying the ratio of barium oxide to silicon dioxide in the sputtering target. The film index varied from 1.48 to 1.62 by varying the barium oxide content from 0 to 40 wt %.

Nd-glass thin-film waveguide is an attractive active medium for Nd thin-film laser for 1064-nm operation. Nd-glass films have been prepared by RF sputtering of barium crown Nd glass on a heated Corning 7059 glass substrate [10]. Clear films were obtained when the target glass was sodium-free. AO 1838 Nd glass (the constituents are SiO_2, NaO_2, K_2O, BaO, Al_2O_3, Li_2O, and Sb_2O_3) yielded brownish sputtered films [11]. The brownish color was attributed to the oxygen deficiency of alkalioxides, whose dissassociation point is much smaller than other oxides. The brownish coloration can be eliminated by annealing at 500°C for 15 h in air. The reported losses of these waveguides were 0.5 dB/cm at λ = 632.8 nm and 0.15 dB/cm at λ = 1064 nm. An attractive feature of these waveguides is a gain of 1 cm^{-1} at λ - 1064 nm when the waveguides is pumped with a dye laser coupled into the waveguide. With a 1-cm^{-1} net gain (including waveguide losses), it is possible to use the film as an active medium for an Nd thin-film laser in either distributed-feedback or Bragg-reflector format.

Aluminum oxide films RF sputtered on single-crystal quartz substrates have been used in waveguide second-harmonic-generation experiments [12]. The measured index of refraction at λ = 546.1 nm was n = 1.663. The waveguide loss was uncharacteristically high, estimated to be about 40 dB/cm.

Ta_2O_5 is a high-index-of-refraction material which is desirable in certain applications requiring high-index overlays, such as waveguide Lunenberg lenses. A film index as high as 2.08 at λ = 632.8 nm was obtained by RF reactive sputtering of Ta metal target in an oxidizing atmosphere, yielding waveguide losses of < 1 dB/cm [13]. If a small amount of N_2 is introduced in the sputtering chamber, the resulting film index can be varied from 1.85 to 2.13 by controlling the N_2/O_2 ratio. The index of refraction can also be varied by the addition of other low-index oxides into the Ta_2O_5 sputtering target. Composite SiO_2-Ta_2O_5 waveguide films have been deposited on Corning Vycor glass substrates from a target consisting of SiO_2 and Ta_2O_5 [14]. Five sputtering targets containing 0, 25, 50, 75, and 100 mol % of Ta_2O_5 were used to deposit films of refractive index ranging from 1.46 to 2.08 at λ = 632.8 nm. Propagation loss of <0.8 dB/cm was obtained for films thicker than 1 μm. The waveguide loss was further reduced by postdeposit annealing in air for 12 h at 450°C. Laser annealing of a 0.445-μm-thick reactively sputtered Ta_2O_5 film on oxidized silicon was found to reduce the losses from 1.3 dB/cm before treatment to 0.4 dB/cm afterward [15]. Nb_2O_5 is also a high-index material which can be deposited by RF reactive sputtering of either Nb metal or Nb_2O_5. Early experiments with Nb_2O_5 film yielded a high waveguide loss, 20 dB/cm at λ = 632.8 nm [16]. X-ray diffraction analysis indicated that the film contained small, randomly oriented crystallites, which gave rise to a high scattering loss. By reducing the substrate temperature during the sputtering process, amorphous films with lower propagation loss can be obtained. The waveguide loss can be further reduced by laser annealing. For 100% O_2 sputtering, a refractive index of 2.297 was measured. In both Ta_2O_5 and Nb_2O_5 cases, the refractive index of the reactive sputtered films was about 94% of the value for the anodic oxide films, probably due to a lower film density. It is important to note that the film density depends on sputtering conditions, which explains the variance in results reported in different laboratories on waveguide properties even when identical sputtering targets are used. CO_2 laser annealing of a 0.518-µm-thick reactively sputtered Nb_2O_5 films was found to reduce the losses from 7.4 dB/cm before annealing to 0.6 dB/cm afterward [15].

Reactive RF sputtering was also used to deposit GeO_2 films on glass slides from a GeO_2 target [17]. The sputtering rate was found to affect attenuation dramatically. Attenuation was found to increase dramatically for sputter rates above 26 Å/min. While the losses were also high from the practical standpoint at sputter rates below 12.7 Å/min ($\sim$18 dB/cm), they were nevertheless reduced to 0.7 dB/cm by annealing at 250°C [17].

RF bias sputtering was also used to fabricate single-mode embedded silica waveguides on silica substrates [18]. Grooves were opened in a silica substrate by photolithography and dry Fr etching. Pure silica was sputtered to fill in the vertical grooves at a rate of 1 µm/h yielding an index difference between the core and the substrate of 0.0043. A cladding layer was sputtered afterward to dover the waveguides. Propagation losses of single-mode waveguides were measured at 0.633 µm to be 1.2 dB/cm [18].

3.2.2 Spin and Dip Coating

Thin-film optical waveguides have been deposited on glass substrates from liquid solutions. Solid films were formed after slow evaporation of the solvents. Similar to the coating of photoresist in the photolithographic process, the liquid films are coated on solid substrates by spinning or dip coating. The final film thickness is determined primarily by the solid content, viscosity, and spinning speed. Various materials have been used, including polyurethane, polystyrene, epoxy, photoresist, and organometallic solutions [19]. The films are usually coated at room temperature and cured at an elevated temperature varying from 50 to 100°C. Waveguide losses of less than 1 dB/cm at λ = 632.8 nm can easily be obtained with most of the above-mentioned materials, except photoresists. Photoresist films have the advantage of waveguide structure definition by photolithographic techniques. However, photoresists, such as Kodak KPR [19] and Shipley AZ1350 [20], are too lossy (>7 dB/cm) to be practical as waveguide materials. The loss can be reduced by removing the photosensitizer, which also reduces the photosensitivity of the waveguide.

Photosensitive polycarbonate polymers were used to form multimode N × N star couplers on glass and plastic substrates [21,22]. A uniform polymer film 20 to 200 µm thick was coated on a glass substrate and ultraviolet (UV) exposed through a mask for 10 min at 2-kW intensity of mercury lamp [21]. After exposure and baking at 90°C to remove the monomer, a lower-index polymer (acrylic emulsion) cladding layer was applied to form a symmetric structure. The transmission losses at 0.85 μm were 0.19 dB/cm, with a wavelength transmission window of 0.48 to 1.1 μm. The same procedure was applied to form a polymer-film guided structure sandwiched between two plastic plates [22]. A six-port multimode star coupler exhibited a fiber-to-fiber insertion loss of 2.6 dB.

Films fabricated from liquid solutions are usually soft and susceptible to chemical and physical attack. The coating technique does not avail good control of film thickness and uniformity as required for integrated optical devices. The ease of application of some of these materials makes them attractive when used as diffusion sources [23] or overlay coatings to reduce the waveguide surface scattering [24]. A spin coating technique is also used to deposit organic films followed by a polymerization process. This technique is classified as a "modification" process and is discussed in Sec. 3.4.

3.2.3 Chemical Vapor Deposition

Silicon oxynitride films deposited on fused silica by chemical vapor deposition (CVD) were considered potentially useful for integrated optical waveguides [25]. SiON is a glassy, amorphous, stable silicon-oxygen-nitrogen polymer of adjustable composition. The index of refraction can be varied between those of deposited SiO_2 (n = 1.455 at λ = 546 nm) and deposited Si_3N_4 (n = 1.98). SiON films were deposited at 850°C in a conventional RF-heated silica tube reactor from a 1-atm ambient typically comprising 0.2 to 0.5% nitric oxide (NO), 0.02 to 0.07% silane (SiH_4), and the remainder nitrogen. The SiON composition was controlled by the NO/SiH_4 concentration ratio. Deposition of 120- to 800-nm-thick waveguide films was achieved in 2 to 10 min. Films of n = 1.48 to 1.54 yielded <4 dB/cm of loss at λ = 632.8 nm. However, higher-refractive-index films were not as successful, as surface cracks developed. By using a low-vapor-pressure CVD technique, excellent-quality films of Si_3N_4 were reported. Silicon wafers were used as substrates, and the Si_3N_4 was separated from the substrate by a thermally grown SiO_2 buffer layer [26]. No evidence of cracks was observed for a film thickness of less than 400 nm. For a film thickness of 321.2 nm, two (TE) waveguide modes with propagation losses of <0.1 dB/cm (TE_0) and 6 dB/cm (TE_1) were observed. Extension of optical fields into the Si substrate was cited as the major loss mechanism, which explains the high loss for the TE_1 mode compared to that of much higher than the TE_0 mode.

Plasma-enhanced CVD was also used to deposit Si_3N_4 films on oxidized silicon [27]. Using a capacitively coupled glow discharge plasma in a reactant gas mixture of SiH_4, NH_3, and N_2, Si_3N_4 films were deposited at a rate of 93.5 Å/min at low substrate temperatures

(200 to 300°C). The measured losses for an Si_3N_4 film 0.935 μm on 1 μm of SiO_2 was 1.14 dB/cm, attributed mostly to scattering.

CVD has been employed to form single-mode [28] and multimode [29] deposited silica waveguides on fused quartz. Single-mode embedded waveguides were fabricated by etching vertical grooves in quartz by reactive sputter etching through a Ti mask and subsequent chemical vapor deposition using vapor-phase $SiCl_4$, BBr_3, and $GeCl_4$, with oxygen as a carrier gas. The deposited particles (SiO_2-B_2O_3-GeO_2) are converted into transparent core glass by sintering. The excess glass is removed by sputter etching, and a cladding silica layer with reduced index is deposited by the same CVD process using a lower $GeCl_4$ flow rate. The waveguide losses were 1.3 dB/cm at 1.25 μm, of which less than 0.1 dB/cm was due to absorption [28]. A similar procedure was used to fabricate embedded-ridge multimode waveguides on quartz [29]. Doped-silica planar film was first deposited and glazed on quartz. A ridge waveguide was formed by reactive sputter etching in a C_2F_6-C_2H_4 gas mixture through an amorphous silicon masking film deposited by RF bias sputtering and sputter etched in $CBrF_3$. Boron-doped silica was then deposited by CVD and glazed to form a lower-index cladding layer. Glazing is done at about 1400°C. The waveguide losses were 1.3 dB/cm at 0.633 μm, attributed mostly to wall roughness as <0.1 dB/cm was caused by absorption.

CVD was also used to improve the surface roughness of sputtered ZnO films [30]. ZnO films 200 to 700 Å thick were sputter deposited first on sapphire. A second layer of ZnO was then deposited by CVD at 975°C at a rate of 0.2 to 1 μm/min in a chamber employing H_2 and O_2 as deoxidizing and oxidizing gases at the source and substrate positions, respectively. N_2, H_2, and H_2O flow was allowed from the source side, and N_2 and O_2 from the substrate side. The film thickness could be controlled from 1 to 70 μm by changing the deposition time and the O_2 and H_2O partial pressures. In a heavily multimode 3.45-μm-thick waveguide supporting eight modes, the loss of the lowest order mode was 0.7 dB/cm. The surface smoothness was much better than that of similar ZnO films grown either by sputtering or CVD.

3.3 EPITAXIAL GROWTH

The epitaxial growth technique allows the fabrication of single-crystalline films for integrated optics, and in particular, active optical devices such as switches, modulators, and scanners. It has been a big challenge to material scientists to refine the growth technique such that low-loss waveguide devices can be constructed. A successful development of low-loss waveguides on GaA or InP material systems will certainly brighten the prospect of fabricating monolithic integrated optical circuits. There are two methods available today for forming single-crystal films: epitaxial growth and modification by diffusion or ion exchange. In spite of the recent intensive research, neither method is satisfactory to form monolithic circuits. More elaborate work is needed to understand and improve the fabrication processes. In certain cases, crystal defects and impurity concentrations are crucial for low-loss waveguide formation.

3.3.1 Epitaxial Growth by RF Sputtering

ZnO films have been prepared by RF sputtering of a ZnO target in O_2-Ar mixtures on various substrate materials. The technology was developed originally for applications in acoustic waves, where ZnO serves as a good transducer medium. Sputtered ZnO is a polycrystalline film with columnar crystallite structure typically 10 nm in diameter. For the films to show electro-optical and electromechanical properties, it is necessary that all the crystallites be oriented in the same direction, so that the associated effects of these crystallites can be accumulated coherently for device applications. ZnO films are known to have a strong tendency to grow with the C axis normal to the substrate surface. A complete C-axis normal orientation with small deviation requires special growth conditions, such as temperature, vapor pressure, growth rate, and so on.

ZnO thin films have been employed as optical waveguides in many integrated optical experiments [31,32]. Early experiments yielded extremely high optical propagation losses (20 to 50 dB/cm) even after gentle polishing of the film surface [33]. This is believed to be due to the columnar structure, which results in a rough surface. Optical scattering losses also arise from voids among crystallites. Although the optical waveguide properties depend strongly on the film deposition process, it is generally recognized that high-optical-quality film (<1 dB/cm) occurs only for nearly epitaxial films. Recently, a nearly epitaxial film exhibiting low optical losses has been observed after annealing the RF sputtered ZnO film with a CO_2 laser [34]. The waveguide loss for the fundamental mode of a three-mode waveguide was reduced to 0.01 dB/cm. The laser annealing process is believed to induce coalescence of neighboring crystallites, thus improving the film density uniformity and orientation.

3.3.2 Epitaxial Growth by Melting

Epitaxial growth by melting (EGM) has been used to grow $6Bi_2O_3$: TiO_3 waveguides on $Bi_{12}GeO_{20}$, and $LiNbO_3$ waveguides on $LiTaO_3$. This is one of the simplest methods of expitaxial single-crystal film growth, achieved by either dipping the substrate into a bulk melt or melting a power of lacquer suspension on the substrate. This technique is attractive for integrated optics for two reasons: First, the waveguide material with index of refraction larger than that of substrate often has a lower melting point. Second, the film grown by EGM always has a transition region at the waveguide-substrate interface, which minimizes the lattice mismatch problem. The refractive index profile of the waveguide is somewhere between exponential and step function.

The growth of various sillenites on bismuth germanate ($Bi_{12}GeO_{20}$) substrate was done by dipping in a supercooled melt (825 to 930°C) at a growth rate of 1 to 2 μm/min [35]. The sillenites used were bismuth gallate ($12Bi_2O_3:Ga_2O_3$) and bismuth titanate ($6Bi_2O_3:TiO_2$).

The growth of $LiNbO_3$ on $LiTaO_3$ substrate was done by spreading $LiNbO_3$ powder over the $LiTaO_3$ substrate, heating the sample to 1300°C to melt the powder, and cooling slowly at a rate of about 20°C/h [36]. The top surface of the as-grown film was rough and required special polishing before waveguiding was observed. An electro-optic modulator was fabricated using this waveguide [37]. The epitaxial-growth-by-melting method was improved by first suspending $LiNbO_3$ powder in a lacquer and then painting the $LiNbO_3$ lacquer on the $LiTaO_3$ substrate [38]. As the sample temperature was raised to 1270°C, the coating first turned black as the organic lacquer was decomposed. It then became transparent and glossy, indicating the melting of $LiNbO_3$. Within a few minutes, the glassy appearance disappeared and a solid solution was formed. An improvement in surface roughness of the final waveguiding layer was reported.

Despite the simplicity of forming high-index single-cyrstalline films, the epitaxial growth-by-melting technique does not appear to produce low-loss waveguides for integrated optical applications.

3.3.3 Liquid-Phase Epitaxy

Liquid-phase epitaxial (LPE) growth of thin films has been successfully developed by the semiconductor industry, and it seems reasonable to apply it to optical waveguide fabrication. However, the LPE technique is still limited to the growth of semiconductor films primarily for the fabrication of optoelectronic devices, such as optical sources and detectors. Limited integration of optoelectronic devices with electronic and integrated optical devices have been reported with good technical progress toward a totally integrated optoelectronic chip.

The first growth of $LiNbO_3$ films on $LiTaO_3$ by the LPE technique was demonstrated using a $Li_2O-V_2O_5$ flux [39]. A C-cut $LiTaO_3$ substrate was dipped into a molten mixture containing 50 mol % Li_2O, 40 mol % V_2O_5 and 10 mol % Nb_2O_5. This mixture was heated to about 1100°C and cooled slowly to the growth temperature of about

850°C. A transparent and colorless $LiNbO_3$ film 3 µm thick grown
epitaxially onto the substrate supported seven TE and TM modes.
Measurements of the modal indices indicated that the film had a uniform index profile with indices $n_0 = 2.288$ and $n_e = 2.191$ at $\lambda =$
632.8 nm. Waveguide losses of 5 and 11 dB/cm were measured for
TM_0 and TE_0 modes, respectively. Different flux systems have been
used. $Li_2B_2O_4$-$Li_2Nb_2O_6$ and Li_2WO_4-$Li_2Nb_2O_6$ flux systems produced films rich in Li; and a K_2WO_4-$Li_2Nb_2O_6$ flux system produced
films rich in Nb [40].

In GaAs, the formation of optical waveguides takes advantage of
the refractive index dependence on Al concentration of $Ga_{1-x}Al_xAs$.
This, coupled with the good lattice match between GaAs and AlAs,
makes it possible to grow good-quality higher-index GaAs buffered
by $Ga_{1-x}Al_xAs$ layers. Heterostructure layers can be grown in a
variety of forms to fabricate surface or buried film waveguides.
Two-dimensional confinement can be achieved by etching, rib formation, strip loading, and so on. Ridge-channel directional coupler
waveguides have been fabricated from liquid-phase epitaxial films of
high-resistivity GaAs or low-resistivity GaAs by ion-beam machining
[41,42] and films of GaAlAs grown on GaAs by groove etching and
growth of a $Ga_{0.7}Al_{0.3}A_3$ low-index buffer layer followed by the
growth of a high-index GaAs waveguide layer [43]. Low scattering
losses were reported due to the smoothing layer, as losses of <4 dB/
cm were measured at 1.15 µm. Efficient electro-optic polarization
modulation was demonstrated in double-heterostructure GaAs-GaAlAs
channel waveguides 0.15 to 0.2 µm thick at 1.15 µm [44].

Single-mode heterostructures, rib waveguide directional couplers,
and switches were also fabricated by the LPE method in GaAs-
$Al_xGa_{1-x}As$ [45]. An isolation layer 6 µm thick of $Al_{0.1}Ga_{0.9}As$
was grown on GaAs followed by the growth of a 0.7-µm-thick layer
of GaAs. Ribs 0.07 µm high by 3 µm wide were then etched into the
waveguide layer by anodization to provide two-dimensional confinement, yielding single-mode operation with propagation losses of about
6 dB/cm at 1.06 µm. Similar waveguide structures were also used to
demonstrate polarization modulation in a 5-µm-wide, 1.2-µm-thick
$Al_xGa_{1-x}As$ rib waveguide grown on (110) GaAs substrate with a
3.8-µm-thick $Al_yGa_{1-y}As$ buffer layer [46]. A loss of <4 dB/cm
was reported, comparable to that grown on the (100) orientation.
In InP, the LPE method was used to fabricate rib GaInAsP on InP
[47]. $Ga_{0.04}In_{0.96}As_{0.1}P_{0.9}$ ($\lambda_{gap} \sim 1$ µm) layers 2.8 µm thick
were grown on Fe-doped (100) InP substrates. Chemical wet etch-
ing was employed to form ribs 6 µm wide, 0.3 µm high in the
GaInAsP layer. The measured losses were about 7 dB/cm at 1.3 µm
and 11 dB/cm at 1.15 µm. The high losses reported were attributed
to rib edge roughness. Lower losses were reported with double-
heterostructure InGaAsP/InP waveguides grown by LPE [48]. Undoped

InGaAsP 1 µm thick was grown on groove-etched (4 to 8 µm wide) undoped (100)n^+-InP by LPE. A 1-µm-thick buffer layer of n^--InP was grown over the waveguide layer, followed by the formation of P^+-InP mesas by dry chemical etching, yielding optical losses of 4 dB/cm at 1.3 µm.

Optical waveguides and magneto-optical waveguide switches have been demonstrated successfully by LPE in gallium and iron garnet films developed for magnetic bubble memory devices. Because of the abrupt interface between the film and substrate, good lattice matching is required for the LPE growth. The garnet family has a good range of lattice parameters and refractive indices to allow a "mix-and-match" set of waveguide structures. Garnet films as grown are smooth, uniform, and pinhole-free, thus yielding low scattering losses. However, the absorption losses vary from 1 to 5 dB/cm, depending on the impurity content of the melt [49]. Magneto-optical switches were demonstrated on a $Eu_3Ge_5O_{12}$ waveguide epitaxially grown on a $Gd_3Sc_2Al_3O_{12}$ substrate [49].

Other waveguide materials fabricated by the LPE method include yttrium aluminum garnet on sapphire [50], and KDP:ADP layers on KDP substrates [51]. Unfortunately, these films are not suitable for integrated optical applications because of either the poor optical quality or water attack of ADP and KDP materials.

3.3.4 Vapor-Phase Epitaxy and Metal-Organic Chemical Vapor Deposition

The vapor-phase epitaxy (VPE) and metal-organic chemical vapor decomposition (MOCVD) are two viable alternative techniques of growing semiconductor films for the fabrication of optoelectronic devices. Unlike the LPE method, the deposited materials are carried by a gas, usually H_2. The deposition rate is slower than that of LPE method, which allows greater control of thickness and doping profile of each deposited layer. Strip-loaded waveguides were fabricated by growing high-index, lightly doped, 5-µm-thick layers of GaAs on a heavily doped n^+-GaAs by VPE, and subsequent chemical etching [52]. Two layers were grown, so that by varying the doping level, a lower-index n^+ layer 2 µm thick is grown on the guiding layer. Chemical etching was employed to form a strip superstrate n^+ layer (5 to 30 µm wide) to achieve two-dimensional confinement. Losses of 4.3 dB/cm were reported for the lowest-order mode at 1.15 µm. Directional coupler switches were also reported following the same procedure of forming lightly doped GaAs layers 2 to 5 µm thick on heavily doped GaAs [53,54]. Layers of SiO_2/metal were deposited on the guiding layer to form channel guides. It should be noted that this is a relatively poor method of two-dimensional confinement due to the small index difference in the horizontal transverse direction. Losses of 6.9 dB/cm were reported at 1.06 μm in 8-μm-wide waveguides [54]. The VPE technique was also used to form oxide-confined waveguides by the lateral epitaxial growth of a single-crystal GaAs over an SiO_2 film [55]. A pyrolytic SiO_2 film 0.3 μm thick is deposited on a single-crystal GaAs. Stripe openings are etched in the SiO_2 and the sample is placed in an $AsCl_3$-H_2-Ga VPE system. Growth nucleates only in the openings in the SiO_2 film, and depending on orientation and growth conditions, the growth rate parallel to the surface is much greater than that perpendicular to it (25:1). Thus the growth proceeds rapidly over the oxide film, yielding a sheet of GaAs 4 μm thick over the SiO_2 film. Channel waveguides are defined by partial chemical etching of the GaAs film, forming a riblike waveguide 6 μm wide. Losses as low as 2.3 dB/cm and 2.7 dB/cm were reported for the TE and TM polarization, respectively, at 1.06 μm [55].

Localized growth of GaAs ridges by VPE through Si_3N_4 openings yielded very low loss channel waveguides (1.5 dB/cm at 1.06 μm [56]). A thick layer of Si_3N_4 is deposited on (100) GaAs, and slots are opened through the nitride film along the (110) direction. A 1-μm- deep groove is chemically etched in the GaAs through the opening to provide a nucleation region for epitaxial growth. A sequence of VPE-grown layers of doped GaAs yielded very smooth side walls in ridges of triangular cross section. The low losses reported are very close to the theoretical 1-dB/cm limit at 1.06 μm.

The MOCVD method was employed to fabricate very low loss ridge GaAs waveguide by growing 1.2-μm-thick GaAlAs and 4.5 to 6.4-μm GaAs layers on n^+-GaAs [57]. The ridge in the GaAs layer was formed by ion milling. Straight-channel as well as curved waveguides with a 10-mm radius of curvature were reported. The propagation losses of a 7-μm-wide straight-channel waveguide with a 1.7-μm ridge step height were 0.5 and 0.2 dB/cm for waveguide thickness of 4.5 and 6.4 μm, respectively, at $\lambda = 1.3$ μm. The same structure was used to fabricate S bends for waveguide routing. A loss of an S bend with a 10-mm radius of curvature was 0.5 dB at 1.3 μm. The losses reported for the above-mentioned waveguides are the lowest reported to date in this material system.

3.3.5 Molecular-Beam Epitaxy

Molecular-beam epitaxy (MBE) is a powerful technique for growing single-crystal films with precise control of stoichiometry, thickness, deposition rate, and dopant concentration. Growth is performed in a high-vacuum chamber where the substrate is kept at an elevated growth temperature. Molecular species are evaporated and directed toward the substrate by separate source chambers. The slow growth rates allow process control down to a film thickness of a few atomic layers. This method has been used to fabricate GaAs double-heterostructure lasers and other devices. MBE was also used to fabricate a GaAs CW-DH laser taper coupler to a passive $Ga_{1-x}Al_xAs$ waveguide layer inside a cavity [58]. The taper coupler approached 100% coupling efficiency and the waveguide loss of the passive layer was about 4 dB/cm.

MBE has also been used to grow single-crystal thin-film $LiNbO_3$ on bulk $LiNbO_3$ and sapphire [59]. Evaporant fluxes of $10^{14}/cm^2$-s for Nb and $10^{15}/cm^2$-s for Li were used with oxygen pressure of 3×10^{-6} torr, giving a flux of $10^{15}/cm^2$-s at the substrate. Under such conditions, uniform composition films close to the bulk material were obtained with a growth rate of 1260 Å/h. The substrate temperature was found to be critical in determining the overgrowth structure. Amorphous films were obtained at ambient temperature, while at 450°C the films were polycrystalline. Above 550°C, single-crystal films were grown. High optical propagation losses were reported at 0.633 μm (15.7 dB/cm), attributed mainly to scattering due to poor crystalline film quality [59].

3.4 MODIFICATIONS

Optical waveguides can also be formed by increasing the index of refraction of the substrate surface through diffusion, ion exchange, ion implantation, polymerization, and so on. The most widely investigated waveguides used in integrated optics are fabricated using one of the methods in this category: diffusion in the case of Ti-diffused $LiNbO_3$ and ion-exchanged glass waveguides.

3.4.1 Out-Diffusion

$LiNbO_3$ and $LiTaO_3$ are among the best waveguide materials for active integrated optical devices due to their high electro-optic, acoustic-optic, and nonlinear opical parameters. The first waveguides on $LiNbO_3$ and $LiTaO_3$ were fabricated by out-diffusion of lithium and oxygen atoms at elevated temperatures [60]. $LiNbO_3$ and $LiTaO_3$ crystals can be grown in a slightly nonstoichiometric form, $(Li_2O)_v(M_2O_5)_{1-v}$, where M may be Nb or Ta and v ranges from 0.48 to 0.50. It was shown experimentally that for a small change of v in $LiNbO_3$ and $LiTaO_3$, the ordinary refractive index (n_0) remains unchanged while the extraordinary refractive index (n_e) increases approximately linearly as v decreases. For $LiNbO_3$, $dn_e/dv = -1.63$, and for $LiTaO_3$, $dn_e/dv = -0.85$. The reduction of Li_2O concentration at the surface caused by out-diffusion forms a high-index guiding layer for the extraordinary waves. Li_2O out-diffused waveguides in $LiNbO_3$ and $LiTaO_3$ have been demonstrated when the crystals were heated at high temperatures (850 to 1200°C) in vacuum or in air for several hours. Out-diffusion in vacuum yielded blackened crystals due to oxygen deficiency. The discoloration could be removed by annealing at high temperature in oxygen. Waveguide layers thus formed had a $\Delta n_e = 10^{-3}$ and a thickness of a few to several hundred micrometers. They usually support a large number of guided modes and are not practical for most applications.

The existence of out-diffusion of Li_2O at high temperature creates a problem for integrated optical waveguides fabricated by in-diffusion. Planar out-diffusion waveguides may cause optical leakage or crosstalk between the useful optical channel waveguides. Several papers have been published on lithium out-diffusion suppression describing techniques, such as annealing the sample in Li_2CO_3 [61] or $LiNbO_3$ powders before or after [63,64], and wetting the incoming gas flow during diffusion [65].

3.4.2 In-Diffusion

Metallic impurity in-diffusion has proven to be the most effective technique of fabricating low-loss waveguides in $LiNbO_3$ and $LiTaO_3$. Metal films such as Ti, Nb, Mn, Fe, Co, and Cu, a few hundred angstroms thick, deposited on the surface and diffused at temperatures in the range 900 to 1100°C yield shallow surface waveguides of varying index change and depth, depending on the parameters used. Of the two materials cited above, $LiNbO_3$ waveguides have become more practical from the fabrication procedure standpoint, due to the high Curie temperature (~ 1250°C). Due to the low Curie temperature of $LiTaO_3$ (~ 600°C), the crystal must be repoled after diffusion. This has so far reduced interest in $LiTaO_3$ despite its superiority to $LiNbO_3$ in relation to optical damage [66,67]. The first demonstration of transition metal in-duffused waveguides in $LiNbO_3$ was reported in 1974 [68]. Since then, the majority of intetrated optical devices, such as couplers, switches, modulators, and mode converters, have been fabricated by Ti diffusion in $LiNbO_3$. An investigation of the formation of optical waveguides in $LiNbO_3$ and $LiTaO_3$ by metal ion diffusion indicated an increase or decrease in the refractive index depending on the valence of the induffused and replaced ions [69]. Divalent ions such as Ni^{2+}, Zn^{2+}, and Mg^{2+} decrease the extraordinary index n_e. The ordinary index n_0, on the other hand, is increased by Ni^{2+} and Zn^{2+} and decreased by Mg^{2+}. Higher-valent ions, such as Fe^{3+}, Cr^{3+}, and Ti^{3+}, increase both the ordinary and extraordinary indices. It appears that lower-valent ions replace Li^+ sites, while higher-valent ions replace Ta^{5+} or Nb^{5+} sites. Results of x-ray photoelectron spectroscopy seem to support this theory [70]. Experimental results indicated that the in-diffused Ti metal in $LiNbO_3$ was all tetravalent (i.e., Ti atoms are fully ionized). There are no electrons in partially filled d-orbitals to absorb the electromagnetic energy at visible wavelengths. This explains the measurement of low optical losses of waveguides fabricated by Ti diffusion into $LiNbO_3$. The spectrum of x-ray photoelectron spectroscopy did not show evidence of Ti^{4+} ions distributed among different sites. Based on these measurements, it is concluded that the Ti ions are bonded chemically in the lattice in the center of the oxygen octahedra. However, it is not certain whether the Ti ions are incorporated chemically into $LiNbO_3$ as substitutional impurities on Nb or equivalent sites. The absorption loss of channel waveguides is measured to be less than 0.3 dB/cm at 632 nm and is decreased to about 0.1 dB/cm at 1300 nm [71]. The predominant loss mechanism is attributed to scattering losses.

X-ray microanalysis has also been employed to determine the refractive-index change and profile of Ti-diffused $LiNbO_3$ waveguides [72]. Several conclusions are drawn from this study:

1. Δn_o and Δn_e are similar at 0.75% Ti concentration. For Ti concentrations smaller than 0.75%, $\Delta n_o > \Delta n_e$, and for Ti concentrations higher than 0.75%, $\Delta n_e > \Delta n_o$.
2. The diffusion profiles were Gaussian, with diffusion depths on the order of a few micrometers. The maximum refractive index changes at the surface of waveguides fabricated at 970°C for 7 h are typically $\Delta n_e = 3.05 \times 10^{-2}$ and $\Delta n_o = 7.7 \times 10^{-3}$ for a Ti film of 800 Å.

While the Ti diffusion in $LiNbO_3$ is carried out at temperatures below the Curie temperature, it was, nevertheless, found that partial domain inversion takes place especially in Z^+-$LiNbO_3$ at temperatures in the range 1000 to 1100°C [73,74]. The Z^--$LiNbO_3$ was found to be less susceptible to domain inversion in this temperature range. The effects of domain inversion on the electro-optic effect in relation to voltage-length product requirements was shown to be dramatic in the Z^+-$LiNbO_3$ orientation [73]. Diffusion at higher temperatures is often performed in order to yield deeper waveguides for operation in the long-wavelength region.

Limited efforts have been made to fabricate waveguides in $LiTaO_3$ by in-diffusion without requiring crystal repoling. Electric field-assisted diffusion of Cu or CuO was shown to yield optical waveguides in $LiTaO_3$ at temperatures below the Curie point [75]. Optical waveguides 10 to 25 μm deep were formed at 550°C with the application of an electric field of 10 V/mm for 15 to 60 min. At such temperatures, no waveguide was obtainable by thermal diffusion along. The electro-diffusion yielded an index increase of approximately 0.005 with a similar index profile in both the ordinary and extraordinary indices. Microcracks were observed on the surface at electrical fields in excess of 30 V/mm.

3.4.3 Ion Exchange

The ion-exchange process is different from diffusion in that ions in
an external source are exchanged with ions in the substrate material [76]. This alteration of the material properties yields a change
in the refractive index of the material. The refractive index can be
increased or decreased as a result of a combination of two effects.
The first one relates to the atomic size of the exchanging ions. If
a small ion such as Li^+ replaces a larger ion such as Na^+ or K^+ in
glass, the glass network will collapse around the smaller ion to produce a more densely packed structure, which usually has a higher
refractive index. Conversely, if a larger ion replaces a smaller ion,
the network expands to a less packed structure, yielding a lower refractive index. The second effect relates to the electronic polarizability of the exchanging ions. If an ion of larger electronic polarizability, such as Tl^+, Cs^+, Ag^+, Rb^+, or K^+, replaces an ion of
smaller polarizability, such as Na^+, an increase in the refractive
index will result, and vice versa.

Sodium ions, present in common glasses such as soda limes and
borosilicates, are known for their high mobility and therefore exchange well with monovalent alkali ions. The extent and rate of exchange depend on the activation energy, mobility, glass composition,
temperature, and various other parameters that govern the diffusion
kinetics. The monovalent alkali ions, which yield an increase in the
refractive index of glass when exchanged with Na^+, include Li^+, Cs^+,
Rb^+, K^+, Ag^+, and Tl^+. Considerable work has been reported on
the use of these ions for the formation of single-mode and multimode waveguides in glass. The general characteristics of the ion-
exchange process associated with these ions is reviewed next.

A. $Li^+ \longleftrightarrow Na^+$ Ion Exchange

The substitution of Li^+ ions for Na^+ ions results in an increase in
the refractive index, because of its smaller ionic size compared to
Na^+, yielding a more densely packed structure. The index increase
is generally small (<0.01), making it suitable for single-mode waveguides but marginally acceptable for multimode waveguides of large
numerical aperture. The diffusion of Li^+ ions in glass is fast due
to its small size, which makes it attractive for producing thick multimode layers in reasonably short times. Planar guided layers in sodium aluminosilicate glasses were produced by Na^+-Li^+ exchange from
a $LiNO_3$ source, with an index change of 0.0018 [77]. Attempts to
bury the guided layers by double diffusion caused surface damage.
A eutectic melt of Li_2SO_4 and K_2SO_4 was also used as a source for
exchanging Li^+ ions with Na^+ ions in soda-lime glass slides at high
temperatures (575°C), yielding thick guiding layers in short diffusion
times [78,79]. The waveguides so produced were 46 to 140 µm thick

with $\Delta n = 0.015$ at the surface and had propagation losses of 1.2 dB/cm.

B. $Cs^+ \longleftrightarrow Na^+$ Ion Exchange

Cs^+ ions can also provide an increase in the refractive index when exchanged with Na^+ ions, due to the gain in electronic polarizability. Planar waveguides produced by Cs^+-Na^+ ion exchange were reported in soda-lime glass from a $CsNO_3$ source at 520°C [80]. Due to its large size, Cs^+ diffuses slowly in glass. Waveguides 8 μm deep were fabricated in 37 h at 520°C. The waveguides so formed had an index increase of 0.03 and an estimated loss of 1 dB/cm. Low loss planar waveguides were also produced by Cs^+-K^+ ion-exchange with and without an electric field in a specially prepared alumina-bor-silicate glass containing 12.5 mol% K_2O [81]. The index could be varied up to 0.043 at 436°C exchange temperature for a few hours yielding waveguides a few microns deep compatible for single-mode operation.

C. $Ag^+ \longleftrightarrow Na^+$ Ion Exchange

The Ag^+-Na^+ ion exchange is the most explored ion-exchange technique. Ag^+ ions from $AgNO_3$ salt exchange very well with Na^+ ions in glass even at low temperatures, typically in the range 220 to 300°C. The strong interdiffusion between Na^+ and Ag^+ ions, coupled with a large index increase obtainable (typically 0.09 in soda-lime glass [82-85], and 0.22 in TiF6 glass [86]) result in highly multimode waveguides for diffusion periods of a few hours. The losses in such waveguides are relatively high (often a few dB/cm) when the silver concentration is high. The refractive index can be adjusted below the saturation level by diluting the silver salts with sodium or potassium salts [84,85]. Another method of controlling the silver content utilizes electrolytic release of silver ions from a silver rod electrode immersed in molten $NaNO3$ [87]. By controlling the current, the silver concentration in the salt and therefore the rate of Ag^+-Na^+ ion exchange can be controlled, yielding the desired index change. Lower silver concentration usually results in a lower loss. The high losses commonly observed in silver ion-exchange waveguides are the result of silver ion reduction in glasses containing impurities, causing staining.

After the silver ions penetrate through the glass, they are reduced to elementary silver if they react with other metastable ions, such as Fe^{2+} and As^{3+}, which serve as electron donors [88]. The low solubility of the elementary silver so formed brings about precipitation of the metal as submicroscopic crystals. It has been found especially in glasses with a high FeO and As_2O_3 content that silver ion exchange yields deep yellow colors, whereas glasses free of these constituents or prepared under strongly oxidizing conditions did not readily accept the silver stain [89]. This suggests the following typical chemical reaction:

$$Ag_2O + 2FeO \longrightarrow Fe_eO_3 + 2Ag$$

$$2Ag_2O + As_2O_3 \longrightarrow As_2O_5 + 4Ag$$

The necessity for a reducing agent to be present in the glass explains the sensitivity with respect to minor constituents such as arsenic, antimony, and ferrous ions. These ions assume the role of electron donors and the Ag^+ ion that of an electron acceptor. The yellowish coloration is produced by aggregation and crystallization. During the ion-exchange process, and during the reduction of silver ion into atoms, the silver atoms have considerable speed. Soon after their formation, they collide, aggregate, and form submicroscopic crystals, responsible for the yellowish staining of silver-treated glasses. Under proper conditions, silver stain can be formed in all silicate glasses. The major constituents are of little influence, while minor constituents such as As_2O_3, Sb_2O_3, or FeO strongly influence the extent of glass coloration [88].

D. $K^+ \longleftrightarrow Na^+$ Ion Exchange

K^+ is another ion that exchanges well with Na^+ ions. This technique proves to be very useful and most promising for applications that require low numerical apertures and single-mode operation. K^+ ions diffuse in glass at a moderate pace, slower than silver but faster than thallium ions. The index increase obtained is in the range 0.006 to 0.009 [81]. This technique has been used to fabricate single mode $1 \times N$ star couplers in glass substrates [90]. Single-mode channel waveguides with typical dimensions of 4×3 μm are formed on glass slides at 350°C during a 30-minute period with a KNO_3 source. The waveguides formed are compatible with single-mode fibers with regards to dimensions and numerical aperture. The propagation losses are typically <0.2 dB/cm at 0.83-μm wavelength. For multimode operation, it is possible to extend the diffusion time and/or apply an electric field to create thick multimode guiding layers. The maximum index increase, however, is typically about 0.009 in soda-lime or borosilicate glasses, which is lower than needed for multimode operation.

E. $Tl^+ \longleftrightarrow Na^+, K^+$ Ion Exchange

Tl^+-Na^+, K^+ ion exchange in glass has been used to produce planar and two-dimensional guiding layers [81,92,93] and multimode star couplers [89]. Tl^+ has the largest electronic polarizability among the monovalent alkali ions, and therefore yields a large change in the refractive index when exchanged with Na^+ or K^+ ions. An index change of 0.1 has been observed, and lower values were obtained by diluting the $TlNO_3$ source with $NaNO_3$ [92]. Tl^+ is a fairly large ion and therefore has very low mobility in glass. For example, no guiding layers in aluminosilicate glass were observed after an ion exchange in $TlNO_3$ melt for a period of 24 h at 250°C [81]. On the other hand, thick guided layers (150 μm deep) were formed in borosilicate crown glass at a much higher temperature (530°C) and longer duration (reaching 72 h) [92]. The application of an electric field considerably shortens the diffusion time. Using this method and a subsequent second exchange process to reduce the index at the surface, excellent buried two-dimensional waveguides with a near circular profile 30 μm in diameter were formed with losses below 0.1 dB/cm [92]. For multimode purposes, this technique proves to be the most feasible among other ion-exchange techniques, due to the ability to control the process inherent in the slow diffusion of Tl^+ ions and the excellent outcome of low losses and compatibility with multimode fibers.

The main problem that inhibits the wider use of Tl^+ ion compounds for ion exchange is its high toxicity. However, when precautions are taken and operation is maintained under proper fume hoods and controlled conditions, hazards can be minimized.

Ion exchange has also been employed for waveguide fabrication in $LiNbO_3$ and $LiTaO_3$ by silver [94,95] and Tl [96] ion exchange. In $LiNbO_3$, optical waveguides were formed in X-cut $LiNbO_3$ by Ag^+-Li^+ ion exchange at 360°C [94]. The extraordinary index was increased by a large amount (∿0.12) when undiluted $AgNO_3$ was used. No increase was observed in the ordinary index. The waveguide losses were >6 dB/cm. Tl^+↔Li^+ ion exchange also yielded a high index increase (0.12 to 0.13) only in the extraordinary index of X-cut $LiNbO_3$ using undiluted $TlNO_3$ at 240°C for few hours [96]. A reconstruction of the index profile from modal properties revealed a profile that closely resembles a step-index distribution.

Ag^+-Li^+ ion exchange was also employed to form waveguides in X-cut $LiTaO_3$ [95]. Here again only the extraordinary index was increased (0.03 to 0.05) at temperatures in the range 240 to 370°C. The $LiTaO_3$ waveguides formed by this method seem to have lower loss than their $LiNbO_3$ counterpart. The $LiTaO_3$ silver-exchanged waveguide losses were reported to be approximately 0.8 to 1.3 dB/cm [95].

The waveguides formed by Tl^+ ion-exchange in $LiNbO_3$ were found to be difficult to reproduce. A careful examination of numerous samples suggested the possibility of a proton exchange taking place rather than an ion-exchange [97]. This was later verified by inducing humidity in the exchange setup which resulted in a high increase in the extraordinary index only under humid environments [98]. This suggests that the waveguide formation was the result of a hydrogen-lithium exchange rather than the heavy silver or thallium ion-exchange.

Proton exchange has also been used as a method of creating waveguides LiNbO$_3$ [97] and LiTaO$_3$ [98]. Very large increases in n_e (about 0.12 for LiNbO$_3$ and 0.14 for LiTaO$_3$) have been obtained in this method by treating X-cut and Z-cut samples in undiluted benzoic acid at low temperatures (110 to 249°C) for a few hours. Low-loss waveguides (0.5 dB/cm) were reported using this method in X-cut LiNbO$_3$ crystals. Treatment of Y-cut LiNbO$_3$ samples resulted in surface deformations, and no waveguides were reported for that crystal orientation.

Similar results were obtained in LiTaO$_3$, although the diffusion depth under similar conditions appears to be smaller than that of LiTaO$_3$, requiring diffusion at higher temperatures [100]. The large index increase obtained in the extraordinary index could be lowered by subsequent annealing, yielding a relaxed, deeper index profile. The index increase could also be controlled by diluting the benzoic acid with lithium-rich solutions such as lithium benzoate [101]. By varying the lithium concentration in the solution, the index increase could be adjusted in the range 0.01 to 0.11. The corporation of Ti indiffusion and proton exchange in LiNbO$_3$ was used to form waveguides with equally effective refractive indices in the ordinary and extraordinary orientations in the guiding layers. The phase matching so obtained is useful for second harmonic generation [102].

F. Ion Implantation

Ion implantation has also been explored for integrated optical waveguide fabrication. One of the problems associated with ion implantation is that the high-energy particles create defects that result in scattering losses. Thermal annealing is usually employed to reduce the damage effect. Channel waveguides on fused quartz were formed by Li$^+$ implantation through a PMMA electron resist mask [101]. The waveguide losses after annealing were about 3 dB/cm.

LiNbO$_3$ implanted with 60-keV Ne or Ar ions showed up to a 10% decrease in refractive index, attributed to polarization effects of the damaged lattice structure [104]. No individual measurement was reported of the effects on the two refractive indices, or the effects of possible annealing damage on the index change. The effect of ion implantation of He, B, N, Ne, and Ar on LiNbO$_3$ was examined in greater detail [105, 106]. Optical waveguides were formed by genating a subsurface low-index layer, 2 to 4 μm below the surface by He implantation. It was noted that the electro-optic coefficient, r_{33}, was reduced by about 60%. In another attempt to maintain the crystallinity of the structure, He ion implantation was used to delineate a channel waveguide structure on a planar Ti-diffused waveguide [107]. Strip waveguides were formed by protecting the strip region with a metallic mask and He^+ implanting the sides at 350 keV at a dose of $10^{16}/cm^2$. Operating the $LiNbO_3$ waveguides in intensity modulator formats indicated that retention of the electro-optic properties of the guiding structures so formed [106]. Light ions (proton, helium, boron) were also used to form waveguides in ZnTe [108]. Waveguide formation in this case is the result of the macroscopic implantation and the microscopic changes in the induced damage effect. Optical losses of 1 to 4 dB/cm were measured.

G. Others

There are many other modification techniques available to fabricate reasonably good optical waveguides of varying quality, including polymerization by photolocking, stress, metal cladding, etching, and laser heating. It is almost impossible to exhaust all the techniques published in the literature. Some of these techniques are of little interest, however, for practical device applications.

3.5 OPTICAL WAVEGUIDE LOSSES

The loss mechanisms of integrated optical waveguides can be classified in two major categories: absorption and scattering. The absorption loss is due to the conversion of the optical energy into phonon energy in the form of heat. The scattering loss is the result of redirection of guided optical light into radiation modes or other guided modes. Depending on the sources of absorption, the absorption loss can be classified in three categories: intrinsic, impurity, and atomic defects. Scattering can also be classified in three categories: intrinsic volume scattering, boundary scattering, and index inhomogeneity.

3.5.1 Absorption

Intrinsic absorption originates from electronic charge transfer of bands in the UV region and atomic vibration and rotation of multiphonon bands in the near infrared (IR). If these bands are sufficiently strong, the absorption tails will extend into the spectral region of interest in integrated optics applications, namely 800 to 1600 nm. In most cases, the IR absorption bands are located beyond 4 μm and are narrow. On the other hand, the UV bands are stronger and potentially more troublesome in the wavelength range of interest.

Impurity absorption arises predominantly from transiton metal ions present in the bulk material or introduced during the waveguide formation process. The absorption loss of these ions varies for different materials and depends on the valence states of impurity ions. In silica-based materials, OH ions, for example, contribute to absorption around 725, 950, 1250, and 1390 nm.

Absorption may also be caused by atomic defects in the crystalline structure, for example, color centers. A color center is a lattice defect that absorbs visible or near UV light. The simplest color center is an F center in alkali halide crystals. The F centers can be formed by heating the crystal in excess alkali vapor or by x-ray irradiation. When excess alkali atoms are added to the alkali halide crystal, it creates negative ion vacancies in the crystal lattice. As a result, the valence electron of an alkali atom becomes bound to the vacant negative-ion site. The optical absorption arises from the electric dipole transition to a bound excited state of the F center.

3.5.2 Scattering

Rayleigh scattering is a major contributor to the intrinsic volume scattering loss. The scattering is due to the frozen-in thermal fluctuation of constituent atoms, which in turn, causes density and refractive index fluctuations. This intrinsic scattering effect is believed to be the lower bound of optical waveguide losses. At present, this effect in commercial optical fibers has been substantially reduced to the fundamental limit of Raleigh scattering. Index inhomogenity is another source of intrinsic scattering. Scattering can be caused by the nucleation and formation of colloidal clusters of impurity ions in reduced form. One example of this scattering is the formation of Ag^+ ion clusters in silver ion-exchanged glass waveguides.

Another dominant loss mechanism is associated with the roughness of waveguide boundaries. In the planar waveguide structure the boundary scattering occurs at both superstrate/waveguide and waveguide/substrate interfaces. In the case of channel waveguide structures, roughness along the channel width, waveguide bend, and curvature will introduce additional scattering losses.

Another loss mechanism associated with electro-optic materials such as $LiNbO_3$ and $LiTaO_3$ waveguides is the photorefractive effect, also known as optical damage. It is caused by charge transfer from optically excited impurities within the band gap to metastable trapping sites [109,108]. This effect manifests itself in refractive index changes in the material upon exposure to visible-light intensities in excess of 1×10^{-4} mW/μm^2. Due to this localized intensity-induced change in the refractive index, the propagating beam diverges gradually as its wavefront becomes distorted by index inhomogeneities. The contribution of this effect to the total loss is not well quantified, as other geometrical imperfections may scatter the light in a similar way. It appears that Ti diffusion does not increase the absorption or level of optical damage from that of the original substrate. With the application of an electric field of few kV/cm, however, the sensitivity to the photorefractive effect is greatly increased in the Ti-diffused layers in comparison to similar fields applied to the bulk crystal [109]. It has also been observed that Ti-in-diffused wave-guides are more susceptible to optical damage than are out-diffused waveguides [110]. Optical damage decreases for longer wavelengths and becomes negligibly small beyond 1 μm [110]. Recently, a technique for suppressing the out-diffusion in Ti-diffused waveguides by carrying the diffusion in humid atmosphere has been found to improve the susceptibility of Ti-diffused waveguides to optical damage as a result of the presence of H^+ ions [112]. In this case, the photo-refractive sensitivity is improved by a factor of 2 to 5 by intro-ducing humidity in the diffusion process.

REFERENCES

1. P. K. Tien, R. Ulrich and R. J. Martin, Modes of propagating light waves in thin deposited semiconductor films, *Appl. Phys. Lett.*, *14*:291 (1969).

2. R. Th. Kersten and W. Rauscher, A low loss thin film optical waveguide for integrated optics made by vacuum deposition, *Opt. Commun.*, *13*:189 (1975).

3. J. L. Vossen and J. J. O'Neill, RF sputtering process, *RCA Rev.*, *29*:149 (1968).

4. J. E. Goel and R. D. Standley, Sputtered glass waveguides for integrated optics, *Bell Syst. Tech. J.*, *48*:3445 (1969).

5. P. D. Davidse and L. I. Maissel, Dielectric thin films through RF sputtering, *J. Appl. Phys.*, *37*:574 (1966).

6. T. Nishimura, Y. Murayana, D. Dota, and H. Matsumaru, Digest of the 3rd Symposium on the Deposition of Thin Films by Sput-tering, Bendix Corp., Rochester, N.Y., pp. 96–106 (1969).

7. C. W. Pitt, Sputtered glass optical waveguides, *Opt. Lett.*, *9*: 401 (1973).

8. B. Chen, Ph.D dissertation, Cornell University, Ithaca, N.Y. (1976).

9. J. E. Goell, Barium silicate films for optical integrated circuits, *Appl. Opt.*, *12*:737 (1973).

10. H. Yajima, S. Kawase, and Y. Sekimoto, Amplification at 106 μm using a Nd-glass thin film waveguide, *Appl. Phys. Lett.*, *21*:407 (1972).

11. B. Chen and C. L. Tang, Nd-glass thin film waveguide, an active medium for Nd thin-film laser, *Appl. Phys. Lett.*, *28*: 435 (1976).

12. B. Chen, C. L. Tang, and J. M. Telle, CW harmonic generation in the UV using a thin-film waveguide on a nonlinear substrate, *Appl. Phys. Lett.*, *25*:495 (1974).

13. H. Terui and M. Kobayashi, Refractive-index-adjustable SiO_2-Ta_2O_5 films for integrated optical circuits, *Appl. Phys. Lett.*, *32*:666 (1978).

14. S. J. Ingrey, W. D. Westwood, Y. C. Cheng, and J. Wei, Variable refractive index and birefringent waveguides by sputtering tantalum in O_2-N_2 Mixture, *Appl. Opt.*, *14*:2194 (1975).

15. S. Dutta, et al., CO_2 laser annealing of Si_3N_4, Nb_2O_5 thin film optical waveguides to achieve scattering loss reduction, *IEEE J. Quantum Electron.*, *QE-18*:800 (1982).

16. S. J. Ingrey and W. D. Westwood, Birefringent waveguides prepared by reactive sputtering of niobium in O_2-N_2 mixtures, *Appl. Opt.*, *15*:607 (1976).

17. A Yin and B. Garside, Low loss GeO_2 optical waveguide fabrication using low deposition rate of sputtering, *Appl. Opt.*, *21*:4 (1982).

18. N. Imoto, N. Shimizu, H. Mori, and M. Ikeda, Sputtered silica waveguides with an embedded three dimensional structure, *IEEE J. Lightwave Technol.*, *LT-1*:289 (1983).

19. R. Ulbrich and H. P. Weber, Solution deposited thin films as passive and active light guides, *Appl. Opt.*, *11*:428 (1972).

20. D. B. Ostrowsky and A. Jacques, Formation of optical waveguides in photo resist films, *Appl. Phys. Lett.*, *18*:556 (1971).

21. T. Kurokawa, N. Takato, and Y. Katayama, Polymer optical circuits for multimode fiber systems, *Appl. Opt.*, *19*:3124 (1980).

22. N. Takato and T. Kurokawa, Polymer waveguide starcoupler, *Appl. Opt.*, *21*:1940 (1982).

23. R. Chen, A new technique for waveguide formation in $LiNbO_3$, *Thin Solid Films*, *64* (1979).

24. S. Dutta, H. Jackson, and J. T. Boyd, Extremely low-loss thin-film optical waveguides utilizing surface coating and laser annealing, *J. Appl. Phys.*, *52*:3873 (1981).

25. M. J. Rand and R. D. Standley, Silicon oxynitride films on fused silica for optical waveguides, *Appl. Opt.*, *11*:2482 (1972).

26. W. Stutius and W. Streifer, Silicon nitride films on silicon for optical waveguides, *Appl. Opt.*, *16*:3218 (1977).

27. S. Sriram, W. Partlow, and S. Lin, Low loss optical waveguides using plasma deposited silicon nitride, *Appl. Opt.*, *22*:3664 (1983).

28. Y. Murakami, M. Ikeda, and I. Izawa, Optical directional coupler using deposited silica waveguides (DS guides), *IEEE J. Quantum Electron.*, *QE-17*:1982 (1981).

29. H. Mori and N. Shimizn, Multimode deposited silica waveguide and its application to an optical branching circuit, *IEEE J. Quantum Electron.*, *QE-18*:776 (1982).

30. T. Shiosaki, O. Ohnishi, Y. Hirokawa, and A. Kawabata, As-grown CVD ZnO optical waveguides on sapphires, *Appl. Phys. Lett.*, *33*:406 (1978).

31. T. Shiosaki, S. Fukuda, K. Sakai, H. Kuroda, and A. Kawabata, Second harmonic generation in as-shuttered ZnO optical waveguide, *Japn. J. Appl. Phys.*, *19*:2391 (1980).

32. S. Zemon, R. R. Alfano, S. L. Shapiro, and E. Conwell, High power effects in nonlinear optical waveguide, *Appl. Phys. Lett.*, *21*:327 (1972).

33. P. K. Tien, Lightwaves in thin films and integrated optics, *Appl. Opt.*, *10*:2395 (1971).

34. S. Dutta, H. E. Jackson, J. T. Boyd, F. S. Hickernell, and R. L. Davis, Scattering loss reduction in ZnO optical waveguides by laser annealing *Appl. Phys. Lett.*, *39*:206 (1981).

35. A. A. Ballman, H. Brown, P. K. Tien, and R. J. Martin, The growth of single crystalline waveguiding thin films of Piezoelectric sillenites, *J. Cryst. Growth*, *20*:251 (1973).

36. S. Miyazawa, Growth of $LiNbO_3$ single crystal film for optical waveguides, *Appl. Phys. Lett.*, *23*:198 (1973).

37. S. Fukunishi, N. Uchinda, S. Miyazawa, and J. Noda, Electro-optic modulation of optical guided waves in $LiNbO_3$ fabricated by EGM method, *Appl. Phys. Lett.*, *24*:424 (1974).

38. P. K. Tien, S. Riva-Sanseverino, R. J., Martin, A. A. Ballman, and H. Brown, Optical waveguide modes in single crystalline $LiNbO_3$-$LiTaO_3$ solid solution films, *Appl. Phys. Lett.*, *24*:503 (1974).

39. S. Miyazawa, S. Fushimi, and S. Kondo, Optical waveguide of $LiNbO_3$ thin film growth by liquid phase epitaxy, *Appl. Phys. Lett.*, *26*:8 (1975).

40. P. K. Tien and A. A. Ballman, Research in optical films for the applications of integrated optics, *J. Vac. Sci. Technol.*, *12*:892 (1975).

41. S. Somekh, E. Garmire, A. Yariv, H. Garvin, and R. Husnperger, *Appl. Opt.*, *13*:327 (1974).

42. S. Somekh, E. Garmire, A. Yariv, H. Garvin, and R. Husnperger, *Appl. Opt.*, *42*:455 (1973).

44. F. K. Reinhart and B. I. Miller, Effecient $GaAs/Ga_{1-x}Al_xAs$ double-heterstructure light modulators, *Appl. Phys. Lett.*, *20*: 36 (1972).

45. J. Shelton, F. Reinhart, and R. Logan, Rib waveguide switches with MOS electrooptic control for monolithic integrated optics in $GaAs\text{-}Al_xGa_{1-x}As$, *Appl. Opt.*, *17*:2548 (1978).

46. F. Reinhart, R. Logan, and W. Sinclair, Electrooptic polarization modulation in multielectrode $Al_xGa_{1-x}As$ rib waveguides, *IEEE J. Quantum Electron.*, *QE-18*:763 (1982).

47. J. Donnelly, N. DeMeo, F. Leonberger, S. Groves, P. Vohl, and F. O'Donnell, Single-mode optical waveguides and phase modulators in the InP material system, *IEEE J. Quantum Electron.*, *QE-21*:1147 (1985).

48. M. Fujiwana, A. Ajisawa, Y. Sugimoto, and Y. Ohta, Giganhertzbandwidth InGaAsP/InP optical modulators with double-hetero waveguides, *Electron. Lett.*, *20*:790 (1984).

49. P. K. Tien, R. J. Martin, S. L. Blank, S. H. Wemple, and L. J. Varnerin, Optical waveguides of single crystal garnet films, *Appl. Phys. Lett.*, *21*:207 (1972).

50. P. K. Tien, R. J. Martin, R. Wolfe, R. C. Le-Craw, and S. L. Blank, Switching and modulation of light in magneto-optic waveguides of garnet films, *Appl. Phys. Lett.*, *21*:394 (1972).

51. V. Ramaswamy, Epitaxial electro-optic mix-crystal $(NH_4)_xK_{1-x}H_2PO_4$ film waveguide, *Appl. Phys. Lett.*, *21*:183 (1972).

52. F. A. Blum, D. W. Shaw, and W. C. Holton, Optical striplines for integrated optical circuits in epitaxial GaAs, *Appl. Phys. Lett.* *25*:116 (1974).

53. J. C. Cambell, F. A. Blum, D. W. Shaw, and K. L. Lawley, GaAs Electro-optic directional-coupler switch, *Appl. Phys. Lett.*, *27*:202 (1975).

54. F. Leonberger and C. Bozler, GaAs directional-coupler switch with stepped Δβ reversal, *Appl. Phys. Lett.*, *31*:223, 313 (1981).

55. F. Leonberger, C. Bozler, R. McClelland, and I. Melngailis, Low loss GaAs optical waveguide formed by lateral epitaxial growth over oxide, *Appl. Phys. Lett.*, *38*:313 (1981).

56. E. Erman, N. Vodjdani, J. Theeten, and J. Cabanie, Low loss waveguides grown on GaAs using localized vapor phase epitaxy, *Appl. Phys. Lett.*, *43*:894 (1983).

57. H. Inoue, K. Hiruma, . Ishida, T. Asai, and H. Matsumura, Low Loss GaAs optical waveguides, *IEEE J. Lightwave Technol.*, *LT-3*:1270 (1985).

58. F. K. Reinhart and A. Y. Cho, Molecular beam epitaxial layer structures for integrated optics, *Opt. Commun.* *18*:79 (1976).

59. Growth of thin film lithium niobate by molecular beam epitaxy, *Electron. Lett.*, *21*:960 (1985).

60. I. P. Kaminow and J. R. Carruthers, Optical waveguiding layers in LiNbO$_3$ and LiTaO$_3$, *Appl. Phys. Lett.*, *22*:326 (1973).

61. S. Miyazawa, R. Guglielmi, and A. Cerenco, A simple technique for suppressing Li$_2$O, Our-diffusion in Ti:LiNbO$_3$ optical waveguide, *Appl. Phys. Lett.*, *31*:570 (1977).

62. B. Chen and A. C. Pastor, Elimination of Li$_2$O out-diffusion waveguide, *Appl. Phys. Lett.*, *31*:742 (1977).

63. T. R. Ranganath and S. Wang, Suppression of Li$_2$O out-diffusion from Ti-diffused LiNbO$_3$ optical waveguides, *Appl. Phys. Lett.*, *30*:376 (1977).

64. R. J. Esdaile, Closed-tube control of out-diffusion during fabrication of optical waveguides in LiNbO$_3$, *Appl. Phys. Lett.*, *33*:733 (1978).

65. J. L. Jackel, V. Ramaswamy, and S. P. Lyman, Elimination of out-diffused surface guiding in titanium diffused LiNbO$_3$, *Appl. Phys. Lett.*, *38*:509 (1981).

66. G. Tangonan, M. Barnoski, J. Lotspeich, and A. Lee, High optical power capabilities of Ti diffused LiTaO$_3$ waveguide modulator, *Appl. Phys. Lett.*, *30*:238 (1977).

67. R. L. Holman, P. Cressman, and J. Revelli, Chemical control of optical damage in lithium niobate, *Appl. Phys. Lett.*, *32*: 280 (1978).

68. R. V. Schmidt and I. P. Kaminow, Metal diffused optical waveguides in LiNbO$_3$, *Appl. Phys. Lett.*, *25*: 458 (1974).

69. J. Noda and H. Iwasaki, Proceedings of the 2nd Meeting on Ferroelectronics and Their Applications, p. 149 (1979).

70. T. P. Pearsall, S. Chiang, and R. V. Schmidt, Study of titanium diffusion in lithium-niobate low-loss optical waveguides by x-ray photoelectron spectroscopy, *J. Appl. Phys.* *47*: 4794 (1976).

71. A. M. Glass, I. P. Kaminow, A. A. Ballman, and D. H. Olsson, Absorption loss and photorefractive-index changes in Ti:LiNbO$_3$ crystals and waveguides, *Appl. Opt.,* $1\,19$: 276 (1980).

72. M. Minakata, S. Saito, M. Shibata, and S. Miyazawa, Precise determination of refractive-index changes in Ti-diffused LiNbO$_3$ optical waveguides, *J. Appl. Phys.*, *49*: 4677 (1978).

73. S. Thaniyavarn, T. Findakly, D. Booher, and J. Moen, Domain inversion effects in Ti-LiNbO$_3$ integrated optical devices, *Appl. Phys. Lett.*, *46*: 933 (1985).

74. S. Miyazawa, Ferroelectric domain inversion in Ti-LiNbO$_3$ optical waveguides, *J. Appl. Phys.*, *50*: 4599 (1979).

75. J. Noda, T. Saku, and N. Uchida, Fabrication of optical waveguiding layer in LiTaO$_3$ by Cu diffusion, *Appl. Phys. Lett.*, *25*: 308 (1974).

76. T. Findakly, Glass waveguides by ion-exchange: a review, *Opt. Eng.*, *24*: 244 (1985).

77. W. G. French and A. D. Pearson, *Ceram. Bull.*, *49*: 974 (1970).

78. G. Chartier, P. Jaussaud, A. DeOliveira, and O. Parriaux, *Electron. Lett.*, *13*: 763 (1977).

79. G. Chartier et al., *Appl. Opt.*, *19*: 1092 (1980).

80. V. Neuman, O. Parriaux, and L. Walpita, *Electron. Lett. 15*: 704 (1974).

81. L. Ross, et al., "Improved substrate glass for planar waveguides by Cs-ion exchange," Topical Meeting on Integrated and Guided-Wave Optics, IGWO '86, Th BB2, Atlanta, Georgia, 1986.

82. T. Gallorenzi et al., *Appl. Opt.*, *13*: 1240 (1973).

83. G. Stewart, C. Millar, P. Laybourn, C. Wikinson, and R. Dela-Rue, *IEEE J. Quantum Electron.*, *QE-13*:192 (1977).

84. G. Stewart, C. Millar, and P. Laybourn, *IEEE J. Quantum Electron.*, *QE-14*:93 (1978).

85. M. Imal, N. Haneda, and Y. Ohtsuka, *IEEE J. Light Wave Technol.*, *LT-1*:611 (1983).

86. J. Coutaz and P. C. Jaussaud, *Appl. Opt.*, *21*:1063 (1981).

87. R. K. Lagu and R. V. Ramaswamy, 7th Topical Meeting on Integrated Optics, Post-deadline paper PD-8 (1984).

88. W. A. Weyl, *Coloured Glasses*, Scholor Press, Ilkley, Yorkshire (1978).

89. R. Zslgmondy, *Dinglers Polytech. J.*, *306*:91 (1981).

90. T. Findakly and B. Chen, *Appl. Phys. Lett.*, *40*:549 (1982).

91. T. Findakly and B. Chen, Topical Meeting on Integrated and Guided Wave Optics, Pacific Grove, California (Jan. 1982).

92. T. Izawa and H. Nakagome, *Appl. Phys. Lett.*, *21*:584 (1972).

93. M. Hatich, D. Chen, and J. Huber, *Appl. Phys. Lett.*, *33*: 997 (1978).

94. M. K. Shah, Optical waveguides in $LiNbO_3$ by ion-exchange technique, *Appl. Phys. Lett.*, *26*:652 (1975).

95. J. Jackel, Optical waveguides in $LiTaO_3$: silver-lithium ion exchange, *Appl. Opt.*, *19*:1966 (1980).

96. J. Jackel, High-Δn optical waveguides in $LiNbO_3$: thallium-lithium ion exchange, *Appl. Phys. Lett.*, *37*:739 (1980).

97. Y. Chen, W. Chang, S. Lan, L. Wielunski, and R. Holman, Characteristics of $LiNbO_3$ waveguides exchanged in $TlNO_3$ solution, *Appl. Phys. Lett.*, *40*:10 (1982).

98. J. Jackel and C. Rice, Variation in waveguides fabricated by immersion of $LiNbO_3$ in $AgNO_3$ and $TlNO_3$: The role of hydrogen, *Appl. Phys. Lett.*, *41*:508 (1982).

99. J. Jackel, C. E. Rice, and J. J. Veselka, Proton exchange for high index waveguides in $LiNbO_3$, *Appl. Phys. Lett.*, *41*:607 (1982).

100. W. Spillman, Jr., N. Sanford, and R. Soref, Optical waveguides in $LiTaO_3$ formed by proton exchange, *Opt. Lett.*, *8*:497 (1983).

101. M. DeMicheli, J. Botinean, S. Neven, P. Sibillot, D. Ostrow-
     sky, and M. Papuchon, Independent control of index and
     profiles in proton-exchanged lithium niobate guides, *Opt.
     Lett.*, *8*:114 (1983).

102. M. DeMicheli, J. Botinean, S. Neven, P. Sibillot, D. Ostrow-
     sky, and M. Papuchon, Independent control of index and
     profiles in proton-exchanged lithium niobate guides, *Opt.
     Lett.*, *8*:116 (1983).

103. J. E. Goell, R. D. Standley, W. M. Gibson and J. W. Rodgers,
     Ion bombardment fabrication of optical waveguides using elec-
     tron resist masks, *Appl. Phys. Lett.*, *21*:72 (1972).

104. D. T. Y. Wei, W. W. Lee, and L. R. Bloom, Large refrac-
     tive index change induced by ion-implantation in lithium nio-
     bate, *Appl. Phys. Lett.*, *25*, 329 (1974).

105. G. Destefanis, P. Townsend, and J. Gailliard, Optical wave-
     guides in $LiNbO_3$ formed by ion implantation of He, *Appl.
     Phys. Lett.*, *32*:293 (1978).

106. G. L. Desterfanis, J. P. Gailliard, E. L. Liegon, and S. Val-
     ette, The formation of waveguides and modulators in $LiNbO_3$
     by ion-implantation, *J. Appl. Phys. 50*:7898 (1979).

107. J. Heibei and E. Voges, Strip waveguides in $LiNbO_3$ fabri-
     cated by combined metal diffusion and ion implantation, *IEEE
     J. Quantum Electron.*, *QE-18*:820 (1982).

108. S. Valette, G. Labrunie, J.-C. Deutsch, and J. Lizet, Planar
     optical waveguides achieved by ion-implantation in zinc tellur-
     ide: general characteristics, *Appl. Opt.*, *16*:1289 (1977).

109. A. M. Glass, I. P. Kaminow, Al A. Ballman, and D. H. Olsen,
     Absorption loss and photorefractive index changes in $LiNbO_3$
     crystals and waveguides, *Appl. Opt.*, *19*:276 (1980).

110. A. M. Glass, I. T. Kaminow, A. A. Ballman, and D. H. Olsen,
     "Absorption Loss and Photorefractive Index Changes in $LiNbO_3$:
     Ti Crystals and Waveguides," paper TuD5, Digest of the Top-
     ical Meeting on Integrated and Guided-Wave Optics, Incline
     Villate, Nev. (Jan. 1980).

111. R. L. Holman, P. J. Cressman, and J. F. Revelli, Chemical
     control of optical damage in $LiNbO_3$, *Appl. Phys. Lett. 32*:280
     (1978).

112. J. L. Jackel, D. H. Olson, and A. M. Glass, Optical damage
     resistance of monovalent ion diffused $LiNbO_3$ and $LiTaO_3$ wave-
     guides, *J. Appl. Phys.*, *52*:4855 (1981).

4

Loss Mechanisms in Planar and Channel Waveguides

DENNIS G. HALL   *The Institute of Optics, University of Rochester, Rochester, New York*

LYNN D. HUTCHESON   *APA Optics, Inc., Blaine, Minnesota*

4.1 INTRODUCTION

The fabrication of low-scatter and low-loss integrated optical circuits is a necessity before one can reduce to practice the many proposed applications. Scattering and waveguide loss can often limit device performance. Since light is guided in materials having dimensions approximate to a wavelength, small imperfections in the waveguide can be detrimental to device performance. In Sec. 4.2 theory and experiments are presented that give the magnitude of the various scattering effects in terms of measurable properties of the waveguide. Scattering caused by waveguide roughness and by refractive index fluctuations is considered.

Another loss mechanism that can be predicted when designing an integrated optic device is radiation loss due to directional changes in the waveguide. This plays an extremely important role in determining the density of components on a single integrated optic substrate. To achieve high component density, waveguides must bend to guide light from one optical component to the next. In Sec. 4.3 losses due to discrete angle bends, curved waveguides, and mode coupling losses are considered theoretically with experimental results. Also, techniques for designing and fabricating bends having low loss are presented.

4.2 SCATTERING IN OPTICAL WAVEGUIDES

The elastic scattering of light by imperfections in optical waveguides
plays a role in determining the ultimate performance level that can
be achieved in an integrated optical system. Energy that resides in
a bound mode of an ideal waveguide remains trapped in that mode
and the guided wave propagates unattenuated along the prescribed
direction. There is no way to fabricate an ideal waveguide, however,
and all real waveguides contain irregularities that lead to scattering.
There are two broad categories of waveguide imperfections: (1) refractive index fluctuations caused by compositional variations within
the volume of the waveguide, and (2) dimensional fluctuations associated with variations in the height and/or width of the waveguide.
One can imagine other potential scattering centers, such as dust
particles that settle on the surface of the waveguide, but this treatment assumes that adequate cleaning procedures have been used and
concerns itself only with the fundamental mechanisms that will be
difficult to avoid.

Either or both of the fluctuations listed above can cause intermode scattering, radiation, or in-plane scattering. The first of
these, intermode scattering, occurs when, for example, light is scattered out of the TE_0 mode into the TE_1 mode of a planar waveguide.
Physically, the scattering mechanism(s) must, in this case, accomplish
two things: It must redistribute the energy to conform to that of
the second mode, and it must make up for the difference in the effective indices of refraction for the two modes (phase matching).
Intermode scattering is easy to observe in multimode optical waveguides, but it can be eliminated, in principle, by using a single-
mode waveguide. It often happens, however, that a single-mode
waveguide is actually a two-mode waveguide: It supports the lowest-
order mode for each of two polarizations (TE_0 and TM_0 for a planar
waveguide). Waveguide components such as lenses and gratings
can often cause intermode scattering between the two polarizations
(often termed mode conversion), but this is less of a concern when
the scattering is caused by small-scale surface roughness or low-
level compositional fluctuations.

Radiation limits device performance primarily through attenuation. The radiation process can be thought of as intermode scattering that takes place between a bound mode and a radiation mode.
The fact that one can often see, at visible wavelengths, a guided
wave as a "streak" in the waveguide (see Fig. 4.1) is direct evidence that the waveguide is losing energy by radiation. In-plane
scattering occurs in planar optical waveguides. Since, by definition,
the scattered light remains within the optical waveguide, it has the
potential to generate a background noise level in discrete detectors
or detector arrays that are part of an integrated optical device.

FIG. 4.1 Photograph of the visible streak generated by the scattering of a TE_0 guided wave propagating in a graded-index glass waveguide. The photo is a two-minute time-exposure.

It is important to be able to estimate the magnitude of the various scattering effects in terms of the measurable properties of the waveguide. Sections 4.2.1 and 4.2.2 consider two basic problems: (1) attenuation caused by surface-roughness-induced radiation from planar optical waveguides, and (2) in-plane scattering caused by refractive index fluctuations and by surface roughness. The specific details of the analyses are omitted in favor of a concise statement of the assumptions and final formulas together with specific examples that serve to illustrate the results. A rather extensive set of references allows the interested reader to pursue the mathematical details in the open literature.

4.2.1 Attenuation in Planar Optical Waveguides

Both surface roughness and refractive index fluctuations in an optical waveguide are capable of ejecting guided radiation out of the
waveguide and into the substrate or cover media that surround the
guiding region. In this section we consider the attenuation that results from radiation losses caused by interface roughness that appears on the top and bottom surfaces of a planar optical waveguide.
Marcuse first considered this problem by applying coupled-mode
theory to a calculation of the amplitudes of the radiation modes generated by a guided-mode in an imperfect symmetric waveguide [1].
The formulas and plots that appear in this section are based on an
alternative theoretical approach that is essentially a perturbation
theory. In this approach, the fields are expanded in a power series
in a small parameter that is proportional to the magnitude of the surface fluctuations. The fields are then obtained by satisfying the
boundary conditions to first order in this small parameter. The
boundary perturbation approach has been used by several authors
for a variety of problems [2–5], and has been shown to lead to
expressions for the scattered fields that are identical to those obtained from coupled-mode theory for the case of a symmetric waveguide [6].

Figure 4.2 shows the basic geometry considered in this section.
A thin film of mean thickness h and refractive index n_f is bounded
by cover and substrate media of refractive indices n_c and n_s, respectively. Both the upper and lower boundaries are assumed to
be rough on a scale that is small compared to either the film thickness or the wavelength of light; the size of the roughness has been
exaggerated in Fig. 4.2. A guided wave propagating in the z direction in such a waveguide radiates light into the cover and substrate
regions with a characteristic intensity distribution. Figure 4.3
shows as an example the radiation patterns measured for both TE_0
and TM_0 guided waves in a Corning 7059 glass waveguide fabricated

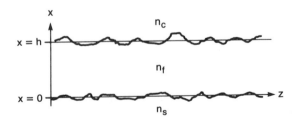

FIG. 4.2 Illustration of the cross-section of a planar waveguide
with rough boundary surfaces. The mean surfaces are at x = 0
and x = h.

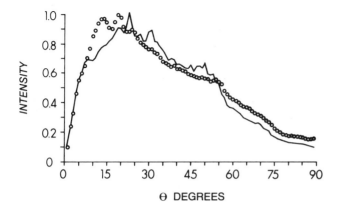

FIG. 4.3 Measured angular distribution of the light scattered out of the TE$_0$ (ooooo) and TM$_0$ (———) modes propagating in a planar glass waveguide. The angle is defined in the text. (From Ref. 4.)

by radio-frequency (RF) sputtering. The intensity patterns were measured in air in the x-z plane (see Fig. 4.2) at wavelength λ = 0.6328 μm and are plotted as functions of the angle θ, where $\theta = 0°$ corresponds to the z direction (forward direction) and $\theta = 90°$ corresponds to the x direction (normal to the mean surface). It is clear that for both polarizations, the radiation patterns are rather broad but are peaked near the forward direction. One usually attempts to observe a guiding streak in a waveguide at normal incidence to the plane of the waveguide (Fig. 4.1 is a case in point), but it is important to realize that this is not the direction for which the radiation is most intense. Data similar to that of Fig. 4.3 have been reported by Suematsu et al. [7] and Gottlieb et al. [8].

The simplest calculation of the guided-wave attenuation constant α involves three basic steps: (1) the use of some type of perturbation theory to obtain the scattered fields, and hence the radiation patterns, for both the substrate and cover media; (2) the choice of a suitable model for the statistical properties of the randomly rough surfaces; and (3) the numerical integration of the radiation patterns over the angle to determine the fraction of the "incident" power that is radiated per unit length of waveguide. Before turning to a statement of the formulas necessary to calculate α, some general comments are in order. It is instructive to examine the way in which α depends on the various waveguide parameters. The major points can be extracted from

$$\alpha_s = \lambda^{-3}(n_f^2 - n_c^2)^2 E_c^2 \int_0^\pi F_1(\theta)g_c(\beta - \xi) \, d\theta$$

$$+ \lambda^{-3}(n_f^2 - n_s^2)^2 E_s^2 \int_0^\pi F_2(\theta)g_s(\beta - \xi) \, d\theta \tag{4.1}$$

where α_s is the attenuation constant due to radiation into the substrate *only* ($\alpha = \alpha_s + \alpha_c$) and θ is measured from the forward (z) direction in each medium. The first term in Eq. (4.1) is due to scattering caused by surface roughness on the cover/film boundary at x = h; the second term is due to scattering caused by surface roughness on the film/substrate boundary at x = 0. $F_1(\theta)$ and $F_2(\theta)$ are complicated functions of angle that also depend on the wavelength λ, the film thickness h, and the refractive indices n_c, n_f, and n_s. E_c and E_s are the values of the modal electric field of the perfect waveguide at x = h and x = 0, respectively. The functions $g_c(\beta - \xi)$ and $g_s(\beta - \xi)$ are each the Fourier transform of the surface roughness autocorrelation function for the surface in question, where β is the propagation constant of the single (TE) guided wave assumed to be supported by the waveguide, and $\xi = (2\pi n_s/\lambda) \cos \theta$.

The λ^{-3} dependence that appears explicitly in Eq. (4.1) is typical of an intrinsically two-dimensional calculation that ignores the y dimension, as has been pointed out by Elson and Bennett in connection with a related surface-scattering problem [9]. The Rayleigh scattering cross sections for a small sphere and a thin wire, for example, display λ^{-4} and λ^{-3} dependences, respectively [10]. A fully three-dimensional calculation of the far-field radiation pattern from a rough waveguide shows the expected λ^{-4} dependence [3], demonstrating that the λ^{-3} dependence is an artifact of the two-dimensional nature of the present as well as previous results. Equation (4.1) contains additional λ dependence that causes the total wavelength dependence to deviate from λ^{-3}, as will be demonstrated shortly.

The functions g_c and g_s contain the statistical information for the randomly rough surfaces. The specific form of these functions is rarely known and it is common practice to choose either an exponential or a Gaussian to model the surface-roughness autocorrelation function. The present discussion assumes the former, for which

$$g(\beta - \xi) = \frac{2\sigma\delta^2}{1 + (\beta - \xi)^2\sigma^2} \tag{4.2}$$

where δ, the root-mean-square (rms) surface roughness, describes the magnitude and σ, the surface autocorrelation length, describes the average wavelength of the surface fluctuations. It is clear, then, from Eqs. (4.1) and (4.2) that the attenuation constant α varies as δ^2, a feature that is common to virtually all surface-scattering theories for which $\delta \ll \lambda$ [11,12].

The strength of the scattering-induced attenuation, according to Eq. (4.1), depends on the surface roughness, the difference in the squares of the refractive indices, and the field at each surface. For most practical geometries, $n_s > n_c$, so Eq. (4.1) predicts that the cover/film interface scatters more strongly than the substrate/film interface. The field strengths and index differences are related according to [13]

$$E_c^2(n_f^2 - n_c^2) = E_s^2(n_f^2 - n_s^2) \tag{4.3}$$

so that scattering from the cover/film interface would seem to exceed that from the substrate/film interface by the factor $(n_f^2 - n_c^2)/(n_f^2 - n_s^2)$ if both surfaces are equally rough. In fact, the functions $F_1(\theta)$ and $F_2(\theta)$ modify this result somewhat, so it is important to examine each case in detail before drawing any final conclusions.

The formulas needed to evaluate the attenuation constant α when scattering from both surfaces is taken into account are complicated, but straightforward. Those that appear below assume that a TE_0 mode propagates in the waveguide and that the scattering from the two surfaces is uncorrelated. The corresponding formulas for correlated scattering appear in Ref. 5, but they are not given here since the attenuation values obtained from these formulas are not very different from those for the uncorrelated case. The expressions for α are

$$\alpha = \int_0^\pi P_{ss}\, d\theta + \int_0^\pi P_{sc}\, d\theta \tag{4.4}$$

where

$$P_{ss} = \frac{n_s^2 k_o}{4\pi}\left(\frac{\mu_o}{\varepsilon_o}\right)^{1/2}\{4\xi_f^2\, B^2 g_c(\beta - \xi) + [2(\xi_f^2 + |\xi_c|^2)$$

$$+ 2(\xi_f^2 - |\xi_c|^2)\cos 2\xi_f h + 8\xi_f\, \mathrm{Im}(\xi_c)\sin \xi_f h \cos \xi_f h]$$

$$\times A^2 g_s(\beta - \xi)\} \times [|-2\xi_f(\xi_s + \xi_c)\cos \xi_f h$$

$$+ 2i(\xi_f^2 + \xi_s\xi_c)\sin \xi_f h|^2]^{-1} \tag{4.5}$$

$$P_{sc} = \frac{n_c^2 k_o}{4\pi} \left(\frac{\mu_o}{\varepsilon_o}\right)^{1/2} \{4\xi_f^2 A^2 g_s(\beta - \xi) + [2(\xi_f^2 + \xi_s^2)$$

$$+ 2(\xi_f^2 - \xi_s^2) \cos 2\xi_f h] B^2 g_c(\beta - \xi)\} [|-2\xi_f(\xi_s + \xi_c)$$

$$\times \cos \xi_f h + 2i(\xi_f^2 + \xi_s \xi_c) \sin \xi_f h|^2]^{-1} \tag{4.6}$$

The various quantities that appear in Eqs. (4.5) and (4.6) are defined below.

$$A = k_o^2 (n_f^2 - n_s^2)^{1/2} \left[\frac{4}{N h_{eff}} \left(\frac{\mu_o}{\varepsilon_o}\right)^{1/2} (n_f^2 - N^2)\right]^{1/2} \tag{4.7}$$

$$B = \frac{-A(n_f^2 - n_c^2)^{1/2}}{(n_f^2 - n_s^2)^{1/2}} \tag{4.8}$$

$$g_c(\beta - \xi) = \frac{2\sigma_c \delta_c^2}{1 + (\beta - \xi)^2 \sigma_c^2} \tag{4.9}$$

$$g_s(\beta - \xi) = \frac{2\sigma_s \delta_s^2}{1 + (\beta - \xi)^2 \sigma_s^2} \tag{4.10}$$

$$\xi_c = (n_c^2 k_o^2 - \xi^2)^{1/2} \tag{4.11}$$

$$\xi_f = (n_f^2 k_o^2 - \xi^2)^{1/2} \tag{4.12}$$

$$\xi_s = (n_s^2 k_o^2 - \xi^2)^{1/2} \tag{4.13}$$

$k_o = 2\pi/\lambda$, $\beta = 2\pi N/\lambda$, the effective thickness h_{eff} was defined in Chapter 2, and $\xi = n_s k_o \cos \theta$ when used in evaluating P_{ss}, but $\xi = n_c k_o \cos \theta$ when used in evaluating P_{sc}. The subscripts c and s have been added to δ and σ in Eqs. (4.9) and (4.10) to make it clear that the surface roughness need not be the same on both surfaces. Equations (4.4)–(4.13) constitute an easily manageable set of equations to describe the attenuation α as a function of all the wave-guide variables. The integration in Eq. (4.4) must, of course, be done numerically.

The attenuation depends on a great many parameters, and so it is worthwhile to examine a number of examples to help build an intuition for the important features. Figure 4.4 shows the wavelength dependence of the attenuation constant α for parameters typical of a glass waveguide deposited on a quartz substrate: $n_c = 1$, $n_f = 1.56$, $n_s = 1.47$, $\sigma = 0.4$ μm, and $\delta_c = \delta_s = 20$ A. The solid curves labeled 1 and 2 correspond to waveguide thicknesses $h = 0.3$ μm and 0.4 μm, respectively. The dashed curve corresponds to the case in which the film thickness is varied to keep the mode size constant with respect to the film thickness; the dotted curve depicts a strict λ^{-3} dependence. It is clear that mode-size variations with λ can cause departures from the simple λ^{-3} expected for Rayleigh scattering in two dimensions. The dashed curve in Fig. 4.4 was obtained by holding h/λ constant at the value that is midway between the cutoff

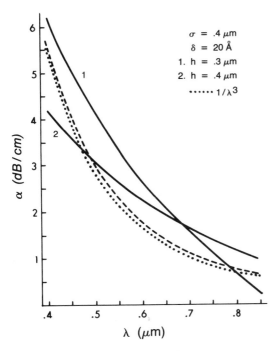

FIG. 4.4 Calculated dependence of attenuation on wavelength. The parameters are: $n_c = 1.0$, $n_s = 1.47$, $n_f = 1.56$. (From Ref. 5.)

values for the TE_0 and TE_1 modes, which has the effect of keeping
the mode size constant with respect to the waveguide. The remaining wavelength dependence follows the λ^{-3} almost perfectly. It is
interesting to note that attenuations of 1 to a few dB/cm arise from
a very slight surface roughness: $\delta/\lambda = 0.003$ for $\lambda = 0.6$ μm and
$\delta/h = 0.005$ for $h = 0.4$ μm. Fortunately, α scales with δ^2 *and*
λ^{-3}, so that a slight decrease in δ, together with an increase in
wavelength, have a significant impact on α.

Figures 4.5 and 4.6 show plots of α versus σ for the same values
of n_s, n_c, n_f, and δ as those of Fig. 4.4, $\lambda = 0.6328$ μm, and several values of the film thickness h. Figure 4.5 gives the result for
the case of uncorrelated scattering from the two waveguide surfaces,
while Fig. 4.6 gives, for comparison, the result for correlated scattering. Since α depends strictly on δ^2, the curves in Figs. 4.5
and 4.6 can easily be scaled to larger or smaller values of δ. Figure 4.7 shows a similar plot for the glass waveguide system of Figs.
4.4–4.6 but for a smooth ($\delta_s = 0$) substrate/film interface and a
rough cover/film interface. The similarity of the magnitudes of α
in Figs. 4.6 and 4.7 shows that the film/cover interface dominates
the scattering losses.

Figures 4.8–4.11 illustrate how the attenuation is influenced by
the choice of materials. Figures 4.8 and 4.9 show plots of α versus
σ for several values of h for parameters that model a GaAs waveguide:
$n_s = 3.4$, $n_f = 3.5$, and $n_c = 1$ and wavelengths $\lambda = 0.6328$ μm (Fig.
4.8) and $\lambda = 0.84$ μm (Fig. 4.9). Figures 4.10 and 4.11 show analogous plots for parameters that attempt to model a Ti-in-diffused
$LiNbO_3$ waveguide: $n_s = 2.0$, $n_f = 2.01$, and $n_c = 1.0$. One should
bear in mind that Figs. 4.8–4.11 are intended primarily to give a

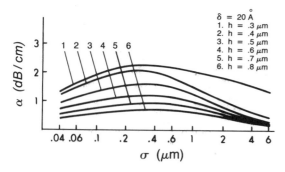

FIG. 4.5 Calculated dependence of attenuation on correlation length
σ for wavelength $\lambda = 0.6328$ μm for the case of uncorrelated scattering surfaces. Parameters are as given in Fig. 4. (From Ref. 5.)

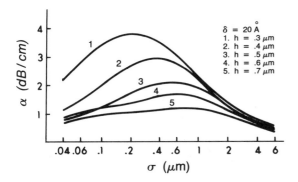

FIG. 4.6 Calculated dependence of attenuation on correlation length σ for wavelength $\lambda = 0.6328$ μm for the case of correlated scattering surfaces. Parameters are as given in Fig. 5. (From Ref. 5.)

feeling for the impact of the choices of n_f, n_s, n_c, and λ and are not intended to represent GaAs and LiNbO$_3$ structures in detail. After all, free carrier absorption is of major concern in GaAs, and Ti-in-diffused LiNbO$_3$ waveguides are graded-index waveguides, not step-index waveguides: neither of these effects is included in the present analysis. In both cases considered here, the scattering losses are dominated by the large index difference at the cover/film boundary: roughness on the substrate/film boundary is typically responsible for less than 5% of the total attenuation for the examples shown here. The rather small values of α in Figs. 4.10

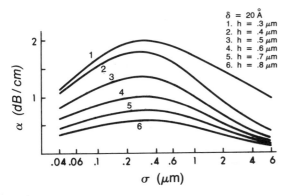

FIG. 4.7 Same as Figs. 5 and 6, but for a smooth substrate/film interface. (From Ref. 5.)

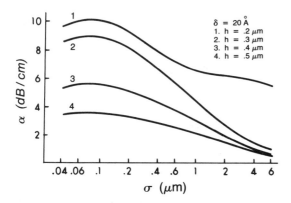

FIG. 4.8 Calculated dependence of attenuation on correlation length σ for wavelength $\lambda = 0.6328$ μm. The parameters are: $n_c = 1.0$, $n_s = 3.4$, and $n_f = 3.5$.

and 4.11 arise because of the very small index difference between the substrate and the waveguiding layer. The small index difference allows the waveguide thickness h to increase and still retain single-mode operation. This tends to spread out the modal field, reducing the values of the electric field at the surfaces. The quantities $E_s^2(n_f^2 - n_s^2)$ and $E_c^2(n_f^2 - n_c^2)$ are reliable indicators of the relative level of attenuation to be expected. For example, for

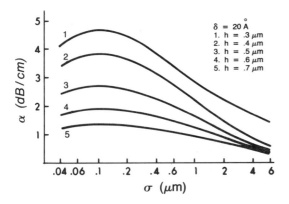

FIG. 4.9 Same as Fig. 8 but for wavelength $\lambda = 0.84$ μm. (From Ref. 5.)

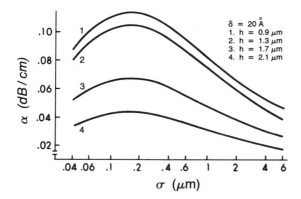

FIG. 4.10 Calculated dependence of attenuation on correlation length σ for wavelength $\lambda = 0.6328$ μm. The parameters are $n_c = 1.0$, $n_s = 2.0$, $n_f = 2.01$. (From Ref. 5.)

h/λ chosen to be midway between the TE_0 and TE_1 cutoff values for each geometry, one finds that $E_s^2(n_f^2 - n_s^2)$ is ~ 160 times larger, and $E_c^2(n_f^2 - n_c^2)$ is roughly 11 times larger, for the glass waveguide parameters of Figs. 4.4–4.7 than for the model-LiNbO$_3$ waveguide parameters of Figs. 4.10 and 4.11.

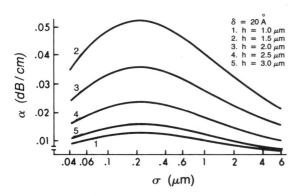

FIG. 4.11 Same as Fig. 10 but for wavelength $\lambda = 0.84$ μm. (From Ref. 5.)

4.2.2 In-Plane Scattering in Planar Optical Waveguides

This section considers in-plane scattering caused by either refractive index fluctuations or surface roughness. The attenuation problem discussed in the preceding section was analyzed by means of a two-dimensional theory that ignores the y direction completely. The in-plane scattering problem, however, requires that all three dimensions be included. Boyd and Anderson [14], in 1978, attempted an intuitive generalization of Marcuse's two-dimensional radiation calculation to three dimensions and applied it to an analysis of in-plane scattering caused by surface roughness in an integrated optical RF spectrum analyzer [15,16]. Since that time, the complete three-dimensional in-plane scattering problem has been solved by two independent methods which happily arrive at the same rsult [17–19]. As was the case in Sec. 4.2.1, this section omits the details of the analysis in favor of a concise statement of the results and examples that illustrate them. The interested reader can consult Ref. 19 for the mathematical detail.

The basic geometry considered appears in Fig. 4.12, which shows a "view from above" of a planar optical waveguide with two identical integrated detectors. One detector (labeled 1) is taken to be on-axis with respect to the incident guided wave (along $\hat{z}$), and the other (labeled 2) collects scattered light generated by the region of width w and length D located in front of the first detector. The theory calculates the ratio of the scattered power collected by detector 2 to the incident power collected by detector 1. When the scattering is caused by surface roughness, this ratio is called R_S. When the scattering is caused by refractive index fluctuations, the ratio is called R_n.

It is necessary to point out, before stating the results, that R_S and R_n each depend on the Fourier transform of an autocorrela-

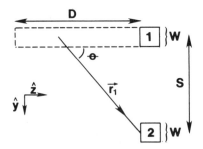

FIG. 4.12 Basic geometry for discussion of the in-plane scattering problem.

tion function. This should come as no surprise since the attenuation constant discussed in the previous section was also expressed in terms of such a function. The autocorrelation functions needed here are different from those of the previous section in two ways. First, both the y- and z-directions play a role in the scattering process, so both the autocorrelation function and its Fourier transform $g(q_y, q_z)$ are two-dimensional. Second, a model must be constructed for both surface roughness and refractive index fluctuations. For simplicity, the present model assumes that only the upper waveguide surface at x = h (see Figure 4.2) is rough. An assumed exponential autocorrelation function gives

$$g_s(q_y, q_z) = \frac{2\pi \delta^2 \sigma^2}{[1 + 4\beta^2 \sigma^2 \sin^2(\theta/2)]^{3/2}} \qquad (4.14)$$

where θ is defined in Figure 4.12, $q_y = \beta \sin \theta$, $q_z = \beta \cos(\theta) - \beta$, β is the (TE) guided-wave's propagation constant, δ is the rms surface roughness, and σ is the surface roughness autocorrelation length, which is assumed to be the same along all directions parallel to the mean surface of the waveguide.

The model for refractive index fluctuations is a simple one that is based on two assumptions: (1) there are no refractive index fluctuations in the cover or substrate media, and (2) the index variations do not depend on x. The second assumption has the effect of modeling the index fluctuations as pillars or cylinders, which may be quite appropriate since many thin films display a columnar growth structure. An assumed exponential form for the index autocorrelation function gives

$$g_n(q_y, q_z) = \frac{2\pi a^2 M^2}{[1 + 4\beta^2 a^2 \sin^2(\theta/2)]^{3/2}} \qquad (4.15)$$

where M is the rms refractive index variation and a is the index autocorrelation length. Note that the subscripts n and s in Eqs. (4.14) and (4.15) denote index (n) and surface (s) effects.

The final formulas for the special case of a step-index waveguide are

$$R_s = \frac{8\pi^3 w \delta^2 \sigma^2 (n_f^2 - N^2)^2}{N h_{eff}^2 \lambda^3} \int_{\theta_{min}}^{\pi/2} \frac{\cos^3 \theta / \sin \theta}{[1 + 4\beta^2 \sigma^2 \sin^2(\theta/2)]^{3/2}} \, d\theta$$

$$(4.16)$$

$$R_n = \frac{8\pi^3 wM^2 a^2 n_f^2 \Gamma^2}{N\lambda^3} \int_{\theta_{min}}^{\pi/2} \frac{\cos^3 \theta/\sin \theta}{[1 + 4\beta^2 a^2 \sin^2(\theta/2)]^{3/2}} \, d\theta \qquad (4.17)$$

where $\tan \theta_{min} = s/D$, N is the effective index of the incident guided wave as given by $\beta = 2\pi N/\lambda$, h_{eff} the effective waveguide thickness, and Γ the confinement factor defined as the ratio of the power carried by the mode within the guiding layer to the total power carried by the mode. Equation (4.17) is sufficiently general that it can be applied as written to either a step-index or a graded-index planar waveguide. The distinction between the two enters only in the evaluation of Γ, which is an easy parameter to model for either structure since by definition, $0 \leqslant \Gamma \leqslant 1$. The specific form of the result that appears in Eq. (4.16) applies only to step-index waveguides. This is not an intrinsic limitation of the theory, however, but arises only because it is necessary to evaluate E_c, the value of the modal electric field at the cover/film interface. For a step index waveguide, one finds that

$$E_c^2 = 4\left(\frac{\mu_o}{\varepsilon_o}\right)^{1/2} \frac{(n_f^2 - N^2)/(n_f^2 - n_c^2)}{Nh_{eff}} \qquad (4.18)$$

The reader can consult Ref. 19 to find Eq. (4.16) expressed directly in terms of E_c^2 in order to modify it for the case of a graded-index planar waveguide.

It is possible to examine the relative importance of refractive index fluctuations and surface roughness for in-plane scattering by defining a certain ratio R_1, a volume-to-surface ratio. In detail, R_1 gives the ratio of the power (per unit guide width) per unit surface area that is scattered into angle θ by index fluctuations to that same quantity scattered into the same angle θ by surface roughness. R_1 is given by

$$R_1 = \left(\frac{n_f h_{eff}\Gamma}{n_f^2 - N^2}\right)^2 \frac{g_n(q_y,q_z)}{g_s(q_y,q_z)} \qquad (4.19)$$

Equations (4.14) and (4.15) and the assumption $a = \sigma$ reduce Eq. (4.19) to

$$R_1(a = \sigma) = \left(\frac{n_f^2}{n_f^2 - N^2}\right)^2 \left[\frac{\Gamma(M/n_f)}{\delta/h_{eff}}\right]^2 \tag{4.20}$$

For the case in which the relative surface roughness δ/h_{eff} is comparable to the relative index variation M/n_f, $R_1(a = \sigma)$ becomes

$$R_1(a = \sigma) = \left(\frac{n_f^2}{n_f^2 - N^2}\right)^2 \Gamma^2 \tag{4.21}$$

which is usually much greater than unity. Figure 4.13 shows a plot of $R_1(a = \sigma)$ from Eq. (4.21) as a function of the waveguide thickness h_o for the fundamental mode in a waveguide specified by the following parameters: $n_s = 1.47$, $n_f = 1.56$, and $n_s = 1$. It is clear that scattering from refractive index fluctuations dominates

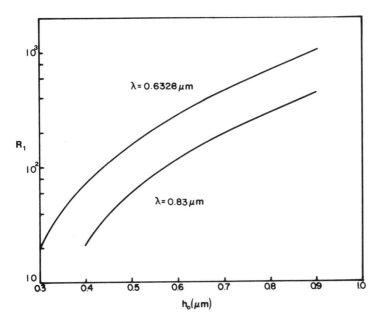

FIG. 4.13 Scattering ratio R_1 as a function of the waveguide thickness for two wavelengths. (From Ref. 19.)

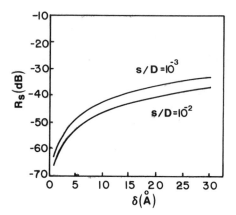

FIG. 4.14 Scattering ratio R_S, the ratio of the scattered power collected by detector 2 to the incident power collected by detector 1,
for the case of surface scattering, as a function of the rms surface
roughness δ. (From Ref. 19.)

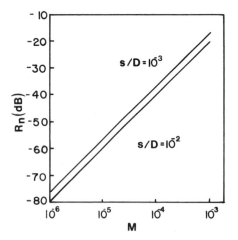

FIG. 4.15 Scattering ratio R_n as a function of M, the rms refractive index fluctuation. R_S and R_n have the same meaning, except
that the former is caused by surface scattering and the latter is
caused by refractive index fluctuations. (From Ref. 19.)

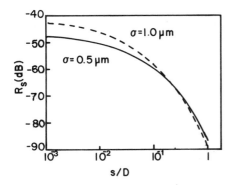

FIG. 4.16 Scattering ratio R_S as a function of the ratio s/D. (From Ref. 19.)

that from surface roughness by at least an order of magnitude, and usually much more, when the relative surface and index fluctuations are comparable.

Figures 4.14–4.17 illustrate the dependence of R_S and R_n, Eqs. (4.16) and (4.17), on a few of the major quantities for the following parameters: $n_S = 1.47$, $n_f = 1.56$, $n_c = 1.0$, $h = 0.9$ μm, $\lambda = 0.84$ μm, $h_{eff} = 1.334$ μm, $N = 1.53$, $\Gamma = 0.9$, and $w = 10$ μm. The integrals were evaluated numerically with a Simpson's rule integration routine. Figure 4.14 shows a plot of R_S as a function of the rms surface roughness δ (in angstroms) for both $s/D = 10^{-3}$ and $s/D = 10^{-2}$. Recall that $\tan \theta_{min} = s/D$ sets the lower limit on

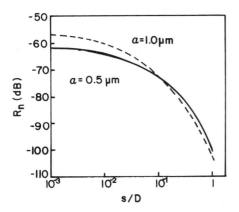

FIG. 4.17 Scattering ratio R_n as a function of the ratio s/D. (From Ref. 19.)

each integral. For D = 1 cm, the two curves in Fig. 4.14 correspond to detector separations of s = 10 μm and s = 100 μm. Figure
4.15 shows the corresponding plot of R_n as a function of M for
s/D = 10^{-2} and s/D = 10^{-3}. Figures 4.14 and 4.15 assume a = σ =
1 μm. Figure 4.16 plots R_s as a function of s/D for correlation
lengths σ = 0.5 μm and σ = 1.0 μm and surface roughness δ = 10 A.
Figure 4.17 plots R_n as a function of s/D for a = 0.5 μm and a =
1.0 μm with M = 10^{-5}. For a detector width w = 10 μm and D =
1 cm, s/D = 10^{-3} corresponds to the case of adjacent detector elements. The in-plane scattering noise in adjacent detectors is below
the 40-dB level for all cases plotted in Figs. 4.16 and 4.17, but the
rate of decrease in R_s and/or R_n with increasing detector separation
s is rather slow.

4.3 LOSSES DUE TO DIRECTIONAL CHANGES

The optical transmission characteristics of optical waveguides having
directional changes play an important role in determining the density
of optical components on a single integrated optic chip. Integrated
optics has not been developed to the same maturity as integrated
electronics. The density of electronic components on a single substrate is quite large and is increasing all the time. The question
of density of optical components on a single chip is being asked.
To achieve high component density, the waveguides must bend to
guide the light from one optical component to the next. In addition
to bending loss affecting component density on a single substrate,
many individual guided-wave components have channel waveguides
with bends such as the Mach-Zender interferometer modulator, the
directional coupler switch, and the Y coupler, to name a few.
    There are two broad categories of loss mechanisms due to wave-
guide bends. The first is a power loss due to curvature of the
waveguide initially presented by Marcatili [20] and Marcuse [21].
The second is the power scattered from the fundamental mode when
incident on a junction between two nonidentical waveguides. The
latter loss mechanism can be due to a straight-to-straight, straight-
to-curved, or curved-to-curved waveguide junction, which is basically due to a mode mismatch.

4.3.1 Physical Description of In-Diffused Channel
        Waveguide

To predict bending losses accurately, an accurate description of the
fundamental mode field of the waveguide and the waveguide parameters is required. Waveguides fabricated by diffusing titanium into
the surface of lithium niobate (LiNbO$_3$) are considered in this chapter. The diffused waveguide complicates the analysis somewhat be-

cause of the asymmetric nature of the waveguide cross section and the inhomogeneous index profile caused by diffusion.

Burns and Hocker [22] and Korotky et al. [23] calculated the dielectric profile of in-diffused channel waveguides to be

$$\varepsilon(x,y) = \varepsilon_o [n_b^2 + 2n_b \; \Delta n \; \varepsilon(x)\varepsilon(y)] \qquad (4.22)$$

where

$$\varepsilon(x) = \frac{1}{2} \left[\text{erfc} \left(\frac{W}{2D} + \frac{x}{D} \right) + \text{erfc} \left(\frac{W}{2D} - \frac{x}{D} \right) \right] \qquad (4.23)$$

where erfc (x) is the complementary error function,

$$\varepsilon(y) = \exp \left[- \left(\frac{y}{D} \right)^2 \right]$$

and

  y = depth direction perpendicular to the air/substrate interface

  x = width direction parallel to the interface

  n_b = refractive index of the bulk substrate

  Δn = refractive index difference between the surface index for one-dimensional diffusion and the refractive index of the substrate

  W = width of the undiffused dopant strip

  T = thickness of the undiffused dopant

  D = diffusion coefficient

The shape in the x direction of the two-dimensional diffused refractive index profile for various values of the ratio of strip width to diffusion depth (W/D) is shown in Fig. 4.18.

Keil and Auracher [24] have shown experimentally that the shape of the fundamental mode of diffused strip waveguides is to a good approximation Gaussian in width and Hermite-Gaussian in depth. An accurate empirical model of the field can be given by [25].

$$\underline{E} = \begin{pmatrix} \hat{x} \\ \hat{y} \end{pmatrix} E(x)E(y)e^{-i\beta z} \qquad (4.24)$$

where

$$E(x) = f(x)e^{-(x/X_o)^2} \tag{4.25}$$

$$E(y) = ye^{-(y-y_o)^2/Y_o^2} \tag{4.26}$$

$$f(x) = \begin{cases} 1 & \text{for the straight guide} \\ 1 + d_1x & \text{for a guide whose axis has constant} \\ & \quad \text{curvature } R_o \end{cases}$$

$$d_1 = \frac{(k_b X_o)^2}{2R_o} \tag{4.27}$$

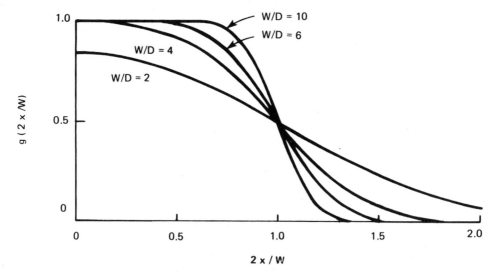

FIG. 4.18 The shape in the x direction of the 2-D diffused index profile for various values of the ratio of strip width to diffusion depth W/D. (From Ref. 22.)

$$k_b = \frac{2\pi}{\lambda} n_b \qquad (4.28)$$

In the curved waveguide the phase change along the axis (see Fig. 4.19) is $\exp(-im\phi)$, where

$$m = \beta R_o \qquad (4.29)$$

In a weakly guiding dielectric waveguide, the total power P_0 guided along the waveguide axis in a mode is [26]

$$P_o = \frac{1}{2}\left(\frac{\varepsilon_b}{\mu}\right)^{1/2} \int\limits_{A_{\infty x}} |\underline{E}|^2 \, dA \qquad (4.30)$$

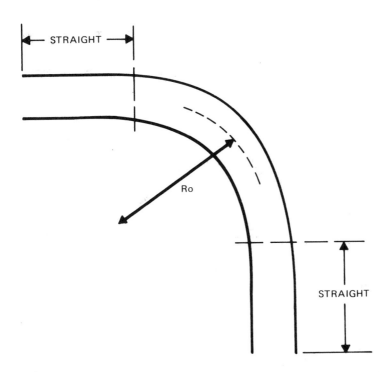

FIG. 4.19 Configuration of curved waveguide with bent axis.

where $A_{\infty x}$ is the infinite transverse cross section. Using the fields defined in Eqs. (4.25) and (4.26) and well-known tables of Gaussian integrals [27],

$$
P_o = \frac{1}{2} \left(\frac{\varepsilon_b}{\mu} \right)^{1/2} P_x P_y \tag{4.31}
$$

with

$$
P_x = \left(\frac{\pi}{2} \right)^{1/2} X_o \left[1 + \left(\frac{d_1 x_o}{2} \right)^2 \right] \tag{4.32}
$$

$$
P_y = \left(\frac{\pi}{8} \right)^{1/2} Y_o^3 \left\{ \left(\frac{y_o^2}{Y_o^2} + 0.25 \right) \left[1 + \mathrm{erf} \left(\frac{\sqrt{2} y_o}{Y_o} \right) \right] \right.
$$

$$
\left. + \frac{y_o}{Y_o \sqrt{2\pi}} \exp \left(\frac{-2 y_o^2}{Y_o^2} \right) \right\} \tag{4.33}
$$

For straight waveguides $R_o = \infty$ and thus $d_1 = 0$.

4.3.2 Scattering Loss at Waveguide Junctions

The scattering loss at waveguide junctions can be divided into two categories. First is the loss at a junction where two straight waveguides are joined together at some discrete angle, in which case there exists a discontinuity in the direction of the waveguide axis. Second, the waveguide junction is continuous (first derivative is continuous), such as occurs when a straight section is joined with a curved section continuously or a curved section is joined continuously with another curved section of opposite curvature. In both the continuous and discontinuous junctions, the loss is due to the mode field profile mismatch on opposite sides of the junction.

In practice, the reason for considering the two different types of junctions is due to the directional change requirement in an integrated optical waveguide. This study considers both the discontinuous directional change with only straight waveguides and the continuous directional change with a combination of straight and curved waveguides.

If we ignore the power scattered back away from a junction between two weakly guiding waveguides, the incident electromagnetic field at the junction is the excitation field for the next section of the guide. Therefore, for single-mode waveguides, to calculate the amount of power coupled from one guide to the next, only one integral needs to be solved for the relative modal amplitude of the fundamental mode of the second guide, a_{00}, excited by the incident mode from the first guide:

$$a_{00} = \frac{1}{2}\left(\frac{\epsilon_b}{\mu}\right)^{1/2} \int_{A_{\infty x}} \underline{E}_{02} \cdot \underline{E}_{01}^* \, dA \qquad (4.34)$$

Where $\underline{E}_{01}$ is the electric field of the fundamental mode of the ith guide normalized to unit power and $A_{\infty x}$ is the infinite cross-sectional area of the junction between the two guides. For an incident field with a total guided power P_{inc}, the power excited in the next guide is

$$P_{ex} = |a_{00}|^2 P_{inc} \qquad (4.35)$$

Because the waveguides are single mode, any power not coupled into the fundamental mode is scattered into the leaky or radiation mode spectrum and rapidly lost from the waveguide core. Therefore, the power lost from the guide to a junction is

$$\Delta P^R = P_{inc}(1 - |a_{00}|^2) \qquad (4.36)$$

For a junction joining two straight, in-diffused, channel waveguides whose axes are inclined at an angle θ_T with respect to each other, and whose axes are curved with radii of curvature R_1 and R_2, the overlap integral Eq. (4.34) can be solved. With the fields described by Eqs. (4.25) and (4.26), Eq. (4.34) is reduced to evaluation of the integral over the width only, because the fields in depth do not change at the junctions but do change in the plane of the waveguide. We obtain from Marcuse [28]

$$|a_{00}|^2 = \frac{2W'^2|(1 + id_1 b)(1 + id_2 b) + (d_1 d_2 / 2)W'^2|^2 \times \exp[-\beta_1^2 W'^2 \sin^2(\theta_T / 2)]}{(X_{01}X_{02})[(1 + (d_1 X_{01}/2)^2][1 + (d_2 X_{02}/2)^2]} \qquad (4.37)$$

where

$$W'^2 = \frac{(X_{01}X_{02})^2}{X_{01}^2 + X_{02}^2} \tag{4.38}$$

$$b = \beta_1 W'^2 \sin \frac{\theta_T}{2} \tag{4.39}$$

and X_{oi} is the width of the fundamental mode fields. β_1 is the longitudinal propagation constant of the first waveguide, from which the mode is incident on the junction and d_1 and d_2 are the field deformations due to curvature of the waveguide axes

$$d_i = \frac{(k_o n_b X_{oi})^2}{2R_o} . \tag{4.40}$$

For a straight waveguide $d_i = 0$ since $R = \infty$. For a curved waveguide d_1 and d_2 may be opposite in sign if the two curved waveguides have opposite curvature.

For two straight identical waveguides ($X_{01} = X_{02}$) inclined at angle θ_T, the power coupling coefficient becomes

$$|a_{00}|^2 = \exp\left(-\beta^2 W'^2 \sin^2 \frac{\theta_T}{2}\right) \tag{4.41}$$

For an identical straight and curved waveguide the power coupling coefficient is

$$|a_{00}|^2 = \frac{1}{1 + (d_2 X_o/2)^2} \tag{4.42}$$

Two identical curved waveguides with opposite curvature yield the power coupling coefficient

$$|a_{00}|^2 = \frac{[1 - (dX_o/2)^2]^2}{[1 + (dX_o/2)^2]^2} \tag{4.43}$$

4.3.3 Curvature-Induced Radiation Loss

There have been basically two different philosophies for calculating curvature-induced radiation losses in waveguides. The first applies coupled-mode theory [29], which spectrally decomposes the radiation field into an orthogonal set of fields to determine the coupling into each component. The second utilizes antenna theory [30] to evaluate the radiation for known current distributions of current sources and is known as the "volume current method." Both methods have been used to calculate bending losses in fibers and dielectric waveguides and has been employed to calculate the radiation loss in diffused $LiNbO_3$ waveguides [31–34].

Applying the volume current method and from Maxwell's equations for source-free media in which the dielectric permitivity tensor is inhomogeneous, $\underline{\underline{\varepsilon}} = \underline{\underline{\varepsilon}}\,(\underline{r})$, we have for time-harmonic fields proportional to $\exp(i\omega t)$,

$$\nabla \times \underline{H} = i\omega\underline{\underline{\varepsilon}}(\underline{r}) \cdot \underline{E} \qquad (4.44)$$

Writing the permitivity tensor in the form

$$\underline{\underline{\varepsilon}}(\underline{r}) = \underline{\underline{\varepsilon}}_b + \underline{\underline{\varepsilon}}(\underline{r}) - \underline{\underline{\varepsilon}}_b = \underline{\underline{\varepsilon}}_b + \delta\underline{\underline{\varepsilon}}(\underline{r}) \qquad (4.45)$$

and substituting Eq. (4.45) into Eq. (4.44), we obtain

$$\nabla \times \underline{H} = i\omega\underline{\underline{\varepsilon}}_b \cdot \underline{E} + i\omega\delta\underline{\underline{\varepsilon}}(\underline{r}) \cdot \underline{E}(\underline{r}) = i\omega\underline{\underline{\varepsilon}}_b \cdot \underline{E} + \underline{J}_e \qquad (4.46)$$

where $\underline{J}_e$ is the psuedovolume current density. In this the imhomogeneous permitivity profile is represented as an induced volume distribution of electric dipoles.

In the present application of the volume current method to radiation from a curved segment of a single-mode waveguide, the bent core of the waveguide is treated as the volume of the antenna. Peterman [35] has shown that the fundamental mode of such a curved structure can be well approximated near the waveguide axis. Peterman [35] showed that for isotropic dielectrics and for a symmetric waveguide, if the electric field of a straight waveguide is Gaussian, the electric field of a bent waveguide is simply that of the straight waveguide modified by a multiplicative factor as shown in Eqs. (4.25) and (4.26). Because the fundamental mode field is confined near the guide axis, the distribution of induced electric dipoles can be determined. The time-averaged radial Poynting vector of the radiation field, S^R, far from the waveguide axis as given by Snyder [36] is

$$S^R = \left(\frac{\mu}{\varepsilon_b}\right)^{1/2} \frac{k_b^2}{32(\pi R)^2} \left| \hat{\underline{R}} \times \int_{core} \underline{J}_e \, \exp(ik_b \hat{\underline{R}} \cdot \underline{R}') \, dV' \right|^2$$

(4.47)

where ε_b is the unperturbed permitivity of the substrate, $\underline{R}'$ the position vector of a source point in the waveguide core, and $\hat{\underline{R}}$ the unit vector in the direction of the observation point in the radiation field a distance R from the origin near the waveguide core. Since we are interested in the total power radiated instead of power density, we will assume that the waveguide is in a homogeneous and isotropic medium.

The total amount of power P^R radiated is

$$P^R = R^2 \int S^R \, d\Omega$$

(4.48)

where $d\Omega$ is an elemental solid angle. The power attenuation coefficient γ from a guide length L is given by

$$\gamma = \frac{P^R}{LP_o}$$

(4.49)

where

$$P(z) = P_0 \exp(-\gamma z)$$

(4.50)

with z an axial distance along the waveguide and P_0 the incident guided power. A complete annular waveguide, as shown in Fig. 4.20, is considered so that the radiation due to curvature of the waveguide axis is isolated. For this configuration

$$L = 2\pi R_o$$

(4.51)

where R_o is the radius of curvature. Assuming a symmetric structure and small bending loss, the radiation field should be assymetric and uniform in the plane of the bend, so that the radiation field and Poynting vector need only be evaluated at one point in this plane. For simplicity choose

$$\phi = \frac{\pi}{2}$$

(4.52)

which yields

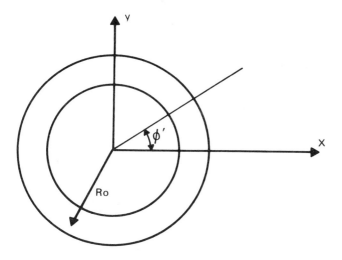

FIG. 4.20 Complete annulus of curved waveguide.

$$P^R = 2\pi R^2 \int S^R \sin\theta \, d\theta \qquad (4.53)$$

The coordinate system for the bent waveguide is shown in Fig. 4.21. From Eqs. (4.47) and (4.53) the power attenuation coefficient γ is calculated to be

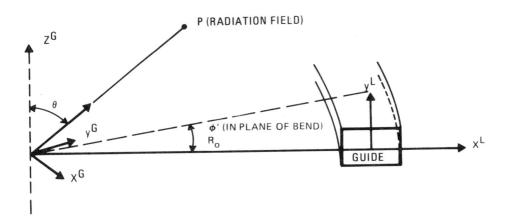

FIG. 4.21 Coordinate system of curved waveguide.

$$\gamma = \left[(\pi Q)^3 R_o \right]^{-1/2} (V^2 Y_1)^2 |F_y|^2 \exp \left[\frac{(QX_o)^2}{2} - \frac{2y_0^2}{Y_o^2 + D^2} \right]$$

$$\times \left[(1 + d_1 X_1) F_x^{(1)} + d_1 F_x^{(2)} \right]^2 \exp \left(-\frac{2Q^3}{3\beta^2} R_o \right)$$

$$\div \left\{ 8X_o Y_o \left[1 + \left(\frac{d_1 X_o}{2} \right)^2 \right] S_y \right\} \qquad (4.54)$$

where

$$V^2 = k_o^2 (n_s^2 - n_b^2) \qquad (4.55)$$

$$Q^2 = \beta^2 - n_b^2 k_o^2 \qquad (4.56)$$

$$Y_1 = \frac{Y_o D}{(Y_o^2 + D^2)^{1/2}} \qquad (4.57)$$

$$d_1 = \frac{(k_o n_b X_o)^2}{2R_o} \qquad (4.58)$$

where X_o and Y_o are the parameters of the fundamental mode field profile. d_1 is the correction term for the field in a curved waveguide described by Eq. (4.26) and D is the diffusion depth. Also,

$$S_y = Y_o^2 \left\{ \frac{y_o^2}{Y_o^2} + 0.25 \left[1 + \mathrm{erf} \frac{\sqrt{2} \, y_o}{Y_o} \right] \right.$$

$$\left. + \frac{y_o}{Y_o} \sqrt{2\pi} \exp \left(\frac{-2y_o^2}{Y_o^2} \right) \right\} \qquad (4.59)$$

$$F_y = Y_1 \left\{ \frac{Y_1 y_o}{Y_o^2} \left[1 + \mathrm{erf} \frac{Y_1 y_o}{Y_o^2} \right] + \frac{\exp[-(Y_1 y_o / Y_o^2)^2]}{\sqrt{\pi}} \right\} \qquad (4.60)$$

Finally,

$$F_x^{(1)} = X_0 \left[\text{erf}\left(\frac{\omega_+ \frac{D}{X_0} x}{X_0}\right) + \text{erf}\left(\frac{\omega_- \frac{D}{X_0} x}{X_0}\right) - \text{erf}\left(\frac{\omega_+}{p}\right) \right.$$

$$\left. - \text{erf}\left(\frac{\omega_-}{p}\right) \right] \tag{4.61}$$

$$F_x^{(2)} = - \sqrt{\pi} \frac{X_0^3}{pD} \sinh \frac{WQX_0^2}{2D^2 p^2} \exp\left[- \frac{(W/2D)^2 + (X_1/D)^2}{p^2} \right] \tag{4.62}$$

where W is the width of the channel before diffusion and

$$p^2 = 1 + \left(\frac{X_0}{D}\right)^2 \tag{4.63}$$

$$X_1 = \frac{QX_0^2}{2} \tag{4.64}$$

$$\omega_\pm = \frac{W}{2D} \pm \frac{X_1}{D} \tag{4.65}$$

Here X_0 is the half-width of the fundamental mode field in the plane of the waveguide. Y_0 and y_0 are weighted Gaussian parameters of the fundamental mode field perpendicular to the plane of the waveguide as defined by

$$E(Y) = Y \exp\left[- \frac{(Y - y_0)^2}{Y_0^2} \right] \tag{4.66}$$

The refractive index at the surface is n_s, n_b is the bulk substrate index, $k_0 = 2\pi/\lambda_0$ (λ_0 is the free-space wavelength). $\beta = n_{eff}k_0$ (n_{eff} is the effective index of the lowest-order mode), D is the diffusion depth, W is the width of the channel before diffusion, and R_0 is the radius of curvature of the curved portion of the waveguide.

4.3.4 Experimental Results

To compare experimental results with theory, as accurate determination of the waveguide parameters and its associated mode parameters

TABLE 4.1 Waveguide Parameters for
Z-Cut TE-Mode LiNbO$_3$

Waveguide parameter	Numerical value
W	3 μm
n_b (ordinary index)	2.2885
n_s	2.2916
n_{eff}	2.2888
D	3.53 μm
X_o	3.56 μm
Y_o	2.08 μm
y_o	0.63 μm

Source: Data from Refs. 37 and 38.

are required. The values for the waveguide parameters in Table
4.1 were obtained by using experimental diffusion measurements of
Burns et al. (39) in conjunction with the theoretical dispersion
curves of Hocker and Burns [40,41].

The waveguides used in these measurements were fabricated by
diffusing 3-μm-wide channels of Ti into Z-cut LiNbO$_3$. They were
diffused for 6 h at 1000°C. All the measurement results in this
section pertains to TE polarized laser light from a HeNe laser emitting at a wavelength of 0.633 μm. For further discussion of results
for TM polarization and other LiNbO$_3$ cuts, see Hutcheson [32]. The
weighted Gaussian mode shape described by Eq. (4.66) and Table
4.1 has the profile shown in Fig. 4.22.

A comparison of theoretical predictions and measurements is discussed for the two cases shown in Fig. 4.23. In Fig. 4.23a is shown
the case for two parallel noncollinear waveguides connected by a
third straight waveguide. In this case, the excess loss is due to
scattering at the two junctions. Figure 4.23b has the same two
waveguides connected by an S shape having a constant curvature.
For the S shape, there exists scattering loss at the two straight-
to-curved waveguides, the curved-to-curved waveguide, and the radiation loss due to curvature of the waveguide. All the measurements
were normalized to exclude absorption and Rayleigh scattering losses
by measuring these losses independently [37]. Results for the junction scattering loss for the corner bend are shown in Fig. 4.24, for

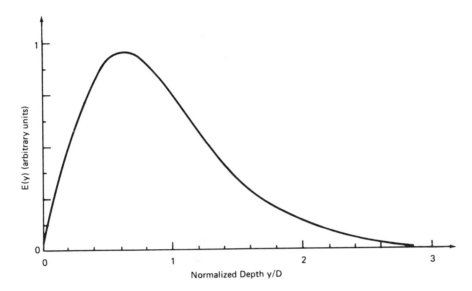

FIG. 4.22 Field profile perpendicular to the surface of the wave-guide for Z-cut LiNbO$_3$ and parameters from Table 4.1.

the straight-to-curved junction in Fig. 4.25, and for the curved-to-curved junction in Fig. 4.26.

To compare the two waveguide configurations shown in Fig. 4.23, the loss is plotted as a function of the ratio of the lateral separation X_S to the longitudinal separation Z_S for both corner and S bends. The results in Fig. 4.27 show in general that for a very small lateral separation, the corner bend approach is better. As the lateral separation increases, the S bend yields the lowest loss; however, the longitudinal separation gets larger.

The results above make a comparison between different types of waveguide transitions, but no attempt was made to measure the effect of waveguide and mode parameters on bend loss. Intuitively, one would think that bend loss can be reduced by increasing the mode confinement. Ramaswamy and Devino [42] showed that fabricating strongly guiding structures reduces loss significantly. The wave-guide structure fabricated in this case consisted of a raised-cosine function S shape proposed by Marcuse [43]. An attenuation of 0.3 dB was achieved for a transition of 0.1/3.0 mm (lateral/longitudinal separation), compared to 2.5 dB for the same transition shown in Fig. 4.27, for which the waveguide parameters were not optimized for mode confinement.

Minford et al. [34] performed a study of bending loss as a function of mode confinement at $\lambda = 1.3$ μm, which has more significance for optical fiber communication systems. The transition region connecting the two offset parallel waveguides was an s-shaped curve specified by

$$y(x) = \frac{h}{l}x - \frac{h}{2\pi} \sin\left(\frac{2\pi}{l}x\right)$$ (4.67)

where l is the longitudinal separation in the x direction and h is the lateral offset in the y direction. The waveguides were fabricated on

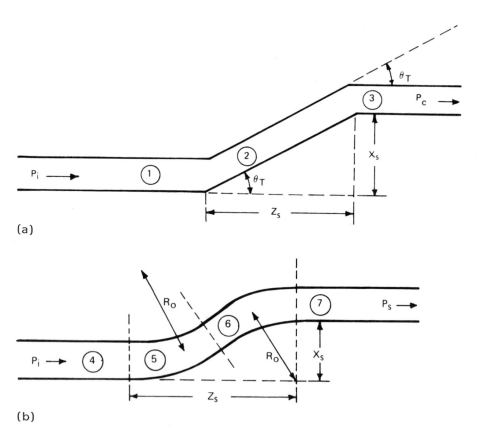

(a)

(b)

FIG. 4.23 Two configurations showing lateral and longitudinal offset; (a) two corner bend approach and (b) S-bend approach.

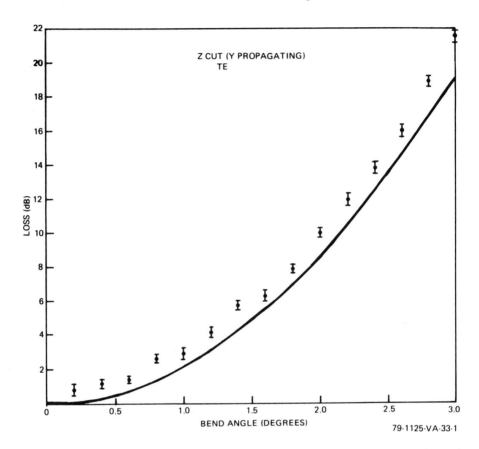

FIG. 4.24 Experimental corner bend scattering loss as a function of bend angle for TE polarized light, for Z-cut, Y-propagating LiNbO$_3$ and compared to theory.

Z-cut LiNbO$_3$ substrates having strip widths of 6 μm. For this experiment, the lateral offset was fixed at 0.1 mm and the loss was measured as a function of longitudinal offset for titanium thicknesses of 740, 850, and 1110 Å. All waveguides were diffused for 6 h at 1050°C. Losses were measured using the output of an Nd-YAG laser having a wavelength of 1.318 μm. Figure 4.28a and b show the effect of bend loss on Ti thickness for TM and TE polarizations, respectively. As can be seen from the figure, the larger Ti thickness yielded lower loss, due to the larger refractive index change, which in turn increased mode confinement. The TM polarization yielded significantly lower losses, which can be attributed to the larger

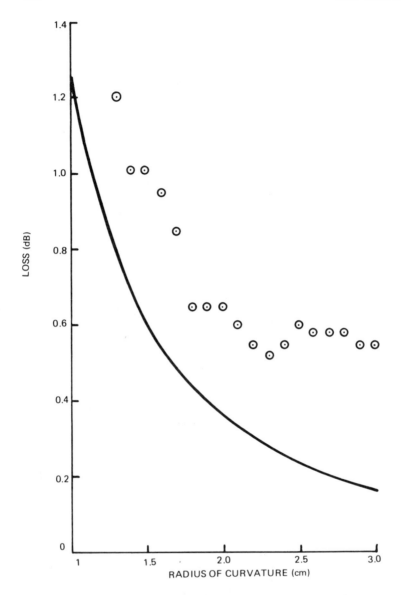

FIG. 4.25 Experimental mode mismatch loss at a single straight to curved junction as a function of radius of curvature and compared to theory. (Same conditions as Fig. 4.24.)

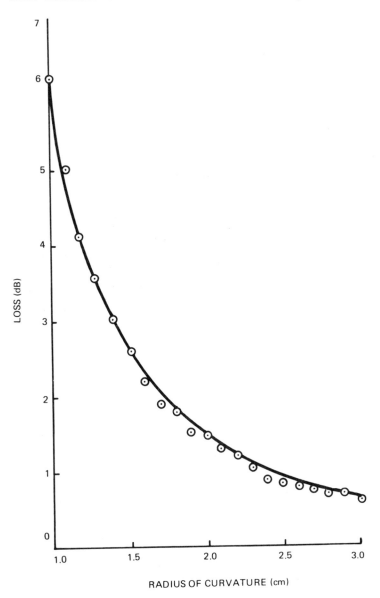

FIG. 4.26 Experimental mode mismatch loss at the curved to curved junction plotted as a function of radius of curvature and compared to theory. (Same conditions as Fig. 4.24.)

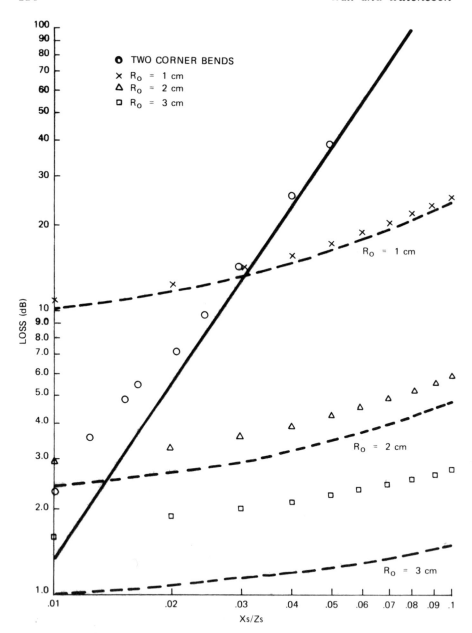

FIG. 4.27 Experimental loss measurements as a function of lateral/
longitudinal ratio for the two waveguide configurations shown in Fig.
4.23.

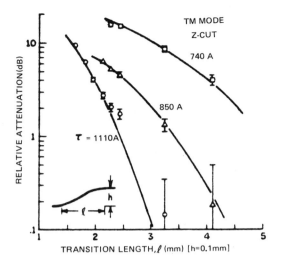

(a)

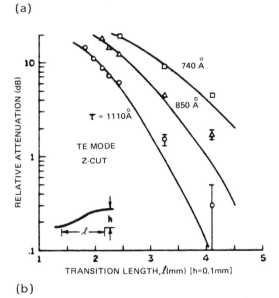

(b)

FIG. 4.28 Bend losses versus transition length (lateral offset = 0.1 mm), for 740, 850 and 1110 Å Ti thickness and (a) TM and (b) TE polarizations. (From Ref. 34.) The solid line is the best fit to Marcatilli's and Miller's model. (From Ref. 45.)

change in refractive index for this polarization [44]. For the
lateral/longitudinal separation ratio of 0.1/3 mm, TE polarization
yielded 1 dB and TM polarization yielded 0.2 dB. The experimental
results are compared to theoretical predictions based on the bend
loss model of Ref. 45.

4.3.5 Bend-Loss-Reduction Techniques

A technique was predicted by Taylor [46] that the loss through a
dielectric waveguide structure of successive closely spaced abrupt
bends will be a strong oscillatory function of the distance L between
each bend due to coherent coupling between the guided and unguided modes. The structure consists of equal-length straight waveguide segments separated by abrupt bends of uniform angle. The
bending loss is shown to be minimized when

$$L = \frac{(2m + 1)\lambda}{2 \, \Delta n} \qquad\qquad (4.68)$$

where L is the waveguide interconnection length, λ the free-space
optical wavelength, m = 0, 1, 2, . . . , and Δn the difference between the effective index of the guided mode and the weighted-
average effective index of the unguided modes excited at the bends.
Physically, light coupled from the guided mode into an unguided
mode at a bend can be completely coupled back into the guided mode
at a succeeding bend if the phase difference between the modes shifts
by $(2m + 1)\pi$ radians.

Waveguide structures of the type illustrated in Fig. 4.29 were
investigated by Johnson and Leonberger [47] to determine the optimum interconnection of length L between successive 1° abrupt bends.
The experimental results for $\lambda = 0.633$ μm with TE and TM polarization are shown in Fig. 4.30. The waveguides were single mode
with a prediffused channel width of 3 μm and the diffusion conditions were chosen to maximize the optical confinement. Oscillatory
behavior in TE optical transmission can be seen with the variation

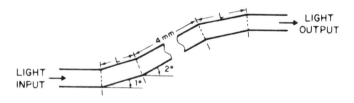

FIG. 4.29 Two pairs of coupled waveguide bends. (From Ref. 47.)

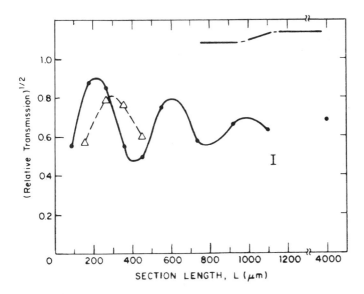

FIG. 4.30 Transmission as a function of L through a single coupled pair for both TE (●) & TM (Δ) polarizations. λ = 0.633 μm. (From Ref. 47.)

becoming damped at large L. The variation in transmission with L is also apparent for TM polarization. These measurements were made on X-cut LiNbO$_3$ with waveguide propagation in the y direction. The devices were tested using end-fire coupling.

Multisection waveguide bend structures of the type shown in Fig. 4.31 were also investigated. This type of structure is necessary to

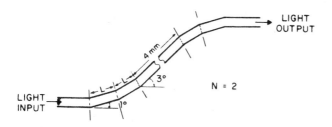

FIG. 4.31 Multisection waveguide bend structure with interconnection length L and consisting of successive 1° bends. A straight 4mm long section isolates two identical, coupled bend structures. (From Ref. 47.)

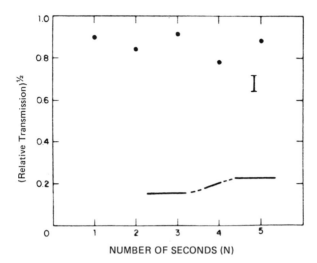

FIG. 4.32 Transmission as a function of N through single multisection bend structures with L = 180 μm. TE polarization λ = 0.633 μm. Oscillation is due to a strong net coupling effect at this interconnection length. (From Ref. 47.)

fabricate large bends. By choosing the interconnection length equal to that for peak transmission in single-section bends, minimum loss can be obtained. Shown in Fig. 4.32 is the TE transmission for N = 1 through N = 5 coupled sections. The interconnection length L = 180 μm, which yielded maximum transmission in the single-section structure, was used in this experiment. As can be seen, optical transmission drops off slowly with N. A multiplicative decrease in transmission with increasing N should not be expected in multisection structures since coupling extends beyond nearest-neighbor bends. A bend structure consisting of 60 successive 1° coupled bends shown in Fig. 4.33 with an effective radius of curvature of 1 cm was also tested. The measured TE transmission was found to be only 6 dB below that of straight waveguides of comparable length on the same substrate. This loss is far lower than the loss of approximately 50 dB expected for a series of 60 isolated 1° bends.

Another technique for reducing bending loss [48] makes use of microprisms of raised index placed in the waveguide path in the bend region, as shown in Fig. 4.34. The concept is called CROWNING (Controlling Radiation from Optical Waveguides by Notching the Index in the Guide) and operates on the principal of equalizing the optical path lengths for the inside and outside edges of the waveguide mode. A planar region of raised index is situated near the

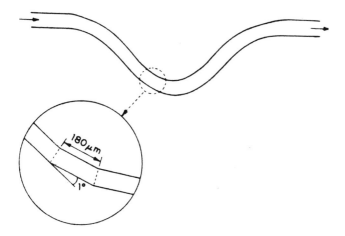

FIG. 4.33 Sixty section bend structure consisting of a succession of coupled 1° bends with L = 180 μm. Bend loss $\stackrel{\sim}{\sim}$ 6 dB; λ = 0.633 μm; TE polarization. (From Ref. 47.)

inner edge of the bend at the base of the triangles to compensate for the tendency of the mode to move toward the outside edge of the bent waveguide.

The CROWNING technique was implemented by Korotky et al. [48] in LiNbO$_3$ waveguides and tested at a wavelength of 1.56 μm. The waveguide fabrication consisted of a two-step diffusion process with the crown structures diffused first. The waveguide and crown dimensions are shown in Fig. 4.34 with an initial waveguide titanium thickness of 950 Å and crown thickness of 1000 Å and diffused for 6 h at 1050°C. The structures were fabricated on Z-cut, y-propagation LiNbO$_3$.

The experiment consisted of measuring losses for the TM mode for constant-radius S bends with and without the crown, as shown

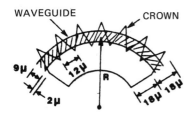

FIG. 4.34 Schematic diagram of CROWNING concept. (From Ref. 48.)

FIG. 4.35 Photomicrograph of S-bend configuration and Crown pattern used to measure loss. (Photo courtesy of AT&T Bell Laboratories, Holmdel, New Jersey.)

in the photograph in Fig. 4.35. A close-up photograph of the transition region for the Ti:LiNbO$_3$ waveguide bend and crown after diffusions is shown in Fig. 4.36. Measurements were made with a fixed lateral offset of 0.1 mm between parallel input and output waveguides. The experimental results plotted in Fig. 4.37 are the measured excess losses incurred for traveling the bends as compared to a straight waveguide. The results are compared to the conventional S-bend configuration [34]. The results show that there is an optimum bend radii of 4 mm for this particular configuration, as predicted by the authors. If the S-bend radius is larger than the fixed design radius, the crown technique becomes less effective due to mode mismatch. Finally, as the S-bend radius becomes much less than the fixed design radius, it is expected that the bend radiation loss will approach that of the conventional S bend.

4.3.6 Other Bending-Loss Results

An experiment was performed by Ramar et al. [49] to isolate all of the integrated optic circuit losses, including waveguide bends, waveguide offsets, metal electrodes, and fiber-to-channel coupling. In this experiment a $\Delta\beta$ reversal switch was fabricated in Z-cut

Ti-diffused $LiNbO_3$. The device was fiber pigtailed to a 0.83-μm-wavelength laser. In this experiment, the loss was measured for a 0.5° angle bend for (1) a 3-μm-wide channel waveguide and 500 Å Ti thickness, (2) a 4-μm-wide channel waveguide and 420 Å Ti thickness, and (3) a 5-μm-wide channel waveguide and 400 Å Ti thickness ness. All samples were diffused for 6 h at 1000°C to yield single-mode operation. The measured bend losses were 1.37, 0.76, and 1 dB, respectively, at a wavelength of 0.83 μm.

All the bending-loss studies discussed in this chapter have been for the dielectric material $LiNbO_3$. Austin [50] and Austin and Flavin [51] studied losses in curved GaAs/AlGaAs rib waveguides. The waveguides were $GaAs/Al_{0.12}Ga_{0.88}As$ and grown by the metal organic chemical vapor deposition technique. Measurements were made for single-mode 2-μm-wide waveguides and 3-μm-wide waveguides

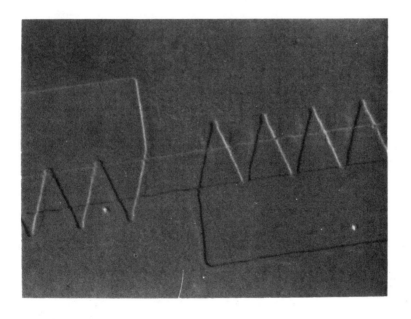

10 μm

FIG. 4.36 Close up photomicrograph of Ti:$LiNbO_3$ waveguides with Crown pattern after diffusion. (Photo courtesy of AT&T Bell Laboratories, Holmdel, New Jersey.)

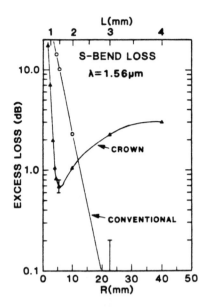

FIG. 4.37 Measured excess loss with and without Crowning technique. (From Ref. 48.)

supporting two modes. The radius of curvature for both cases was 0.3 mm and measured loss was 0.6 and 1.3 dB/rad, respectively. The channel waveguides were defined by ion-beam milling, and for these measurements both scattering from rough edges and radiation loss due to bends contributed to the loss.

REFERENCES

1. D. Marcuse, Mode conversion caused by surface imperfections of a dielectric slab waveguide, *Bell Syst. Tech. J.*, *48*:185 (1969).

2. T. L. Tsai and H. S. Tuan, Reflection and scattering by a single groove in integrated optics, *IEEE J. Quantum Electron.*, *QE-10*:326 (1974).

3. D. G. Hall, Scattering of optical guided waves by waveguide surface roughness: a three-dimensional treatment, *Opt. Lett.*, *6*:601 (1981).

4. E. Bradley and D. G. Hall, Out-of-plane scattering from glass waveguides: comparison of theory and experiment, *Opt. Lett.*, *7*:235 (1982).

5. G. H. Ames and D. G. Hall, Attenuation in planar optical wave-guides: comparison of theory and experiment, *IEEE J. Quantum Electron.*, *QE-19*:845 (1983).

6. D. G. Hall, Comparison of two approaches to the waveguide scattering problem, *Appl. Opt.*, *19*:1732 (1980).

7. Y. Suematsu, K. Furuya, M. Hakuta, and K. Chiba, Farfield radiation pattern caused by random wall distortion of dielectric waveguides and determination of correlation length, *Electron. Commun. Jpn.*, *56-C*:62 (1973).

8. M. Gottlieb, G. B. Brandt, and J. J. Conroy, Out-of-plane scattering in optical waveguides, *IEEE Trans. Circuits Syst.*, *CAS-26*:1029 (1979).

9. J. M. Elson and J. M. Bennett, Relation between the angular dependence of scattering and the statistical properties of optical surfaces, *J. Opt. Soc. Am.*, *69*:31 (1979).

10. W. Panofsky and M. Phillips, *Classical Electricity and Magnetism*, Addison-Wesley, Reading, Mass., Chap. 12 (1955).

11. J. M. Elson and J. M. Bennett, Vector scattering theory, *Opt. Eng.*, *18*:116 (1979).

12. P. K. Tien, Light waves in thin films and integrated optics, *Appl. Opt.*, *10*:2395 (1971).

13. H. Kogelnik, Theory of dielectric waveguides, Topics in Applied Physics, Vol. 7, *Integrated Optics* (T. Tamir, ed.), Springer-Verlag, New York (1979).

14. J. T. Boyd and D. B. Anderson, Effect of waveguide optical scattering on the integrated optical spectrum analyzer dynamic range, *IEEE J. Quantum Electron.*, *QE-14*:437 (1978).

15. M. C. Hamilton, D. A. Wille, and W. J. Micheli, "An Integrated Optical RF Spectrum Analyzer," Proceedings of the IEEE 1976 Ultrasonics Symposium, Institute of Electrical and Electronics Engineers, New York (1976).

16. D. Mergerian, E. C. Malarkey, R. P. Pautienus, J. C. Bradley, G. E. Marx, L. D. Hutcheson, and A. L. Kellner, Operational integrated optical R.F. spectrum analyzer, *Appl. Opt.*, *19*:3033 (1980).

17. R. A. Modavis, In-plane scattering in planar optical waveguides, M.S. Thesis, The Institute of Optics, University of Rochester, Rochester, N.Y. (1983).

18. R. A. Modavis and D. G. Hall, In-plane scattering in planar optical waveguides, *Opt. Lett.* *9*:96 (1984).

19. D. G. Hall, In-plane scattering in planar optical waveguides: refractive index fluctuations and surface roughness, *J. Opt. Soc. Am.*, *A2*: 747 (1985).

20. E. A. J. Marcatili, Bends in optical dielectric guides, *Bell Syst. Tech. J.*, *48*: 2013 (1969).

21. D. Marcuse, Bending losses of the asymmetric slab waveguide, *Bell Syst. Tech. J.*, *50*: 2551 (1971).

22. W. K. Burns and G. B. Hocker, End fire coupling between optical fibers and diffused channel waveguides, *Appl. Opt.*, *16*: 2048 (1977).

23. S. K. Korotky, W. J. Minford, L. L. Buhl, M. D. Divino, and R. C. Alferness, Mode size and method for estimating the propogation constant of single-mode Ti:LiNbO$_3$ strip waveguides, *IEEE J. Quantum Electron.* *QE-18*:1796 (1982).

24. R. Keil and F. Auracher, Coupling of single-mode Ti-diffused LiNbO$_3$ waveguides to single mode fibers, *Opt. Commun.*, *30*: 23 (1979).

25. I. A. White, L. D. Hutcheson, and J. J. Burke, End-fire coupling between optical fibers and diffused channel waveguides: comment, *Appl. Opt.*, *18*: 2362 (1979).

26. A. W. Snyder, Excitation and scattering of modes on a dielectric or optical fiber, *IEEE Trans. Microwave Theory Tech.*, *MTT-17*: 1138 (1969).

27. I. S. Gradsteyn and I. M. Ryzhik, *Tables of Integrals Series and Products*, Academic Press, New York (1965).

28. D. Marcuse, Radiation losses of parabolic index slabs and fibers with bent axes, *Appl. Opt.*, *17*: 755 (1978).

29. D. Marcuse, Curvature loss formula for optical fibers, *J. Opt. Soc. Am.*, *66*: 216 (1976).

30. I. A. White, Radiation from bends in optical waveguides: the volume-current method, *Microwav., Opt. and Acoust.*, *3*: 185 (1979).

31. I. A. White, L. D. Hutcheson, and J. J. Burke, Modal fields and curvature losses in Ti:diffused LiNbO$_3$ waveguides, *SPIE Proc.*, *239*: 74 (1980).

32. L. D. Hutcheson, Losses in titanium—diffused lithium nobate channel waveguides due to directional changes, Ph.D. dissertation, Optical Sciences Center, University of Arizona, Tucson, Ariz. (1980).

33. L. D., Hutcheson, Design criteria for integrated optical devices, *SPIE Proc.*, *317*: 32 (1981).

34. W. J. Minford, S. K. Korotky, and R. C. Alferness, Low-loss Ti:LiNbO$_3$ waveguide bends at $\lambda = 1.3$ µm, *IEEE J. Quantum Electron.*, *QE-18*:1802 (1982).

35. K. Peterman, Theory of microbending loss in monomode fibers with arbitrary refractive index profile, *Arch. Elektr. Uberti*, *30*:337 (1976).

36. A. W. Snyder, Radiation loss due to variations of radius on dielectric or optical fibers, *IEEE Trans. Microwave Theory Tech.*, *MTT-18*:608 (1970).

37. L. D. Hutcheson, I. A. White, and J. J. Burke, Comparison of bending losses in integrated optical circuits, *Opt. Lett.*, *5*:276 (1980).

38. L. D. Hutcheson, I. A. White, and J. J. Burke, "Losses in Diffused LiNbO$_3$ Waveguides Caused by Directional Changes," Paper WB2, Digest of the Topical Meeting on Integrated and Guided-Wave Optics, Incline Village, Nev. (Jan. 1980).

39. W. K. Burns, P. H. Klein, E. J. West, and L. E. Plew, Ti diffusion in Ti:LiNbO$_3$ planar and channel optical waveguides, *J. Appl. Phys.*, *50*:6175 (1979).

40. G. B. Hocker and W. K. Burns, Modes in diffused optical waveguides of arbitrary index profile, *IEEE J. Quantum Electron.*, *QE-11*:270 (1975).

41. G. B. Hocker and W. K. Burns, Mode dispersion in diffused channel waveguides by the effective index method, *Appl. Opt.*, *16*:113 (1977).

42. V. Ramaswamy and M. D. Divino, "Low-loss Bends for Integrated Optics," Proceedings of the Conference on Lasers and Electrooptics, Washington, D.C. Paper THP1 (1981).

43. D. Marcuse, Length optimization of an S-shaped transition between offset optical waveguides, *Appl. Opt.*, *17*:763 (1978).

44. R. V. Schmit and I. P. Kaminov, Metal-diffused optical waveguides in LiNbO$_3$, *Appl. Phys. Lett.*, *25*:458 (1974).

45. E. A. J. Marcatili and S. E. Miller, Improved relations describing directional control in electromagnetic wave guidance, *Bell Syst. Tech. J.*, *48*:2161 (1969).

46. H. F. Taylor, Losses at corner bends in dielectric waveguides, *Appl. Opt.*, *16*:711 (1977).

47. L. M. Johnson and F. J. Leonberger, Low-loss LiNbO$_3$ waveguide bends with coherent coupling, *Opt. Lett.*, *8*:111 (1983).

48. S. K. Korotky, E. A. J. Marcatili, J. J. Veselka, and R. H. Bosworth, Greatly reduced losses for small-radius bends in Ti:LiNbO$_3$ waveguides, *Appl. Phys. Lett.*, *48*:92 (1986).

49. O. G. Ramer, C. Nelson, and C. Mohr, Experimental integrated optic circuit losses and fiber pigtailing of chips, *IEEE J. Quantum Electron.*, *QE-17*:970 (1981).

50. M. W. Austin, GaAs/GaAlAs curved rib waveguides, *IEEE J. Quantum Electron.*, *QE-18*:795 (1982).

51. M. W. Austin and P. G. Flavin, Small-radii curved rib waveguides in GaAs/GaAlAs using electron-beam lithography, *IEEE J. Lightwave Technology*, *LT-1*:236 (1983).

5

Sources and Detectors

VIRGINIA M. ROBBINS and GREGORY E. STILLMAN   *Center for Compound Semiconductor Microelectronics, University of Illinois at Urbana-Champaign, Urbana, Illinois*

5.1 INTRODUCTION

Sources and detectors are, of course, important elements in any useful integrated optical circuits. Integrated optical structures can provide enhanced performance over discrete sources and detectors for specific applications, including fiber optical communications, space communications, optical disks for computer memories, and optical disks for home entertainment systems. Because of this, there has been research in a number of areas to try and realize the potential of integrated sources and detectors.

For fiber optical communication systems there are many instances in which semiconductor light-emitting diodes (LEDs) are satisfactory sources. However, for integrated optical applications extremely large bandwidths are generally of interest. In addition, many electrooptical effects require single-frequency, single-mode laser emission. In those cases, semiconductor lasers are the only suitable sources.

The characteristics that are important for integrated optics, as well as for other applications of semiconductor lasers, include room-temperature operation, good temperature stability, linear L-I characteristics, single-mode operation, and suitable high-frequency modulation characteristics. In Sec. 5.2 the various types of semiconductor laser structures that have been developed to satisfy these

requirements are described briefly, and then refinements of these
structures for integrated optical applications are discussed. Next,
integrated optical devices that have produced unprecedented semiconductor laser performance are summarized.

The detector characteristics of importance include high responsivity at the wavelength of operation, low-noise performance, and
high-frequency operation. Several types of detectors are discussed
in Sec. 5.3, together with the factors that influence the foregoing
parameters in these devices. Some of the integrated detector structures that have been developed are described.

5.2 SOURCES

Semiconductor injection lasers, independently invented by four
groups of researchers [1–4], are the sources that make fiber optics
and integrated optics feasible, but it was not until 1970 that
continuous-wave (CW) room-temperature operation of semiconductor
injection lasers was achieved using heterostructures [5,6]. Refinements of the heterostructure used by these workers are still used
today. A schematic diagram of the $Al_xGa_{1-x}As$-$GaAs$-$Al_xGa_{1-x}As$
double-heterojunction p^+nn^+ laser structure is shown in Fig. 5.1a,
together with a simplified energy-band diagram (b), the corresponding variation of the refractive index throughout the structure (c),
and the intensity distribution of the fundamental transverse mode
with a fraction Γ of the radiant energy guided within the active region shown by the crosshatched area (d). The changes in bandgap
between the cladding and active layers in this structure provide barriers in the valence and conduction bands that under forward bias
confine the injected carriers to the active layer. The active layer
is made thin (<1.0 μm) so that high carrier concentrations are obtained for moderate injection currents. The radiation resulting from
the recombination of electrons and holes in this region is guided
between the facet mirrors of the Fabry-Perot cavity by the planar
slab waveguide formed by the low-index cladding layers and the
high-index active layer. The radiant energy is confined in the plane
parallel to the p-n junction, as shown in Fig. 5.1d. This structure
would have been grown epitaxially on an n^+-GaAs substrate, which
is not shown in the figure.

5.2.1 Wavelength Threshold Current

The wavelength of the laser emission is coarsely determined by the
energy gap of the active layer and more finely by the gain and loss
within the optical cavity. For reliable and efficient operation of
heterostructure laser devices, it is important that the heterointerfaces

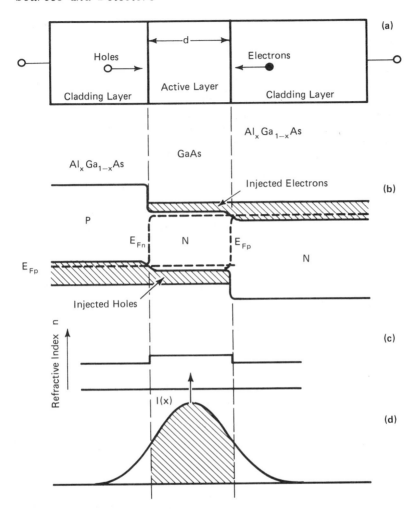

FIG. 5.1 (a) P⁺NN⁺ AlGaAs-GaAs double heterostructure laser structure; (b) energy band diagram under forward bias; (c) index of refraction variation with position; (d) light intensity profile in the
structure. The fraction of the radiant energy confined in the active
region of the laser structures is indicated by Γ.

be free of defects that result in nonradiative recombination centers. Practically, this requires a close lattice match between adjacent layers. The AlGaAs alloy has nearly the same lattice constant as GaAs, so that high-quality, defect-free heterostructures can readily be formed in this material system. The carrier confinement in the active region depends exponentially on the height of the potential steps at the interface between the cladding and active layers, ΔE_C and ΔE_V, respectively; the energy gap difference between the cladding and the active region must be sufficient to provide the required confinement. Thus the useful wavelength range for the AlGaAs-GaAs double heterostructures is about 0.9 to 0.65 µm. Other wavelengths can be covered with different lattice-matched semiconductor alloys, and the lattice-matched GaInAsP/InP quaternary alloy system permits the formation of double-heterostructure lasers in the wavelength range 1.1 to 1.6 µm, which is particularly important for low-loss fiber optic applications.

The thickness of the active layer d influences both the carrier confinement and the optical confinement in the double heterostructure. Figure 5.1d shows the radiant energy for the lowest-order mode in the plane parallel to the p-n junction. For active-layer thicknesses of less than about 1 µm, only the lowest-order mode is guided in the slab waveguide, but for this case a significant part of the energy is guided in the lossy cladding layers. The thinner the active layer, the greater the energy that is transported in the cladding layers (i.e., the smaller d is, the smaller Γ is). Thus as d is decreased, the threshold current first decreases because of increased carrier confinement and the ease of obtaining population inversion, but as d is decreased still further, Γ decreases, the optical losses increase, and the threshold current begins to increase. This variation is shown schematically in Fig. 5.2. The lowest threshold currents are obtained for active-layer thicknesses of less than 1 µm for both GaAlAs/GaAs and GaInAsP/InP lasers. The thickness for lowest threshold current density depends on the compositions of the cladding and active layers for either alloy system through the variation of the energy-band discontinuities ΔE_C and ΔE_V and the steps in the refractive index, Δn, and is smallest for the largest barriers and step Δn because of the better confinement of the carriers and the radiation to the active-layer thickness for large ΔE_C and ΔE_V, and Δn, respectively.

5.2.2 Transverse Mode Control

In actuality, the modal distribution in the plane parallel to the p-n junction and the heterobarriers must also be considered as well as that perpendicular to the plane of the junction; that is, both horizontal (or lateral) and vertical transverse modes must be considered.

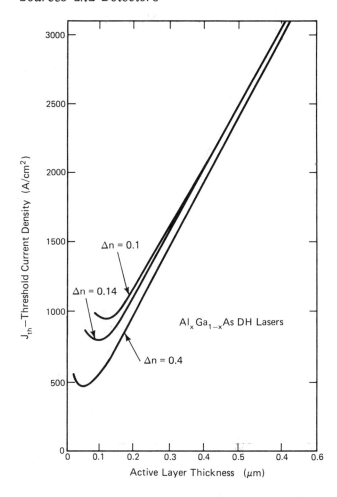

FIG. 5.2 Variation of threshold current density with active layer thickness d for different index steps Δn.

(The vertical transverse modes are often simply referred to as transverse modes.)

A. Gain-Guided Stripe Geometry Lasers

The thickness of the active layer d which gives the minimum threshold current density will also produce an output beam that has considerable divergence, so this may have to be considered in particular applications. For the broad-area double heterostructure shown schematically in Fig. 5.1, there are many possible "horizontal" modes, and in addition the threshold current will be very large even though the threshold current density is low. The large number of horizontal modes in broad-area structures leads to many-lobed far-field patterns in the laser emission which change with temperature or drive current. A technique that has been used to obtain lower threshold currents and better horizontal transverse mode control is to form "stripe" geometry laser structures. In this type of double heterostructure, the current through the junction is confined to a narrow stripe. The injected carriers in this stripe produce an increase in the index of refraction in the active region under the stripe, which is also the region where the optical gain is produced. This structure reduces the total threshold current, minimizes the power dissipation and heat-sinking requirements, and through a process called gain guiding, weakly confines the light emission in the horizontal transverse direction. This structure also minimizes multiple lasing filaments and permits single horizontal transverse mode operation over a narrow range of currents. There are a number of different techniques that have been used for the fabrication of stripe geometry double-heterojunction lasers, and Fig. 5.3 shows the cross sections of AlGaAs/GaAs stripe geometry lasers formed by (a) oxide masking, (b) proton bombardment, and (c) shallow diffusion.

   While the stripe geometry structures described above have many desirable features, they all have one undesirable characteristic that is critical for integrated optics applications. The emission intensity versus drive current (L-I) characteristics for these structures often show distinct nonlinearities that in the extreme, as shown by the upper curve in Fig. 5.4, result in an actual decrease in the emitted power with increasing current over a narrow range. These "kinks" in the L-I characteristic are now known to be due to changes in the horizontal transverse mode structure which result from changes in the "gain guiding" with drive current.

B. Index-Guided Stripe Geometry Lasers (TM)

Because of the importance of single-transverse-horizontal-mode operation in fiber optical communications, a number of schemes have been developed to confine the light in the lateral direction in double-heterostructure lasers. Most of these schemes produce a high refractive index in the three-dimensional region where the light is to be guided. The gain-guided laser also does this, but because in this case the amount of refractive index change is dependent on the injection current, the guiding conditions change with the laser operating conditions. These new structures provide a high-index waveguide in which the propagating light is internally reflected at the waveguide boundaries in the lateral direction just as it is in the vertical direction in a slab waveguide.

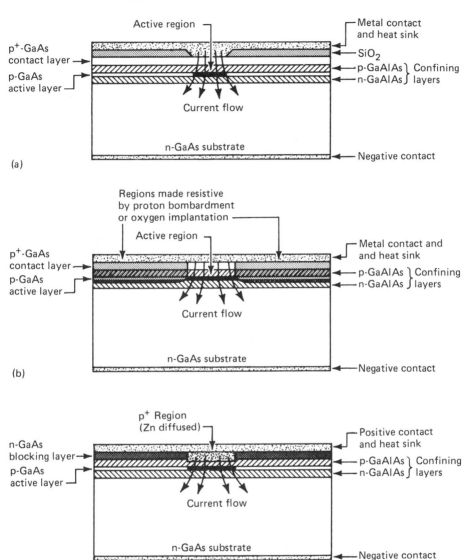

FIG. 5.3 Stripe geometry laser structures: (a) oxide masked stripe; (b) proton bombarded stripe; (c) deep Zn diffused stripe.

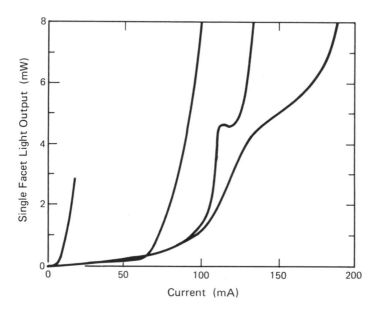

FIG. 5.4 L-I characteristics for gain-guided stripe geometry lasers showing non-linear behavior and "kinks".

Four different structures that have been used to obtain single-transverse-mode index-guided double-heterostructure lasers are shown in Fig. 5.5. The buried heterostructure shown schematically in this figure provides a high-index three-dimensional waveguide and three-dimensional confinement of the injected carriers as well. In practical devices this structure is often combined with the oxide-stripe structure to minimize the requirements on the resistivity of the GaAlAs layers that are regrown around the etched mesa during fabrication of this structure. The output power obtainable from this structure is limited because of optical damage that results due to the small emitting area (the power density threshold of optical damage in GaAs lasers is about 11 mW/μm^2). In the small structures required for single-transverse-mode operation, this damage threshold limits the optical power to about 1 mW for reliable devices. Low threshold currents, good linearity, and high-frequency response have been obtained with this structure, but the fabrication process is complex and the reproducibility and yield for devices using this structure are not high.

The selectively diffused structure shown in Fig. 5.5 is much easier to fabricate and provides lateral transverse index guiding because of the high doping in the diffused area. The channeled-substrate laser in Fig. 5.5 uses a different mechanism to obtain the

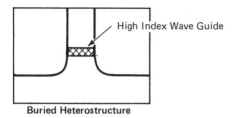

Buried Heterostructure

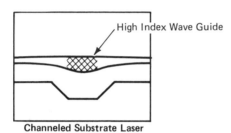

Selectively Diffused Structure

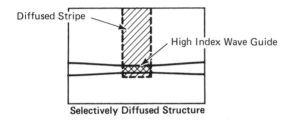

Channeled Substrate Laser

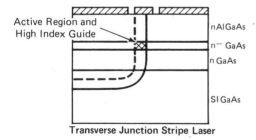

Transverse Junction Stripe Laser

FIG. 5.5 Index-guided double heterostructure stripe geometry laser structures.

required index change in the lateral transverse direction essential
for single-transverse-mode operation. The channel etched in the
substrate produces thickness variation in the subsequent epitaxial
layers and the thickness variation results in an effective index of
refraction in the waveguide which is highest in the region shown
as the high-index waveguide. Still another single-lateral-transverse-
mode laser structure is the transverse-junction stripe laser shown in
Fig. 5.5 [7,8]. In one form of this structure, the transverse junction
is formed by zinc diffusion, and carriers are injected transversely when
the p^+-zinc-diffused layer on the left is biased positively with respect to the n-type contact on the right. The heterojunctions confine the injected carriers to the n-GaAs active region when the injection is highest because of the lower forward voltage drop of this
junction compared to the GaAlAs transverse p-n junctions. The
lateral transverse waveguiding occurs because the doping profile introduced by the zinc diffusion into the heavily doped n-type material produces a high index of refraction in the active region indicated. This structure has been shown to be capable of single-
longitudinal-mode operation as well as single-transverse-mode operation with low threshold currents and good reliability. Other structures, such as bent-guide or "crank" stripe lasers [9] and mis-
aligned Fabry-Perot mirrors [10], have been devised to introduce
higher losses for the higher-order lateral transverse modes.

5.2.3 Longitudinal Mode Control

The structures described above and shown in Fig. 5.5 all operate
in a single transverse mode, but with the exception of the channeled
substrates laser and the transverse-junction stripe laser structure,
all operate in a number of longitudinal modes. This multimode operation results in a series of peaks in the emission spectra separated
by a few angstroms, dependent on the mode separation of the Fabry-
Perot cavity. For most integrated optics applications, it is important that the laser sources have both transverse and longitudinal
single-mode operation. The transverse-junction stripe (TJS) lasers
and the channeled-substrate lasers seem to operate naturally in a
single longitudinal mode. In the TJS structure the active region is
heavily doped p-type, and Scifres et al. [11] explained the single-
longitudinal-mode operation in terms of enhanced diffusion of the
carriers in the p^+ active layers.

A different means of obtaining single-longitudinal-mode operation
is to make the cavity short enough that only one mode falls within
the gain spectrum of the laser. Extremely short semiconductor
cavities will lase [12], but for semiconductor injection lasers, it is
impractical to make the cavities short enough to obtain reliable
single-longitudinal-mode operation.

Another means of obtaining single-longitudinal-mode operation with a cavity that has only one mode within the gain band of the laser is to use a distributed feedback (DFB) [13] or distributed Bragg refector (DBR) [14]. A diagram of a GaAlAs DFB laser is shown in Fig. 5.6. The fabrication of such a structure requires a number of complex steps, including precision lithography to define a grating with a submicron period, epitaxial growth over the grating, and so on. The transverse mode control of DFB lasers is just as important as for the structures described previously, and the same techniques can be used. For the DFB laser structure shown in Fig. 5.6, lateral transverse mode control is obtained using the channeled-substrate structure. This device operates in a single longitudinal and single transverse mode over a wide range of temperatures and drive currents, with an output power of between 5 and 10 mW.

5.2.4 Modulation Speed

The simplest way to modulate the power output of a semiconductor laser is to modulate the drive current. When a current pulse is applied to a semiconductor laser there is generally a delay of a few nanoseconds while the electron and hole populations build up to the threshold level for stimulated emission. Once the device starts to lase there is usually a decaying relaxation oscillation because of the coupling between the excess electron and hole populations and the photon flux [16]. These relaxation oscillations are minimized and/or

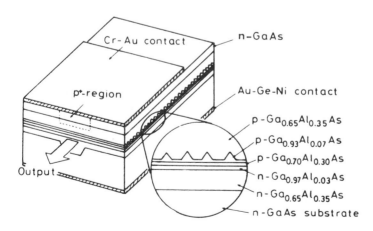

FIG. 5.6 Schematic structure of a channeled-substrate planar distributed feedback laser. (From Ref. 15.)

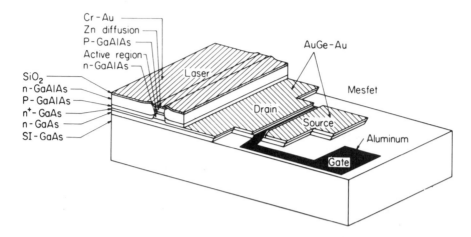

FIG. 5.7 Monolithic integration of a buried heterostructure laser with a MESFET. (From Ref. 18.)

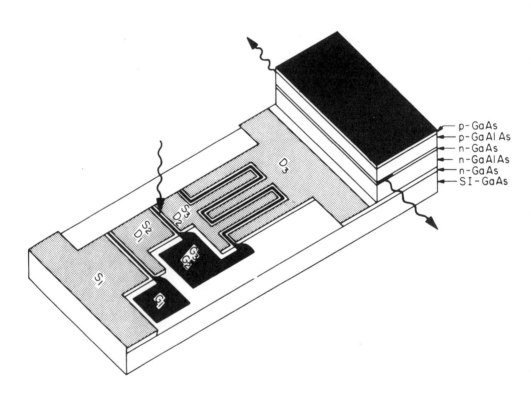

FIG. 5.8 An GaAs/GaAlAs integrated optical repeater. (From Ref. 14.)

suppressed in the narrow-stripe single-transverse-mode laser structures described above, so that the resonance in the frequency response is suppressed, and modulation well into the gigahertz range can be obtained.

5.2.5 Integration of Sources

There are many reasons that the integration of optical sources with other optical and electrical components is desirable. Yariv [17] has described the beginnings of integrated optoelectronic circuits and presented several examples of integration. The continued development of integrated circuits should provide less expensive and more reliable sources that can be modulated at high frequencies due to the reduced parasitic inductances and capacitances compared to the hybrid approach. Figure 5.7 is an example of the monolithic integration of a buried heterostructure GaAs/GaAlAs laser with a MESFET for high-frequency modulation capabilities. An integrated optoelectronic circuit that provides the function of an optical repeater is shown in Fig. 5.8, where the incident signal is detected, amplified, and then used to modulate the laser to reemit a stronger optical signal.

5.3 DETECTORS

The integration of detectors with electronic components has concentrated on the GaAs/AlGaAs and the InP/GaInAsP materials systems. Devices that take advantage of the unique properties of heterostructures can be fabricated with these materials. Also, because they are direct bandgap materials, they are suitable for lasers and high-speed detectors. The bandgap of the InP/GaInAsP quaternary covers the wavelength region that is of interest for low-loss, low-dispersion fiber optic applications (1.3 to 1.6 µm). This material is therefore suitable for long-distance communication links. The GaAs/AlGaAs material system operates at 0.85 µm, which limits the bit rate and link length in optical fiber systems due to dispersion and losses. However, since GaAs technology is more developed and integrated electronic circuits have been fabricated in GaAs, this material system may be advantageous for high-speed local networks.

In Sec. 5.3.1 the different types of detectors that have potential for use in integrated optoelectronic applications and the general performance characteristics of each type of detector are discussed. The theoretical and experimental performance of these detectors in optical receivers are compared in Sec. 5.3.2. Finally, some examples of how these detectors have been integrated with other components and the performance of those integrated optoelectronic circuits that have been fabricated to date are given in Sec. 5.3.3.

5.3.1 Types of Detectors

Primarily, three types of detectors have received attention for integration applications. These are PIN diodes, avalanche photodiodes
(APDs), and photoconductive detectors. Each will be discussed in
some detail.

A. PIN Diodes

A schematic diagram of a PIN photodiode is shown in Fig. 5.9. The
diode is operated in the reverse bias with the depletion region extending through the intrinsic region. Electron-hole pairs are created
when light of bandgap energy or higher is incident on the diode.
Those pairs that are generated in the depletion region or in the
contact region within a diffusion length of the depletion region are
swept across the depletion region by the electric field and collected.
If all the electron-hole pairs generated by the absorbed light are
collected, the quantum efficiency of the diode is simply

$$\eta = (1 - R)(1 - e^{-\alpha w}) \tag{5.1}$$

where R is the reflectivity, α the absorption coefficient, and w the
thickness of the material where the light is absorbed. In a practical device, R can be made small by the use of an antireflection coating for the wavelength of operation. Therefore, good quantum efficiency requires that the device be thick enough for all the light to
be absorbed. The absorption coefficient for 1.5-µm light in GaInAs

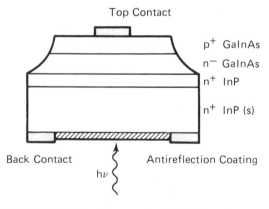

FIG. 5.9 Structure of a GaInAs pin mesa diode.

is approximately 10^4 cm^{-1}, so a device with an absorption region 3 µm thick would have an internal quantum efficiency of 95%.

The frequency response of a PIN diode can be limited by three factors: (1) the time it takes for carriers generated outside the depletion region to diffuse to the depletion region, (2) the time it takes for carriers to traverse the depletion region, and (3) the RC time constant determined by the junction capacitance and the equivalent resistance of the diode and the external circuit. Since the diffusion of carriers is a relatively slow process it is desirable to have the carriers generated in the depletion region. As noted above, a 95% internal quantum efficiency requires a 3-µm absorption region. The reverse-bias voltage required to obtain a depletion region this wide depends on the doping concentrations of the n and p sides of the junction and for an abrupt junction is given by the expression

$$W = \frac{\sqrt{2\varepsilon(V_0 + V_R)}}{qN_B} \qquad (5.2)$$

where ε is the dielectric constant, V_0 the built-in junction voltage, V_R the applied reverse-bias voltage, and N_B the reduced carrier concentration, $N_B = N_A N_D / N_A + N_D$. A narrower depletion region reduces the transit time of the carriers but increases the junction capacitance as well as decreasing the quantum efficiency. The design of the diodes in other materials depends on the wavelength to be detected and the materials that are being used.

A PIN detector is subject to several sources of noise that degrade its performance. The current that flows in a PIN diode is the result of three effects: the current due to the optical signal I_P, the current due to background radiation I_B, and the dark current, I_D, that is due to surface leakage, tunneling [19-21], and the thermal generation of electron-hole pairs in and within a diffusion length of the depletion region. All of these currents are generated randomly and contribute to shot noise, which can be expressed as

$$\langle i_s^2 \rangle = 2q(I_P + I_B + I_D)B \qquad (5.3)$$

where B is the bandwidth of the system. In addition, a thermal noise known as Johnson noise contributes to the total noise in the system. This is due to the shunt resistance of the diode combined with the total input resistance of the following preamplifier stage. The thermal noise may be expressed as

$$\langle i_T^2 \rangle = 4kT_{eff} \frac{1}{R_{eff}} B \qquad (5.4)$$

where T_{eff} is an effective temperature related to the noise figure of the amplifier [$T_{eff} = T(10^{NF/10} - 1)$] and R_{eff} is the parallel combination of the detector and the preamplifier input resistances.

All of the constraints above must be taken into account when designing a PIN detector. A diagram that summarizes the design rules for an GaInAs PIN is shown in Fig. 5.10. The dashed line indicates the breakdown voltage of the diodes, while the shaded region above this shows the voltages at which the tunneling current becomes unacceptably high. This diagram indicates that to obtain a sufficiently wide depletion region for good quantum efficiency while avoiding high leakage currents due to tunneling, it is necessary to have a doping level of less than about 8×10^{15} cm^{-3}. A device with this doping level would operate with a depletion width of about 2 µm, corresponding to a transit time of 30 ps [23].

B. Avalanche Photodiodes

For applications where the preamplifier noise dominates the detector noise, improved performance can be obtained using an avalanche

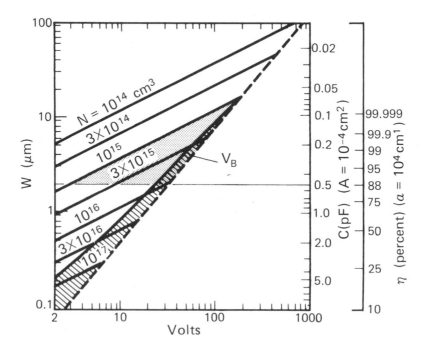

FIG. 5.10 Design constraints for GaInAs pin detectors. (From Ref. 22.)

photodiode (APD). Like the PIN detector, the APD is a reverse-biased p-n junction. When in operation, the electric field is high enough that carriers are accelerated to energies sufficient to excite electrons from the valence band into the conduction band in a process known as impact ionization, thereby creating electron-hole pairs. These carriers may then create additional electron-hole pairs, and this process, called carrier multiplication or avalanche gain, continues until all the carriers are collected at the edges of the depletion region. In this way, a single photo-generated carrier can create many more carriers. The avalanche process is essentially an internal gain mechanism and this gain increases with applied voltage. The voltage at which the multiplication becomes arbitrarily large is called the breakdown voltage of the diode.

The frequency response of an avalanche photodiode at low values of multiplication is essentially the same as that for an equivalent PIN diode. For a hypothetical avalanche photodiode in which only the injected carrier type can cause impact ionization, at high values of multiplication the transit-time-limited frequency response is degraded by at most a factor of 2. However, the high values of multiplication when both electrons and holes can cause impact ionization, the frequency response is degraded by the time it takes for all the carriers created by the multiplication process to be collected, and this may be more severe. This results in a finite gain-bandwidth product that is fixed for a given material and device structure.

The sources of shot noise in an APD are the same as in a PIN diode, but the photocurrent and the bulk dark current shot noise are both multiplied. Unfortunately, because the avalanche process is itself a random process, there is also excess noise generated by the multiplication process. The mean-square shot noise current after multiplication can be described by

$$\langle i_s^2 \rangle = 2q[(I_P F_P + I_B F_B + I_{DB} F_D)M^2 + I_{DS}] \tag{5.5}$$

where F_P, F_B, and F_D are excess noise factors for the photocurrent the background current, and the bulk dark current, respectively, and M is the multiplication or gain [24]. The dark current is composed of two components: the bulk dark current, which flows through the high-field region and is multiplied, and the surface leakage current, which is not multiplied. In some diodes, the surface leakage current may be orders of magnitude greater than the bulk current. The thermal noise of the APD circuit is the same as that for the PIN diode. As a result, even though the multiplication process introduces excess noise into the detection process, if the noise in the preamplifier circuit was the dominant noise source the overall signal-to-noise ratio may be improved.

As can be seen in Eq. (5.5), the bulk leakage current is multiplied and contributes excess notes to the total noise of the device.
As a result, the leakage current degrades the performance of APDs
more severely than in PINs. It was shown that the use of GaInAs
APDs would not result in improved performance over GaInAs PINs
because of the high leakage currents in this small-bandgap material
[25]. To obtain a detector that operates at 1.5 μm and yet avoid
the problem of high leakage current, diodes were fabricated that had
a GaInAs absorption layer but had the p-n junction, and therefore
the high-field region, in InP. This device, the SAM (separate absorption and multiplication) APD [26–28], is shown schematically in
Fig. 5.11a. For electric field values $\lesssim$ 1.5 × 10^5 V/cm in the GaInAs,

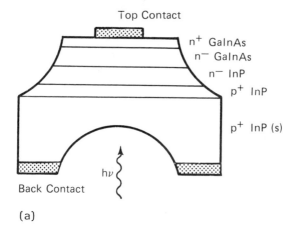

(a)

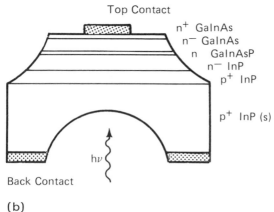

(b)

FIG. 5.11 (a) Structure of a SAM APD. (b) SAM structure with a
grading layer to improve frequency response.

this device exhibited negligible tunneling currents. However, carrier accumulation due to the valence-band discontinuity at the hetero-interface and subsequent thermionic emission of holes resulted in very slow response times for these devices [22]. The inclusion of an inter-mediate-bandgap layer between the InP and GaInAs, as shown in Fig. 5.11b, significantly improves the response time [29]. It has recently been shown, however, that there can still be residual trapping even with the inclusion of the intermediate layer [30].

C. Photoconductive Detectors

In contrast to the diodes discussed in Sec. B, this type of detector consists of a single layer of semiconductor between two ohmic contacts. The type of photoconductors considered here are referred to as intrinsic photoconductors, meaning that the incident optical signal excites electrons from the valence band to the conduction band. This is in contrast to extrinsic photoconductors, where the transitions are from an impurity level to the conduction or valence band. When the optical signal creates an electron-hole pair, the conductivity of the device is increased and the signal is detected as an increased current flow under constant-voltage bias or as a decreased voltage drop under constant-current bias.

Optically generated carriers contribute to current flow until they recombine. If the minority-carrier lifetime is greater than the majority-carrier transit time between the contacts, many carriers can pass between the contacts before recombination occurs. This results in a photoconductive gain which is given by

$$\Gamma = \frac{\tau_{eff}(v_n + v_p)}{l} \qquad (5.6)$$

where τ_{eff} is an effective carrier lifetime which contains terms for bulk, surface, and contact recombination; l is the channel length, and v is the carrier velocity [31].

The bandwidth of a photoconductor is given by

$$B = \frac{1}{2\pi\tau_{eff}} \qquad (5.7)$$

In a long device, τ_{eff} is dominated by the bulk recombination rate; in a short device, the transit time of the minority carrier becomes smaller and dominates. However, as noted above, a long combination time is desirable to obtain high gain. So, as was the case for APDs, there is a trade-off between gain and bandwidth and the gain-bandwidth product is fixed for a given material and detector configuration.

For a photoconductive detector the noise present in the detector is a combination of the thermal noise of the conductor and the generation-recombination noise due to the random generation and recombination of carriers in the conduction channel. For a detector of average conductivity G, the thermal noise is given by the usual expression

$$\langle i_T^2 \rangle = 4kTGB \tag{5.8}$$

The generation-recombination noise is given by

$$\langle i_{gr}^2 \rangle = \frac{4qI_0GB}{1 + \omega^2\tau_{eff}^2} \tag{5.9}$$

where G here is the photoconductive gain, I_0 the photo-generated current, B the bandwidth, and τ_{eff} the effective lifetime of the carriers. The mean-square noise is twice that of shot noise for diodes because there are contributions to the noise from both the generation and the recombination of the carriers.

In practical photoconductors, the device must be large enough that the light from an optical fiber can be efficiently coupled into it. To do this without limiting the gain of the device requires an inter-digitated geometry. A schematic diagram of such a device is shown in Fig. 5.12. This device had an average resistance of about 400 Ω. The dominant noise source in a photoconductive detector is the thermal noise of the conducting channel [33]. To reduce this noise it would be necessary to increase the resistance of the channel. To do so by altering the geometrical factors of length or width would not increase the total detector sensitivity because it would decrease the coupling efficiency. Increasing the resistance by decreasing the channel thickness would decrease the quantum efficiency because of incomplete absorption of the incident radiation. Therefore, to obtain the highest resistance and hence the lowest thermal noise possible without sacrificing gain or quantum efficiency, it is necessary to use material of the lowest carrier concentration available.

5.3.2 Integrated Optical Receiver Considerations

The area of prime interest for the integration of a detector involves combining it with a low-noise preamplifier to obtain an integrated optical receiver. All of the detectors described in Sec. 5.3.1 have been used in hybrid or discrete component optical receivers. The detectors discussed in this section use $Ga_{0.47}In_{0.53}As$ light absorption layers and so are suitable for long-wavelength fiber optic

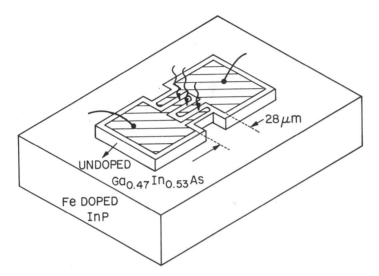

FIG. 5.12 Interdigitated photoconductive GaInAs detector. (From Ref. 32.)

communication systems. The performance of devices designed to operate at shorter wavelengths will be different from that given below because of differences in material parameters. From these considerations of discrete optical receivers we can determine what advantages an integrated receiver would have and what detectors should be considered for integration for a particular application.

A. FET Performance

In an optical receiver the performance is determined not only by the detector but by the preamplifier circuit as well. An equivalent circuit of the front end of an optical receiver is shown in Fig. 5.13. As noted above, the load resistor of the amplifier circuit contributes thermal noise to the system. Most receivers use a low-noise FET in the front end of the amplifier circuit. The FET also contributes noise in addition to the detector noise and therefore degrades the sensitivity of the receiver. The noise sources in an FET can be modeled as a shunt current noise source that is the result of thermal noise of the load or feedback resistor of the circuit and the shot noise of the gate leakage current,

$$\frac{d}{df} \langle i^2 \rangle_{shunt} = \frac{4kT}{R_L} + 2qI_{gate} \qquad (5.10)$$

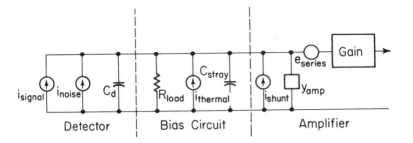

FIG. 5.13 Equivalent circuit for the front end of an optical receiver. (From Ref. 34.)

and a series voltage noise source that is the result of channel noise in the FET,

$$\frac{d}{df} \langle e^2_{series} \rangle = \frac{4kT\,\Gamma}{g_m} \tag{5.11}$$

where Γ is the noise figure of the FET and g_m is the transconductance [34]. The total circuit noise is given by

$$\langle i^2 \rangle_{circuit} = \left[\frac{4kT}{R_L} \left(1 + \frac{\Gamma}{g_m R_L} \right) + 2qI_{gate} \right] I_2 B \tag{5.12}$$

$$+ 4kT\Gamma \frac{(2\pi C_T)^2}{g_m} I_3 B^3$$

Here C_T is the sum of the detector, circuit, gate-source, and gate-drain capacitances. The values of I_2 and I_3 depend on the shape of the input and output pulses, and B is the bit rate. Note that the different noise sources vary differently with the bandwidth or bit rate of receiver operation. Therefore, the dominant noise source will vary with bit rate.

B. Sensitivity of Optical Receivers

In a PIN/FET receiver, for example, the noise at low bit rates is due to the shot noise of the FET gate leakage current and the PIN diode reverse leakage current, whereas at high bit rates the FET channel noise is dominant. In a photoconductive receiver, however, the dominant noise source is the Johnson noise of the photoconductive channel until bit rates greater than 4 Gbit/s, when FET channel noise becomes dominant. In a receiver utilizing an APD and a FET amplifier, the gain of the APD is adjusted to an optimum value where the excess noise due to the multiplication process is balanced against the gain-independent noise sources in the preamplifier circuit.

A summary of both the theoretical and experimentally obtained sensitivities of different types of nonintegrated receivers is given in Fig. 5.14. The calculations in this figure are for receivers using the same FET in the amplifier for all cases. For the PIN receiver the calculations were done for front-end capacitances of 1.5 and 0.5 pF. The 0.5-pF value corresponds to the best value that has been reported to date and the 1.5-pF value corresponds to a typical value. The calculations for the APD receiver used a capacitance of 1.5 pF and assumed that the APD had a sufficient bandwidth for the bit rates used. The data points in this figure are the best experimental values obtained so far. The data for the PIN receiver agree fairly well with the theory. The discrepancies between the calculated and the experimental values for the other receivers arise for the following reasons. The difference between the theoretical and the experimental APD sensitivity is due to high dark currents, large circuit capacitances, and the fact that equalization was necessary because of the slow response of the detector used. The main reason for the discrepancies for the photoconductive receiver data is the less than 100% coupling of the light into the detector due to the area covered by the contacts.

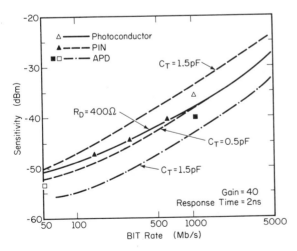

FIG. 5.14 Theoretical performance for different detectors as well as best experimental results to date. (From Ref. 33.)

C. Integrated Receiver Goals

Based on the data and calculations in Sec. B, the following conclusions can be made. The integration of the detector and the preamplifier into an optical receiver will result in the reduction of the parasitic capacitance of the circuit. For the case of the PIN receiver this would mean that 0.5 pF or lower capacitance could be obtained routinely, compared to the 0.5-pF example given above, which was the best result obtained to date with hybrid circuits. This would result in an improvement in the sensitivity of the receiver to the level where it was comparable to the photoconductive receiver. The photoconductive receiver would gain little sensitivity from integration because it is limited by the noise of the photoconductive channel rather than by circuit parasitics. However, the simplicity of the device would make it easier to integrate than a PIN diode, and therefore reasons of cost and reliability might make an integrated photoconductive receiver desirable in some applications. The highest sensitivity could in principle be obtained with an integrated APD receiver. However, the complexities of the structure make it difficult to fabricate optimized devices even in discrete form. In addition, the restrictions for optimized device performance are in conflict with the design requirements of a planar APD. Therefore, at present the problems associated with APDs outweigh the gains that might be obtained with their use in integrated optoelectronic circuits.

5.3.3 Examples of Detector Integration

A. GaInAs PIN/FET

As discussed in Sec. 5.3.2, one area that has received attention is the integration of a PIN diode and an FET preamplifier to form an optical receiver. Integration of these two components offers the advantage of minimizing the imput capacitance at the preamplifier. The work that has been done in this area has utilized $Ga_{0.47}In_{0.53}As$ grown on InP substrates. This material allows the absorption of light out to 1.65 μm. In addition, the high-field transport properties of this material indicate that high FET transconductances may be possible.

A diagram of the PIN/FET photoreceiver is shown in Fig. 5.15. The transistor is a junction field-effect transistor and the photodiode is a part of the FET gate electrode. The device was fabricated by first growing an undoped 1-μm layer of lattice-matched GaInAs on a semi-insulating <100> InP:Fe substrate. The unintentional doping in this case is n type with a carrier concentration of 10^{16} cm^{-3}. The p layer is formed by a Zn diffusion to create a diode region the size of an optical fiber. This preliminary device exhibited a current gain of about 30. However, this device was not

pin/FET Photo Receiver

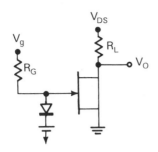

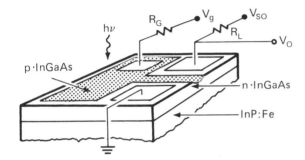

FIG. 5.15 Schematic of GaInAs pin/FET integrated photo receiver. (From Ref. 35.)

optimized either in the structure or in the material used. Calculations show that to optimize the photodiode performance by minimizing the capacitance, while staying well below the breakdown voltage to minimize dark current, a carrier concentration below 5×10^{15} cm^{-3} is needed. However, to obtain a g_m in an GaInAs FET comparable to those obtained in GaAs FETs, the channel doping should be in the range 1.0×10^{17} cm^{-3}. It has been proposed that this apparent incompatability could be avoided by growing a lightly doped buffer layer beneath the FET channel and fabricating the diode junction in this buffer layer [36]. However, this structure has not been realized, nor have device dimensions been optimized to date.

B. Wavelength Demultiplexing Detector

An interesting device that takes advantage of the range of bandgaps that are obtainable with the GaInAsP system is shown in Fig. 5.16. This is a two-wavelength demultiplexing photodetector. The Q_1

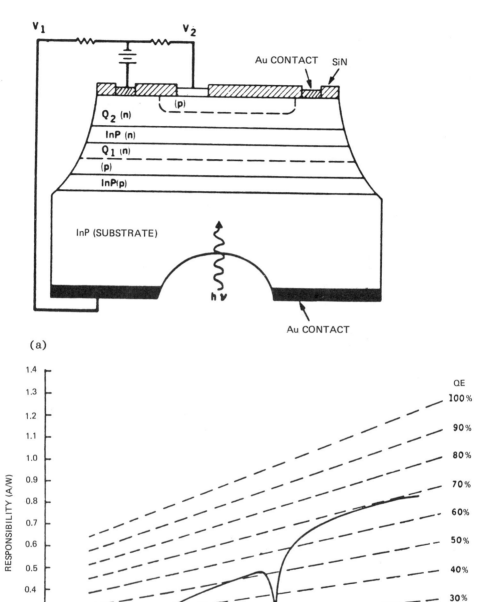

FIG. 5.16 (a) Structure and (b) responsivity curve of wavelength demultiplexing device. (From Ref. 37.)

layer, is composed of $Ga_{0.30}In_{0.70}As_{0.66}P_{0.34}$ and detects photons in the range 0.9 to 1.25 μm. The Q_2 layer is the ternary $Ga_{0.47}In_{0.53}As$ and extends the wavelength response to 1.6 μm. The external quantum efficiency of Q_1 is 50% at 1.15 μm and at 1.3 μm, Q_2 has a quantum efficiency of 65%. The crosstalk on the short-wavelength side is -30 dB at 1.15 μm and is -43 dB on the long-wavelength side at 1.3 μm. The use of a device such as this could increase the information capacity of an optical fiber since two wavelengths of light could be used.

C. GaAs Optoelectronic Receiver

As mentioned earlier, the technology of GaAs processing is more developed than that of the other III-V materials. Consequently, the most sophisticated optoelectronic integrated circuits to date have been fabricated in this material system. An optoelectronic receiver has been fabricated that consists of a GaAs PIN photodiode and a transimpedance preamplifier that incorporates six GaAs depletion-mode MESFETs and five Schottky diodes [38]. The general method used to fabricate this circuit is applicable to other optoelectronic circuits. The advantages of this method include the fact that it is a planar method and yet does not require selective epitaxy. The processing sequence is shown in Fig. 5.17. A well is etched in the substrate and then a highly doped n layer followed by a lightly doped layer are grown over the entire surface, forming the contact and 1 regions of the PIN diode. The well region is then masked and the epitaxial layers are etched down to the substrate over the remaining wafer until the surface is planar. In this way there are epitaxial layers in the well suitable for the fabrication of the PIN diodes, but the FETs can be made in the area where the substrate is exposed. The p region of the photodiode and the FETs and Schottky diodes are then formed by selective ion implantation. This method allows for the optimization of the optical and electronic devices independently of each other. The photoreceiver fabricating using this method had a responsivity of 0.3 A/W at 820 nm and a bandwidth of 30 MHz that was limited by the photodiode capacitance.

D. Photoconductive Receiver

The integration of a photoconductor with FETs to obtain a photoreceiver has also been achieved [39]. This circuit is illustrated in Fig. 5.18. Two identical PCDs were connected in series to provide stability against common-mode signals. The signal from the photodetectors was fed into one of the MESFETs while the other gate was used for automatic gain control. This circuit was fabricated on an InP semi-insulating substrates. The FETs were double-heterostructure MESFETs fabricated in $Al_{0.48}In_{0.52}As/Ga_{0.53}In_{0.47}As/Al_{0.48}In_{0.52}As$ epitaxial layers, and the photoconductors were fabricated in the

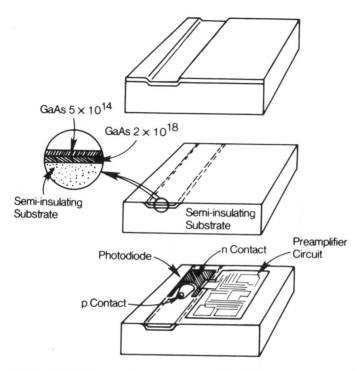

GaAs 5×10^{14}

GaAs 2×10^{18}

Semi-insulating
Substrate

Semi-insulating
Substrate

Photodiode

n Contact

Preamplifier
Circuit

p Contact

FIG. 5.17 Processing sequence for GaAs optoelectronic receiver.
(From Ref. 38.)

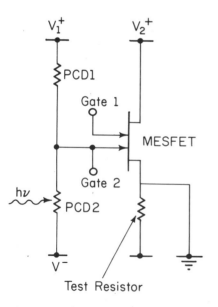

V_1^+

V_2^+

PCD1

Gate 1

MESFET

Gate 2

hν

PCD2

Test Resistor

V^-

FIG. 5.18 Equivalent circuit for GaInAs photoconductive integrated
receiver. (From Ref. 39.)

GaInAs after the AlInAs layer was etched away. This photoreceiver has a responsivity of 0.15 A/W and the calculated RC time constant was 70 ps. No frequency response measurements were reported for this device.

5.4 CONCLUSIONS

From the discussion, the advantages of integrated optoelectronics can be seen. The results that have been obtained in this area have been very promising. However, it is also clear that there are many areas where further work must be done to achieve the full potential of these devices.

REFERENCES

1. N. Holonyak, Jr., and S. F. Bevacqua, *Appl. Phys. Lett.*, *1*: 82 (1962).

2. R. N. Hall, G. E. Fenner, J. D. Kingsley, T. J. Soltys, and R. O. Carison, *Phys. Rev. Lett.*, *9*: 366 (1962).

3. M. I. Nathan, W. P. Dunke, G. Burrus, F. H. Dill, and G. J. Lesher, *Appl. Phys. Lett.*, *1*: 62 (1962).

4. T. M. Quist, R. H. Rediker, R. J. Keyes, W. E. Kraz, B. Lax, A. L. McWhorter, and H. J. Zeiger, *Appl. Phys. Lett.*, *1*: 91 (1962).

5. Zh. I. Alferov, M. Andreev, D. Z. Garbuzov, Yu. V. Zhilyaev, E. P. Morozov, E. L. Portnoi, and V. G. Triofim, *Sov. Phys. Semicond. Triofim*, *4*: 1753 (1971).

6. I. Hayashi, M. B. Panish, P. W. Foy, and S. Sumski, *Appl. Phys. Lett.*, *17*: 109 (1970).

7. H. Namizaki, *IEEE J. Quantum Electron.*, *QE-11*: 427 (1975).

8. H. Kumabe, T. Tanaka, H. Nakamizaki, M. Ishii, and W. Sasaki, *Appl. Phys. Lett.*, *33*: 38 (1978).

9. N. Matsumoto, *IEEE J. Quantum Electron.*, *QE-13*: 560 (1977).

10. B. Frescura, C. Hwang, H. Luechinger, and J. Ripper, *Appl. Phys. Lett.*, *31*: 770 (1977).

11. D. F. Scifres, R. D. Burnham, and W. Streifer, *Appl. Phys. Lett. 14*: 112 (1977).

12. G. E. Stillman, M. D. Sirkis, J. A. Rossi, M. R. Johnson, and N. Holonyak, Jr., *Appl. Phys. Lett.*, *9*: 268 (1966).

13. H. Kogelnik and C. V. Shank, *J. Appl. Phys.*, *43*:2327 (1972).

14. A. Yariv and M. Nakamura, *IEEE J. Quantum Electron.*, *QE-13*: 574 (1977).

15. T. Kuroda, S. Yamashita, M. Nakamura, and J. Umeda, *Appl. Phys. Lett.*, *33*:173 (1978).

16. M. J. Adams, *Opto-electronics*, *5*:201 (1973).

17. A. Yariv, *IEEE Trans. Electron Devices*, *ED-31*:1656 (1984).

18. I. Ury, K. Lau, N. Bar Chaim, and A. Yariv, *Appl. Phys. Lett.*, *41*:126 (1982).

19. S. R. Forrest, R. F. Leheny, R. E. Nahory, and M. A. Pollack, *Appl. Phys. Lett.*, *37*:322 (1980).

20. Y. Takanashi, M. Hawashima, and Y. Korikashi, *Jpn. J. Appl. Phys.*, *19*:693 (1980).

21. H. Ando, H. Kanbe, M. Ito, and T. Kaneda, *Jpn. J. Appl. Phys. 19*:1277 (1980).

22. S. R. Forrest, *Fiberoptic Technology*, 81 (1982).

23. T. H. Windhorn, L. W. Cook, and G. E. Stillman, Proceedings of the International Electron Devices Meeting, pp. 641–644 (1981).

24. G. E. Stillman and C. M. Wolfe, *Semiconductors and Semimetals*, Vol. 12 (R. K. Willardson and A. C. Beer, eds.), Academic Press, New York, p. 291 (1977).

25. Takanashi and Horikoshi *Jpn. J. Appl. Phys.*, *20*:1915 (1981).

26. K. Nishida, K. Taguchi, and Y. Matsumoto, *Appl. Phys. Lett.*, *35*:251 (1979).

27. H. Kanbe, N. Susa, H. Nakagome, and H. Ando, *Electron. Lett.*, *16*:163 (1980).

28. N. Susa, H. Nakagome, O. Mikami, H. Ando, and H. Kanbe, *IEEE J. Quantum Electron.*, *QE-16*:684 (1980).

29. J. C. Campbell, A. G. Dentai, W. S. Holden and B. L. Kasper, *Electron. Lett.*, *17*:818 (1983).

30. J. C. Campbell, W. S. Holden, G. J. Qua, and A. G. Dentai, *IEEE J. Quantum Electron.*, *QE-21*:1743.

31. H. Beneking, *IEEE Trans. Electron. Devices*, *ED-29*:1420 (1982).

32. C. Y. Chen, B. L. Kasper, H. M. Cox, J. K. Plourde, *Appl. Phys. Lett.*, *46*:379 (1984).

33. S. R. Forrest, *IEEE Electron Devices Lett.*, *EDL-5*:536 (1984).

34. R. G. Smith and S. D. Personick, *Topics in Applied Physics, Semiconductor Devices for Optical Communication* (H. Kressel, ed.), Springer-Verlag, Berlin, p. 89 (1982).

35. R. F. Leheny, R. E. Nahory, M. A. Pollack, A. A. Ballman, E. D. Beebe, J. C. DeWinter, and R. J. Martin, *Electron. Lett.*, *16*:353 (1980).

36. R. F. Leheny, A. S. H. Liao, B. Tell, and L. Mayer, *Proc. SPIE*, *408*:115 (1983).

37. J. C. Campbell, A. G. Dentai, T. P. Lee, and C. A. Burrus, Digest of the Topical Meeting on Integrated and Guided-Wave Optics, Incline Village, Nev., pp. WD3-1-WD3-3 (1980).

38. R. M. Kolbas, J. Abrokwah, J. K. Casney, D. H. Bradshaw, B. R. Elmer, and J. R. Biard, *Appl. Phys. Lett.*, *43*:821 (1983).

39. J. Barnard, H. Ohno, C. E. C. Wood, and L. F. Eastman, *IEEE Electron Devices Lett.*, *EDL-2*:7 (1981).

40. S. R. Forrest, O. K. Kim, and R. G. Smith, *Appl. Phys. Lett.*, *41*:95 (1982).

41. H. Namejaki, *IEEE J. Quantum Electron.*, *QE-11*:427 (1975).

6

Ti:LiNbO₃ Integrated Optic Technology

Considerations, and Capabilities Fundamentals, Design

STEVEN K. KOROTKY and RODNEY C. ALFERNESS   *AT & T Bell Laboratories, Holmdel, New Jersey*

6.1 INTRODUCTION

6.1.1 Background and Scope

The Ti:LiNbO₃ waveguide technology is at a stage when its penetration into general use appears to be imminent. Yet there are few places where an introductory description of the many facets of this particular technology can be found. Here we have gathered together a description of the key ingredients of this technology to provide an overview for those just encountering the field. In doing so, a conscious effort has been made to draw together both fabrication and device issues.

A. Ti:LiNbO₃ Integrated Optic Technology

Definition and Description. At the most basic level, the Ti: LiNbO₃ integrated optic technology is a methodology for guiding and processing light. The action of waveguiding is attained by using small amounts of titanium (Ti) to dope a lithium niobate (LiNbO₃) crystal and locally increase the index of refraction. It is an electro-optic technology, meaning that the control of the optical signal is mediated through an electrically induced change in the optical characteristics of the crystal. For this reason, as well as the reliance

on semiconductor-based devices to provide the functions of optical
gain and optical detection, the technology is necessarily a hybrid
one. The technology is also to be distinguished from an all-optical
approach, in which the goal is eventually to implement the control
optically without the use of intermediate electronics.

With the Ti:LiNbO$_3$ technology, essentially all aspects of light
can be controlled. These characteristics include the phase, amplitude, polarization, frequency, and direction of propagation of the
optical wave and their temporal dependence.

Goals and Applications. The goal of the Ti:LiNbO$_3$ waveguide
technology is to provide devices for controlling and processing the
characteristics of light in a manner permitting a significant level of
integration. It was of this vision that the term *integrated optics*
was fashioned [1]. The major areas of application are optical communication, signal processing, and sensing.

Advantages. The advantage of this approach is the economy of
scale and the level of functionality in analogy to integrated electronic circuitry. The ability to integrate many devices on a single substrate not only reduces the cost per function, but adds significantly
to the stability and robustness of the optical circuit. Additionally,
because the dimensions are small, the voltages required to perform
a given operation can be low and the speeds can be high. Since
the devices are voltage driven, and lithium niobate is an insulating
crystal, they do not continuously draw power, but expend energy
only when changing states. The Ti:LiNbO$_3$ waveguides also share
the unique feature of optical fibers: the very large intrinsic bandwidth over which optical information can be transmitted.

Status. Within the last five years, Ti:LiNbO$_3$ waveguide devices
have begun to make the transition from laboratory curiosities to
practical systems alternatives. In addition to continued efforts to
push the capabilities of devices and to formulate new device concepts, efforts to develop the technology for application have been
initiated. The level of integration over the last few years has been
characterized by an exponential trend, with the number of interconnected elements on a single chip increasing from a few to 16 to
64 [2,3]. A variety of experimental systems using Ti:LiNbO$_3$ devices have been reported and the first prototype samples are commercially available.

Competition. In general, optical technologies compete with the
electronics industry to meet the needs of information processing demands. For communication purposes, optics has in its favor a
broad-bandwidth, low-loss medium for transmission—the optical
fiber. To take full advantage of the enormous bandwidth, however,

requires a supporting optical technology that interfaces to the fiber and electronics. Integrated optics aims at fulfilling this need as well as providing improved performance and functionality.

The Ti:LiNbO$_3$ integrated optic technology is the most mature among the alternatives. At present it offers the highest level of functionality and performance, and consequently it is the closest to meeting the near-term needs of increased capacity. It is fully expected that at some time in the future higher levels of integration will be made possible by implementing all functions in semiconductors. At a later time still, an optical-optical or truly photonic technology may evolve.

B. Outline of Chapter

In this chapter it is our purpose to introduce the fundamentals, design considerations, and capabilities of the Ti:LiNbO$_3$ waveguide technology. We review the fabrication and design of waveguides in this material, with emphasis on issues pertaining to the attainment of high figures of merit in the areas of insertion loss, operating voltage, and speed. As one important category of devices, we examine in closer detail the design of high-speed modulators and switches and their application to optical communication.

The reader should bear in mind that in many instances the references cited here are representative and not necessarily complete. The scientists and engineers who have made significant contributions to the field over the past 20 years number far more than those represented on these pages. We dedicate this chapter to one such person, William Silva, a crystal grower, whom we have had the pleasure to know.

6.1.2 Waveguiding and Electro-Optics

A. Optical Waveguide

The physical principles of waveguiding in Ti:LiNbO$_3$ waveguides are identical to those of optical fibers and are discussed in Chapter 1 as well as in other texts devoted to integrated optics [4,5]. The essential ingredient is a region of raised index of refraction which is used to guide light in a manner related to total internal reflection. The principal difference is the planar nature of the Ti:LiNbO$_3$ technology—the crystalline substrate serving the function of a cladding material in which many "core" regions can be fabricated and easily interconnected. This correspondence is not only of pedantic interest, but is of practical significance as the evolutionary trend toward single-mode optical fibers has been a driving force for extending and perfecting the Ti:LiNbO$_3$ channel waveguide technology.

B. Electro-Optic Effect

The linear electro-optic effect, or Pockels effect, is a phenomenon
whereby the index of refraction of a material is changed in proportion
to an applied electric field. In a crystalline material, the dielectric
characteristics are described by second-rank tensors and changes in
the index are described by third-rank tensors. Through microscopic
changes in the medium under the application of an electric field,
the coefficients of the dielectric tensor are modified and correspond
to changes in the index of refraction for a given polarization as well
as possible conversion to the orthogonal polarization. These changes
provide a means to implement a host of optical operations and control devices (see, e.g., Ref. 6). By modifying the index of refraction, the phase of the optical signal is controlled and this serves as
the basis of devices such as phase, amplitude, and intensity modulators. Via mode conversion the state of polarization can be controlled and devices such as wavelength filters and frequency shifters
can be designed.

6.2 Ti:LiNbO$_3$ WAVEGUIDES AND FABRICATION

6.2.1 Basic Fabrication Steps

The first optical waveguides formed by the doping of lithium niobate
(LiNbO$_3$) with titanium (Ti) were fabricated 12 years ago by Schmidt
and Kaminow [7]. These early waveguides were multimode planar
structures, but channel-type guides supporting a single mode were
demonstrated soon after [8]. The basic fabrication steps employed
to process Ti:LiNbO$_3$ devices today are very similar to those of the
pioneering works. Significant differences are mainly the result of
the introduction of tighter control over the fabrication parameters
and the diffusion environment and are a direct consequence of the
requirement of device reproducibility and uniformity. Before considering specific aspects we summarize the basic steps in fabricating
channel waveguides into lithium niobate by means of titanium
diffusion.

A. Fabrication Parameters

An important feature of titanium-in-diffusion waveguide fabrication,
in addition to being convenient and reproducible, is the independent
control of the important waveguide parameters: waveguide width and
depth and the waveguide-substrate index difference. The waveguide
width is controlled photolithographically following the procedure illustrated in Fig. 6.1. Photoresist is spun on the oriented, polished,
and cleaned lithium niobate crystal. The desired waveguide width
and path are delineated by opening a window in the photoresist using

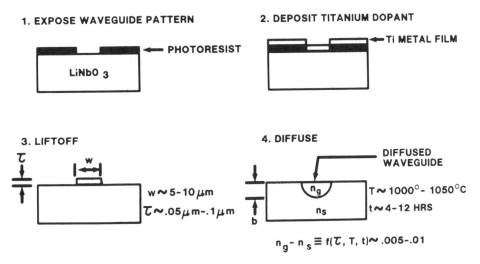

FIG. 6.1 Waveguide fabrication procedure.

ultraviolet exposure through a mask. All other factors being constant, the titanium thickness provides control over the index change. Titanium, of carefully controlled thickness, is deposited over the entire crystal surface by evaporation (either electron-beam or resistively heater induced) or radio-frequency (RF) sputtering. The titanium to form the channel is obtained by dissolving the photoresist to "lift off" the undesired metal. Alternatively, etching techniques may be used to transfer the waveguide pattern to the titanium. The waveguide depth is controlled by the time and temperature of the diffusion.

B. Diffusion Environment

The diffusion is carried out in a furnace, with typical temperatures ranging from 980 to 1050°C. Typical diffusion times are 4 to 12 h depending on the wavelength of operation and the mode size desired. A variety of diffusion atmospheres have been explored to fabricate Ti:LiNbO₃ waveguides. Historically, the heating and bake cycle has been performed in an atmosphere of flowing argon bubbled through a water column. This technique was originally employed to limit the susceptibility to optical damage. However, later is was realized that it also reduces out-diffusion of Li_2O [9]. Such outdiffusion can result in an unwanted planar waveguide because the Li deficiency causes an increase in the extraordinary index of lithium niobate. Cool down is generally carried out in flowing oxygen to reoxidize the crystal.

Another technique used to reduce out-diffusion is to perform diffusion in an overpressure of Li_2O. This can be achieved by placing a powdered source of Li_2O in the oven. Both closed and open tubes have been used [10,11]. An atmosphere of dry O_2 is often used [12].

C. Buffer Layer

After the diffusion, an insulating buffer layer, typically SiO_2, is deposited over the waveguides. This layer reduces the optical attenuation (loading) of light polarized perpendicular to the crystal surface (TM polarization) that would otherwise result from induced currents in an overlaid control electrode. The necessity of the buffer layer depends on the optical polarization and electrode placement.

The optimum buffer layer to isolate the TM optical mode from an overlaid electrode—required, for example to use the maximum r_{33} coefficient in Z-cut lithium niobate—remains an area of investigation. Early buffer layers made by RF-sputtered films provided adequate reduction of electrode loading loss. However, devices made with these buffer layers exhibited severe drift problems when voltage was applied to the overlaid electrodes. Thermal annealing was found to reduce the drift somewhat but was quite nonreproducible. The problem seemed to be that the films were not sufficiently insulating and defects trapped charge at the $SiO_2/LiNbO_3$ interface. Drift problems in these devices could be dramatically reduced by etching away the buffer layer in the gap region between the electrodes [13]. Devices with stable device characteristics (i.e., no apparent short-term drift) have been demonstrated with SiO_2 films made by chemical vapor deposition [14].

D. End-Face Preparation

For end-fire or fiber coupling a carefully polished end face is essential. The end face should be perpendicular to the waveguide to ensure low coupling loss. A cover plate is clamped or epoxied to the crystal with very little gap (<1 μm) to avoid rounding of the end surface and to provide strength during polishing. Typical polishing is performed with 1- and 1/4-μm grit. To ensure low (<−30 dB) optical feedback to lasers, the end faces are antireflection coated for optimum performance [15]. Fiber attachment is typically accomplished by careful alignment and bonding using a UV-cured epoxy. The device is then packaged for system use [16].

6.2.2 General Characteristics of $LiNbO_3$

A. Attributes

Of all known crystals that may potentially find application in the waveguide processing of optical signals, lithium niobate is by far the most thoroughly studied. Long before waveguiding was demonstrated in this material, it was already well established for optical applications. Today, lithium niobate waveguide devices play a central role in state-of-the-art experiments in optical fiber communication, signal processing, and sensing [17–19].

There are several very basic reasons why $LiNbO_3$ is suited for use in these areas. For all practical purposes it is transparent in the visible and near-infrared region of the optical spectrum. Large single-domain crystals of optical quality are also routinely grown. In addition, the electro-optic figure of merit is among the largest of known materials. Finally, waveguides are relatively easily fabricated in $LiNbO_3$ using standard processing methods.

B. Crystalline Properties

The study of the material science aspects of lithium niobate has a long history (see, e.g., Ref. 20). Tabulated characteristics of physical characteristics are collected and discussed in several reviews and monographs [21–23]. It has been determined that lithium niobate has a rhombohedral crystalline structure and its symmetry is that of the 3m point group [24,25]. Consequently, $LiNbO_3$ is a uniaxial crystal with the symmetry axis defined as the z axis, or sometimes c axis, and referred to as the optic axis. It is a birefringent crystal and light propagating in the crystal along the z axis with polarization in the x y plane is an ordinary wave. The z axis, or extraordinary axis, is the fast axis. Crystals can be grown over a range of compositions, $(Li_2O)_x(Nb_2O_5)_{1-x}$. The congruent composition—that composition where the solid and liquid in contact are in equilibrium—is preferred for device applications because of the high degree of uniformity attained during growth. This composition is shifted slightly from the stochiometric composition (x = 0.5) and occurs for a mole% of Li_2O corresponding to x = 0.486 [26]. The melting point of congruent $LiNbO_3$ is about 1243°C [27].

C. Growth and Poling Methods

Crystals are grown using the Czochralski pulling technique as reported by Ballman [28]. Methods for growing and poling single-domain crystals have been described by Nassau et al. [29,30]. Today, single-crystal boules over 8 cm in diameter and 10 cm in length are drawn from the melt and poled. We note that $LiNbO_3$ has a relatively high Curie temperature of 1150°C. Consequently, its characteristics are quite stable at room temperature.

D. Electro-Optic Coefficients

The ferroelectric nature of $LiNbO_3$ was reported in 1949 by Matthias and Remeika [31]. In the mid-1960s, work on crystal growth facilitated electro-optic investigations of the material [32]. The linear electro-optic effect or Pockels effect in $LiNbO_3$ is attributed to a subatomic scale displacement of the Li ion with respect to the oxygen lattice under an applied field [33,34]. The induced polarization, $\underline{\Delta P}$, for $LiNbO_3$ in terms of the optical field, $\underline{E}^\circ$, and the applied electric field, $\underline{E}^a$, can be written as

$$\Delta P_i = \sum_j \Delta\varepsilon_{ij} E_j^0 \tag{6.1a}$$

with

$$\Delta\varepsilon_{ij} \propto \begin{bmatrix} -r_{22}E_y^a + r_{13}E_z^a & -r_{22}E_z^a & r_{51}E_z^a \\ -r_{22}E_z^a & r_{22}E_y^a + r_{13}E_z^a & r_{51}E_y^a \\ r_{51}E_z^a & r_{51}E_y^a & r_{33}E_z^a \end{bmatrix} \tag{6.1b}$$

The diagonal terms in this matrix represent electro-optically induced changes of the phase of the propagating optical signal, and the off-diagonal terms represent polarization conversion. Values for the electro-optic coefficients, r_{ij}, have been collected from the literature and tabulated by Kaminow [35]. The largest of the diagonal coefficients is r_{33}, with a value of about 31×10^{-12} m/V. This coefficient is about three times larger than the second largest diagonal coefficient, r_{13}. Polarization conversion is possible using the r_{51} and r_{22} coefficients. The r_{51} coefficient is comparable in magnitude to the r_{33} coefficient and is about 10 times larger than r_{22}. For this reason r_{51} is most often used in polarization controllers, although broadband optical performance is more easily achieved using the r_{22} coefficient.

A relevant electro-optic figure of merit for a material is $n^3 r$, where n is the index of refraction. $LiNbO_3$ has an index of refraction just slightly greater than $n = 2$ and exhibits a birefringence of about 0.1. Values for the congruent composition have been measured over the transparent region of the spectrum [36]. We note that room-temperature measurements [37] of the bulk indices of refraction for the ordinary and extraordinary axes of $LiNbO_3$ are summarized by $n_0(\lambda) = 2.195 + 0.037/\lambda^2$ (μm) and $n_e(\lambda) = 2.122 + 0.031/\lambda^2$ (μm). The temperature dependence of these indices has also been investigated [38].

6.2.3 Approach to Ti:LiNbO$_3$ Waveguide Design

A. Motivation

In designing a Ti:LiNbO$_3$ integrated-optic circuit, the goal is to provide the desired optical function with a performance meeting or exceeding the required specification. The performance of the optical circuit depends on the quality of the individual building block components and interconnections, and is ultimately related to the modal characteristics of the waveguides from which it is fabricated. In turn, it is the fabrication parameters and diffusion conditions over which the designer has control. Our purpose in this section is to provide the reader with a basic understanding of how a working relationship between device performance and these latter variables may be established. Through the feedback made possible, the optimal values of parameters can be determined and tolerances can be set.

By way of illustration, we consider the high-speed directional coupler switch. The device (Fig. 6.2) consists of a directional coupler formed from two identical waveguides brought into proximity and of a traveling-wave electrode, which can be used to electro-optically control the coupler state. The operating principles are discussed in more detail in Sec. 6.3. Briefly, the coupler is normally designed with an interaction length corresponding to one coupling length. In this case, the coupler is in the crossover state without application of voltage to the electrode. The coupler can be switched to the straight-through state when the proper voltage is applied to the electrode. Such a device finds application in optical modulation and spatial switching.

The specification of the performance of the directional coupler switch/modulator would certainly include the wavelength of operation, the optical insertion loss, the modulation bandwidth, and the switching

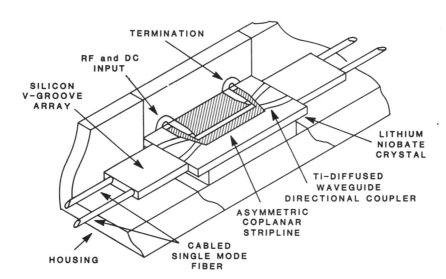

FIG. 6.2 Traveling-wave directional coupler switch.

voltage. In addition, a complete specification would include the modulation efficiency for encoding applications and the crosstalk and voltage tolerances for switching applications. Quantities such as the optical and electrical reflectivities and the size of the device are also relevant concerns. These device characteristics are usually not independent of each other, however, but are correlated through their common dependence on the physical characteristics of the mode of the waveguide.

As one example, we suppose that the fiber constituting the optical system into which the modulator will be installed is specified. In this situation, the optical insertion loss will include a contribution that is a function of the overlap of the optical mode distribution of the waveguide with that of the fiber. Since the fiber mode is typically made large, to ease the splicing tolerances, this provides a reason to argue for a large waveguide mode size. On the other hand, the switching voltage depends on the electric field strength and density that is developed by the electrode within the crystal in the region of the waveguide mode. This electric field is increased as the gap between the electrodes is reduced, but is used most efficiently when the gap is of the size of the waveguide mode. Thus the desire for a low switching voltage would tend to favor a small waveguide mode.

A trade-off must obviously be made. However, the situation may be further complicated by limits on the drive voltage available from broadband electronic amplifiers or power dissipation. In this manner, the optical insertion loss, the drive voltage, and the bandwidth are interrelated and depend on the waveguide mode size. Clearly, the designer requires the capability to correlate the waveguide mode size with the fabrication conditions to develop an effective device solution for a given system application and set of constraints.

The example above illustrates how one characteristic, the waveguide mode size, can influence the overall device performance. A more complete set of modal characteristics for Ti:LiNbO$_3$ waveguides, which would permit a comprehensive analysis of the device behavior, are the two-dimensional optical field distribution, $\psi(x,y)$, of the mode; the propagation constant of the optical mode, β; and the coupling coefficient, κ, describing the rate of power transfer between two waveguides in proximity. These modal characteristics will influence the fiber-to-waveguide coupling loss, the electro-optic efficiency, the waveguide propagation and bending loss, and the coupling length of the directional coupler. Other devices may be dependent on a given characteristic to a greater or lesser degree.

Below we describe the general steps that are used to formulate a design aid that permits predictions of device performance given the set of fabrication parameters and diffusion conditions. In Sec. 6.5 we describe in more detail the issues involved in optimizing high-speed device performance.

B.   Modal Characteristics of Ti:LiNbO₃ Waveguides

Algorithm for Design Aid.   A chart summarizing the steps for formulating a design aid for Ti:LiNbO₃ devices is given below.

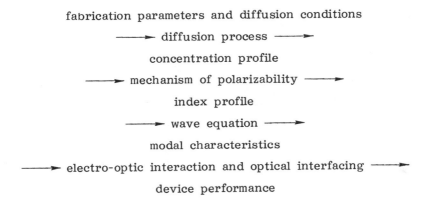

fabrication parameters and diffusion conditions

⟶ diffusion process ⟶

concentration profile

⟶ mechanism of polarizability ⟶

index profile

⟶ wave equation ⟶

modal characteristics

⟶ electro-optic interaction and optical interfacing ⟶

device performance

As indicated in the chart and discussed earlier, the goal of a comprehensive design aid is to provide a connection between the controllable device fabrication parameters and the device performance.   The relevant physical processes that affect this relationship are the diffusion defect chemistry, the mechanism for index change, the physics of electromagnetic wave propagation, and the electro-optic interaction and optical interconnections.   Researchers continue to refine our understanding and control of each of these contributing factors through more precise experiments and theoretical models. Even so, the problem at hand may be addressed with a wide spectrum of attacks of varying levels of sophistication, accuracy, and computational effort.   The choice depends on the specific application and stage of the design.   Consequently, here our intention is not to review the variety of approaches, but to illustrate relevant concepts and typical behavior by particular example.   In practice, sufficient accuracy for surveying trends and developing prototypes has been attained without recourse to strictly numerical solutions.   It is the more analytic approach that provides an intuitive understanding of the process and which we elaborate on here.

Diffusion of Ti into LiNbO₃.   The process by which a lithium niobate crystal is doped with titanium to form waveguides is a combination of chemistry and classical diffusion.   According to present understanding, the titanium incorporation occurs in three identifiable stages [39].   The first stage takes place during the period when the substrate is being heated from room temperature to the chosen diffusion temperature (ca. 1000°C).   At about 500°C the deposited titanium film begins to oxidize.   The oxidizing reaction is complete by the time the diffusion soak temperature is reached, leaving a polycrystalline TiO_2 compound on the surface in place of the metal.

The second stage is characterized by the formation of a complex source compound at the surface, without significant penetration of titanium into the substrate. Material scientists believe this source compound to be $(Li_{0.25}Nb_{0.75}O_2)_{1-x}(TiO_2)_x$, with $x = 0.59$ [40]. It is formed by the movement of the mobile Li species into the TiO_2 layer, and the subsequent counter diffusion of Nb from the $LiNbO_3$ into the volume of the TiO_2 layer and of Ti from the TiO_2 layer into a similar volume in the $LiNbO_3$ crystal.

The final stage of the doping process is essentially the classical diffusion of the limited supply of Ti from the source compound into the crystal volume. As such, the extent of Ti diffusion depends on the diffusion coefficient and on the time and temperature of the heat treatment. The latter parameters are accurately and easily controlled and provide a convenient method of waveguide adjustment. The diffusion coefficient is normally not an explicit parameter. At the same time, however, it must not be ignored as an immutable constant. This is a consequence of the dependence of the diffusion coefficient on the defect density of vacancies that are, in turn, a function of the stoichiometry (i.e., composition) of the $LiNbO_3$ substrate [41,42]. Experiments have been carried out to quantify the compositional dependence of the diffusion constant and have shown it to vary by over a factor of 4 over the range 48 to 50 mol % Li_2O composition [42]. No difference in the diffusion rates along the different crystalline axes that would be intrinsic to the structure of the crystal have been observed, however. For reasons of waveguide reproducibility, it is therefore important to control the starting composition of the substrate. These same results also have implications concerning the control of the diffusion environment. Unchecked loss of Li_2O to out-diffusion during the heat treatment, for example, can radically alter the rate of lateral surface diffusion versus vertical diffusion into the bulk [42,43]. As mentioned previously, a variety of approaches to control the diffusion atmosphere are currently employed.

By far, the modal characteristics of the waveguides are most strongly influenced by the third stage of the titanium incorporation because of the substantially larger crystal volume that it affects. Thus any model of the waveguide characteristics begins with a model of the classical diffusion process for a laterally distributed source. Assuming that suitable means have been taken to ensure the stability of the Li_2O content of the crystal, we can consider the diffusion constant as given. Many experiments to measure the diffusion profile and rates for planar diffusion of Ti into $LiNbO_3$ have been carried out [7,42-48]. All are described well by classical diffusion theory [49].

In the case of diffusion for the purpose of waveguide formation. the diffusion time is long compared to that to deplete the source compound. In fact, the region of the polycrystalline source compound crystalizes to $LiNbO_3$ from below as a consequence of Ti loss. The theory of diffusion under this circumstance predicts that the diffusion profile has a Gaussian dependence as a function of the depth (y) from the crystal surface. For a planar diffusion, the concentration of Ti within the $LiNbO_3$, relative to pure metallic titanium, is then approximated by

$$C(y) = \frac{2}{\sqrt{\pi}} \frac{\tau}{D} \exp \left(- \frac{y}{D} \right)^2 \qquad (6.2)$$

where τ is the thickness of the deposited titanium layer before oxidation and **D** is the diffusion depth. The diffusion depth is expressed as $D = \sqrt{4Dt}$ with $D = D_0 \exp(-T_A/T)$. Here D is the diffusivity, D_0 the diffusion constant, t is the diffusion time, T the diffusion temperature, and T_A the characteristic activation temperature.

The diffusion rates depend on the stoichiometry of the $LiNbO_3$ crystal and the diffusion environment. Therefore, the values of D_0 and T_A may differ from laboratory to laboratory and must be adjusted accordingly. Measured values for the diffusivity, D, for the congruent $LiNbO_3$ composition (ca. 48.6 mol % Li_2O) range from 0.5×10^{-12} cm²/s for T = 1000°C to 1×10^{-12} cm²/s for T = 1050°C, and 2×10^{-12} cm²/s for T = 1100°C. The corresponding values for D_0 and T_A that describe the diffusivity in this temperature region are $D_0 = 2.5 \times 10^{-4}$ cm²/s and $T_A = 2.55 \times 10^4$ K. It is also worth pointing out that the density of deposited titanium is also dependent on the method of deposition. Electron-beam deposition of the titanium yields a density very close to that of bulk metallic titanium, whereas deposition via resistive thermal evaporation tends to yield a density lower by approximately 25%. This is presumably a consequence of the incorporation of oxygen, and therefore the measured thickness, τ, must be devalued appropriately.

Channel waveguides are fabricated by photolithographically defining a stripe of width W from the deposited Ti (Fig. 6.1). The fact that there is no inherent anisotropy observed for diffusion along the different axes of the $LiNbO_3$ crystal greatly simplifies the theoretical analysis of the diffusion from this strip. Under isotropic conditions, the strip may be considered as a continuous set of infinitesimally wide line sources and we may integrate their contributions to obtain the concentration at a specified point [50]. In this case, the theoretical concentration profile is separable in the lateral and vertical coordinates, x and y, and can be written as

$$C(x,y) = C_0 f(x) g(y) \qquad (6.3a)$$

with

$$g(y) = \exp \left(- \frac{y}{D}\right)^2 \tag{6.3b}$$

$$f(x) = \frac{1}{2\,\mathrm{erf}(W/2D)} \left\{ \mathrm{erf}\left[\frac{1}{D}\left(x + \frac{W}{2}\right)\right] - \mathrm{erf}\left[\frac{1}{D}\left(x - \frac{W}{2}\right)\right] \right\} \tag{6.3c}$$

$$C_0 = \frac{2}{\sqrt{\pi}} \frac{\tau}{D} \,\mathrm{erf}\left(\frac{W}{2D}\right) \tag{6.3d}$$

where erf is the error function.

To give a general impression of the dimensions that are involved, we note that the value of the initial metal strip width, W, for single-mode Ti:LiNbO$_3$ waveguides is empirically 5 to 6λ, where λ is the free-space wavelength of operation. The ratio of the strip width to the diffusion depth is W/D $\sim$ 2 to 4, depending on the application. The metal thickness is roughly 2.5 to 3% of D. We emphasize that although these relations identify the region of the multidimensional parameter space that is relevant to single-mode waveguide devices, the characteristics of the waveguide may be significantly altered when a given parameter is changed by as little as 10% and the others remain fixed.

Dependence of Index on Ti Concentration. The index change produced by the doping of the LiNbO$_3$ crystal with Ti has been measured by several groups [44–46,48,51,52]. It has been observed that the index change for the extraordinary axis is in direct proportion to the Ti concentration, and the index change for the ordinary axes is a sublinear function of concentration. In Fig. 6.3 we summarize the results of Minakata et al. [45] for 0.63-μm wavelength. As can be seen, the index change for the extraordinary axis is 1×10^{-2} for a Ti concentration of 1.5% of the bulk metallic Ti density. The constant of proportionality, b, between index change and percent Ti concentration is approximately 0.7 at $\lambda = 0.63$ μm. The index change exhibits a small wavelength dependence [52] (i.e., dispersion) that can be approximated by b(λ) = $0.552 + 0.065/\lambda^2$ (μm). Researchers have not conclusively identified the microscopic mechanism of the index change associated with the incorporation of Ti into the crystal, but the dominant contribution is attributed to induced strain in combination with the photoelastic effect [46].

With the knowledge of how the index change depends on Ti concentration, the index distribution can be constructed from the concentration profile. Since the index change is proportional to concentration to a good approximation, we can write

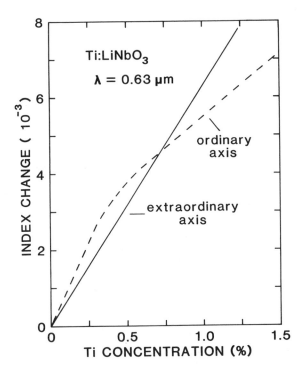

FIG. 6.3 Index change versus titanium concentration. (Derived from data of Ref. 45.)

$$\Delta n(x,y) = \Delta n_0 f(x)g(y) \qquad\qquad (6.4a)$$

Having enforced the conservation of the number of Ti atoms, the peak index change Δn_0 is given by

$$\Delta n_0 = bC_0 = \frac{2}{\sqrt{\pi}} \ \text{erf} \left(\frac{W}{2D}\right) \frac{b\tau}{D} \qquad\qquad (6.4b)$$

Wave Equation and Solutions. The propagation of light along the Ti:LiNbO₃ dielectric waveguide is described by the vector wave equation that can be derived from Maxwell's equations. Exact solutions for the corresponding eigenmode problem can be constructed only for the simplest of waveguide geometries. Consequently, it is customary to use the fact that the index change associated with the formation of single-mode waveguides is small to simplify the wave equation to a manageable form. To a very good approximation the problem can be reduced to a scalar wave equation for the electric field components [53]. The wave equation to be solved is thus

$$\nabla^2 \underline{E} = \mu_0 \varepsilon(x,y) \frac{\partial^2}{\partial t^2} \underline{E} \qquad (6.5)$$

where μ_0 is the free-space permeability and $\varepsilon(x,y)$ denotes the permittivity of the structure at optical frequencies. In this approximation, the propagating mode is essentially TEM in nature; that is, the major field components are orthogonal to the direction of propagation. In reality, the fields have small projections along the propagation axis, but for most practical purposes these may be neglected without our error [53].

Considering the wave to be of frequency ω with polarization in the vertical (y) direction and propagating in the z direction with a constant lateral field profile, the wave equation takes the form

$$\left(\frac{\partial^2}{\partial x^2} + \frac{\partial^2}{\partial y^2} \right) E_y(x,y) + N^2(x,y)k^2 E_y(x,y) = \beta^2 E_y(x,y) \qquad (6.6)$$

where β is the propagation constant, n the index of refraction given by $n^2(x,y) = \varepsilon(x,y)/\varepsilon_0$, and $k^2 = \mu_0\varepsilon_0\omega^2$, with ε_0 being the free-space permittivity.* Because we are working with small index changes, it is convenient to represent $n(x,y)$ as the sum of the bulk index, n_B, and the index difference, $\Delta n(x,y)$, induced by doping. Mathematically, we write $n(x,y) = n_B + \Delta n(x,y)$ and expand $n^2(x,y)$ as $n^2(x,y) = n_B{}^2 + 2n_B\Delta n(x,y)$. Similarly, the eigenvalue β is written as $\beta = (n_B + \Delta N)k$, where ΔN is referred to as the effective index difference of the guided mode and is the new eigenvalue of the problem. The wave equation to be solved for the mode profile $\psi(x,y)$ is then

$$\left[\frac{\partial^2}{\partial x^2} + \frac{\partial^2}{\partial y^2} + 2n_B k^2 \Delta n(x,y) \right] \psi(x,y) = 2n_B k^2 \Delta N \psi(x,y) \qquad (6.7)$$

There are no completely analytic solutions to the foregoing eigenvector problem for the index distribution of interest. If a purely analytic solution is desired, the problem must be simplified to the point where too few details of the Ti:LiNbO$_3$ waveguide remain to permit as specific comparison with experiment. The most accurate

---

*(Note: here the coordinates x, y, z do not necessarily refer to the crystallographic axes.)

method of solution is in principle a numerical one, such as, for example, a finite element iteration technique [54]. These methods require large amounts of computer memory and time and therefore are used to examine particular details rather than to survey general trends. Intermediate to the strictly analytical and numerical techniques are approaches such as the effective index method [50, 52,55,56] and the variational method [57–59]. One difference between these two methods is that the effective index method may be applied to modes of arbitrary order, whereas the variational method is rigorously applicable only to the fundamental mode, which is symmetric, and to the lowest-order antisymmetric mode. The restriction on the variational method does not represent a limitation for single-mode waveguides. The variational method does, however, permit the analysis to proceed further analytically. Consequently, it is more instructional and requires less computational effort. We use it here to convey the behavior of the Ti:LiNbO₃ waveguides.

The variational principle is a consequence of the ability to express the eigenvalue of the linear second-order differential operator

$$O = \left[\frac{\partial^2}{\partial x^2} + \frac{\partial^2}{\partial y^2} + 2n_B k^2 \, \Delta n(x,y) \right] \qquad (6.8)$$

as an expectation value of that same operator, that is,

$$2n_B k^2 \, \Delta N = \int\!\!\int \psi^*(x,y) O \psi(x,y) \, dx \, dy \equiv F \qquad (6.9)$$

The variational principle then rests on the fact that the expectation value of O is largest when $\psi_0(x,y)$ is the exact solution for the fundamental mode of the corresponding wave equation. If $\psi_0^\alpha(x,y)$ is an approximate parametric solution of the wave equation depending on a parameter α, the approximate solution best represents the exact solution when

$$\frac{\partial F_0(\alpha)}{\partial \alpha} = 0 \qquad (6.10)$$

In addition, if the error in the trial waveform compared to the exact solution is of order ξ, the error in the calculated eigenvalue is of order ξ^2.

To a very good approximation the measured mode profiles of single-mode Ti:LiNbO₃ waveguides can be represented by a field distribution separable in x and y and having a lateral shape that is Gaussian and a vertical profile that is Hermite-Gaussian. In many applications it is the size of the mode in the orthogonal directions and the effective index difference that sufficiently characterize the eigenmode. A reasonable trial solution for $\psi_0(x,y)$ is then

$$\psi_0^{w,d}(x,y) = \psi_w(x)\phi_d(y) \tag{6.11a}$$

with

$$\psi_w(x) = \frac{1}{\sqrt{(w/2)\sqrt{\pi}}} \exp\left[-\frac{1}{2}\left(\frac{x}{w/2}\right)^2\right] \tag{6.11b}$$

$$\phi_d(y) = \frac{2}{\sqrt{d\sqrt{\pi}}} \frac{y}{d} \exp\left[-\frac{1}{2}\left(\frac{y}{d}\right)^2\right] \tag{6.11c}$$

Here d and w are the mode-size parameters for depth and width to be determined, together with ΔN, using the variational method. The expectation value of O using the trial solution above can be calculated analytically [58] and is

$$2n_B k^2 \Delta N = \frac{-3}{2(D/W)^2(d/D)^2} - \frac{4}{2(w/W)^2} \tag{6.12}$$

$$+ [2n_B b\tau Wk^2]\left(\frac{2}{\sqrt{\pi}}\right)\left(\frac{W}{D}\right)\left[1 + \left(\frac{d}{D}\right)^2\right]^{-3/2}$$

$$\times \operatorname{erf}\left[\frac{1}{\sqrt{(w/W)^2 + r(D/W)^2}}\right]$$

An interesting point clearly brought out in the variational approach is that the effective index difference, ΔN, is determined by a competition between two terms, one being a measure of the average curvature of the mode profile, the other being a measure of the normalized mode overlap with the index distribution. The contribution of the former term to the functional $F_0(w,d)$ is maximized for a large mode size, while the contribution of the latter term is maximized for a small mode size.

General Trends. To illustrate the behavior of single-mode Ti: LiNbO$_3$ waveguides, calculated mode sizes and the corresponding effective index difference for diffusion parameters typical of 1.3-μm wavelength are plotted in Figs. 6.4 and 6.5. The following trends are empirically observed and are reflected in the calculations. First, considering mode sizes as a function of strip width, the mode size is very large when the strip width is small and guiding is weak.

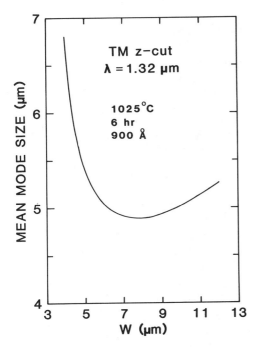

FIG. 6.4 Calculated Ti:LiNbO$_3$ mean mode size, $\sqrt{W_x, W_y}$, where $W_{x,y}$ is full width at 1/2 intensity.

Increasing the strip width reduces the size of the mode as the increasing cross section of the doped region pulls the mode within the waveguide. Once within the waveguide, the mode size reaches a minimum and with further increase of the strip width the mode size begins to increase to follow the waveguide size. For strip widths larger than that required to achieve the minimum size, the guide moves close to the multimode regime. The optimal strip width dimension is characterized by a relative insensitivity of the mode size to variations of the strip width.

Control of the diffusion depth permits a method of controlling the mode size once the strip width is chosen. The mode width is relatively insensitive to changes of the diffusion depth, especially when the metal thickness has been adjusted in concert. The vertical mode size, however, conveniently increases with increasing diffusion depth. The choice of metal thickness for fixed W and D is dictated by the desire to operate close to the cutoff of the second-order mode to ensure low propagation and bending losses of the fundamental mode.

In Fig. 6.5 the dependence of ΔN on τ is illustrated and shows
a clear threshold behavior. As the thickness is increased from zero,
the first effect is to begin to reduce the size of the mode, which is
much larger than the waveguide size. Until the mode is drawn within the waveguide, the effective index difference is very small, because the overlap is poor. Once the mode size is roughly the waveguide size, there is little advantage to reduce the mode size further,
as the overlap is good. At this point, changes in the metal thickness are directly reflected in increases in the effective index difference. It is because of this threshold behavior that the tolerance on
metal thickness is particularly stringent compared to other parameters. Although the required metal thickness is roughly 1000 Å at
1.3 µm, the single-mode region of good confinement extends over a
range of less than 200 Å. In this region, the mode size hardly
changes; rather, by comparison, the rate of the decaying tails of the
mode profile change radically. These tails and therefore the metal
thickness strongly influence the inter-waveguide coupling coefficient
and the bending loss.

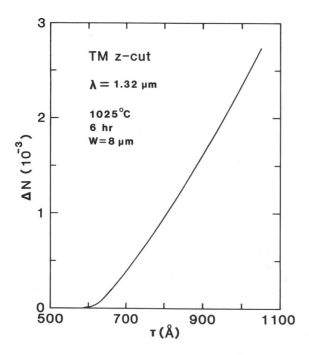

FIG. 6.5 Calculated Ti:LiNbO$_3$ effective index difference.

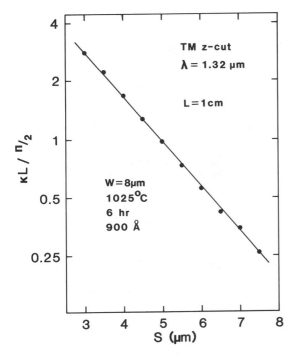

FIG. 6.6 Inter-waveguide coupling coefficient. Here, S is the separation of the inner edges of the waveguides.

Although the Gaussian × Hermite-Gaussian waveforms provide an accurate determination of the mode size and effective index difference, the functional form of the rapidly decaying tails requires modification to predict the inter-waveguide coupling coefficient. Experimentally, the coupling coefficient is an exponential function of the inter-waveguide gap. It is usually expressed as

$$\kappa(s) = \kappa_0 e^{-\gamma s} \qquad (6.13)$$

where κ is the coupling coefficient per unit length and s is the inter-waveguide separation. The variables κ_0 and γ depend on the waveguide fabrication parameters and can be determined by fitting measurements as illustrated in Fig. 6.6. The functional form e^{-x} for the waveguide tail is thus empirically and theoretically more appropriate outside the 1/e radius.

The theoretical dependence of the coupling coefficient on waveguide parameters for Ti:LiNbO₃ structures has been analyzed using

the effective index method [52], the variational technique [59], and
the beam propagation method [60]. Analytic results have also been
derived for idealized rectangular channel waveguides [53].

6.3 ELECTRO-OPTIC WAVEGUIDE MODULATORS
AND SWITCHES

6.3.1 Waveguide Devices for Electro-Optic Control
of Light

An indication of the strength and flexibility of the Ti:LiNbO$_3$ technology is the large family of devices that have been demonstrated
with it. These include such devices as the electro-optic polarization
controler [61], frequency shifter [62], tunable wavelength filter
[63], polarization multiplexer/demultiplexer [64], and tunable wave-
length-selective coupler [65]. It is not the intention here to survey all these devices. Rather, to illustrate technological issues we
consider devices to implement high-speed switching and modulation.
These functions are currently the most likely to be first implemented
with practical integrated optic devices. In fact, research systems
experiments employing such devices are described later in the chapter. We examine the more mature devices to perform the switching
and modulation functions, namely the directional coupler switch and
the Y-branch interferometric modulator. They are instructive examples to discuss basic design issues such as drive voltage, modulation bandwidth, and insertion loss. In addition, they are the devices most cited to compare the performance of different guided-wave
technologies.

6.3.2 Interferometric and Directional
Coupler Structures

A. Interferometric Modulators and Switches

Two versions of interferometric-type modulators are shown in Fig.
6.7. The most popular and easiest to implement is the Y-branch
interferometer, in which the 3-dB (50/50) splitter and combiner is
a symmetric Y branch (Fig. 6.7). Light entering in the single-
mode waveguide is equally divided into the two waveguides with
zero relative phase difference at the junction. From here, the
guided light enters the two arms of the interferometer. The waveguide arms are sufficiently separated that there is no coupling between them. If no voltage is applied to the electrode, the two
beams arrive at the second Y junction in phase and are combined
onto the output single-mode waveguide. Other than small radiation
losses, the output beam is equal in intensity to the input. However, if a π phase difference is introduced between the two beams

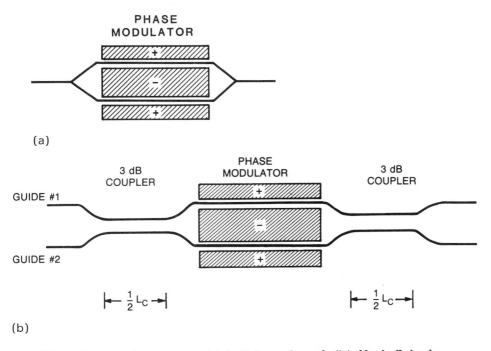

FIG. 6.7 Interferometers: (a) Y-branch and (b) Mach-Zehnder.

via the electro-optic, the combined beam has a lateral amplitude profile of odd spatial symmetry. Such a profile is similar to that of a second-order mode and is not supported by the output single-mode waveguide. The light is consequently forced to radiate into the substrate and is lost. In this way, the device operates as an electrically driven optical on/off modulator.

The same interferometric principal can be used to make an optical switch by replacing the Y-branch splitter and combiner with two 2 × 2 directional couplers. Such a four-port or Mach-Zehnder interferometer is shown in Fig. 6.7b. The couplers are fabricated or electro-optically adjusted to provide 3-dB splitting. Again the interferometer arms are separated to eliminate coupling in the active region. For input into the upper waveguide and without voltage applied to the interferometer, the light arrives at the second coupler and the coupling process is simply continued, resulting in transfer to the lower waveguide. However by introducing a π phase shift in the interferometer the light incident in the upper waveguide also exists from the upper waveguide. The result is a 2 × 2 switch for which, in principle, both switch states can be electrically achieved.

Assuming perfect 3-dB splitting and combining for the Mach-Zehnder and Y-branch interferometers, the transfer efficiency or intensity transmittance is given by

$$\eta = \cos^2\left(\frac{\Delta \beta L}{2}\right) \qquad\qquad (6.14)$$

where $\Delta\beta = (2\pi/\lambda)\,\Delta N_{e0}$, ΔN_{e0} is the electro-optically induced index difference between the two interferometer arms, and λ is the free-space wavelength. The induced phase mismatch is proportional to the applied voltage as discussed in Sec. B. The modulation versus voltage response of the interferometer is thus strictly periodic. A reduction in the on/off ratio or isolation results if the splitter and combiner are not perfect 3-dB couplers.

B.   Switched Directional Coupler

The switched directional coupler is a convenient optical circuit device for both on/off modulation and switching. In addition, variations of this structure have been used for wavelength [65] and polarization [64] filtering. The switched directional coupler based on adjustable phase mismatch, shown schematically in Fig. 6.8, consists of a pair of identical channel waveguides placed in proximity for an interaction length L. The evanescent tail of the optical mode profile of each waveguide extends into the other and results in a local coupling coefficient, κ, between the two waveguides. In addition to the coupling coefficient, the transfer efficiency of light between the waveguide depends on the difference in propagation constants $\Delta\beta = (2\pi/\lambda)|N_2 - N_1|$ of the two waveguides. The crossover efficiency, η, can be derived from the coupled-mode equations [66], which describe the operation of this and several other integrated optic devices, and is

$$\eta = \frac{\sin^2 \kappa L[1 + (\delta/\kappa)^2]^{1/2}}{1 + (\delta/\kappa)^2} \qquad\qquad (6.15)$$

where $\delta = \Delta\beta/2$.

Complete transfer ($\eta = 1$), defined as the crossover state, occurs only if $\delta = 0$ and $\kappa L = m\pi/2$, m being an odd integer. However, the crossover efficiency can be made zero (i.e., all the output light is in the incident waveguide or straight-through state) for arbitrary values of κL by applying an appropriate phase-mismatch value. For $\kappa L = m\pi/2$ the switch condition is

$$\Delta\beta L = \sqrt{2m + 1}\,\pi \qquad\qquad (6.16)$$

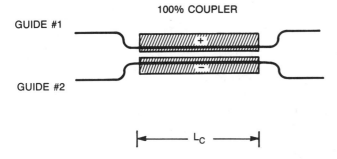

FIG. 6.8 Switched directional coupler.

Therefore, to minimize the voltage for a given length, that length should correspond to one coupling length, i.e., m = 1.

With two accessible output ports, the directional coupler is ideally suited for electrically controlled signal routing applications. However, it obviously can also be used as a simple on/off modulator. In the former application, where both switch states must be realized with very low crosstalk (typically, < -30 dB), the critical criterion placed on κL to achieve complete crossover implies stringent control over the waveguide fabrication parameters. Although this may be achievable, complete electrical control over both switch states for much more flexible fabrication tolerance can be achieved by using a split-electrode design—referred to as the $\Delta\beta$-reversal electrode [67]—shown in Fig. 6.9. In the ideal reversed-$\Delta\beta$ configuration, voltages of equal magnitude and opposite polarity are applied to the pair of electrode sections. For the two-section device shown in Fig. 6.9, the cross and bar states can be achieved provided that the total path length corresponds to between one and three coupling lengths—an easily achieved condition.

The transfer efficiency for the reversed $\Delta\beta$ coupler of N sections (N even) can be written simply as [6,67]

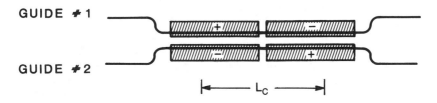

FIG. 6.9 Reversed-$\Delta\beta$ directional coupler switch.

$$\eta_{\Delta\beta} = \sin^2\kappa_{eff}L \tag{6.17}$$

where $\kappa_{eff} = (N/L)\sin^{-1}\sqrt{\eta}$, and η, is the crossover efficiency for one section of length L/N, as given by Eq. (6.15). Here it is assumed that both the cross and bar states are attained with opposite polarity voltage on the two sections. To minimize the switch voltage for a given device length under this condition, it is best to match the number of sections to the number of coupling lengths. In so doing there is no reduction in the effectiveness of the applied mismatch, unlike the case of a multiple-coupling-length device with uniform $\Delta\beta$.

C. Differences and Similarities

Because the Y-branch interferometer and directional coupler are basic structures for optical integrated circuits, it is instructive to analyze their differences and similarities. Most apparent is that the Y-branch interferometer provides simple on/off modulation, while the directional coupler provides the added feature of spatial switching. When switching is required, the Mach-Zehnder structure can also be used but with added complexity.

Another important difference is that the interferometer exhibits a periodic intensity transmittance versus phase difference (voltage); whereas that of the directional coupler is a passband theoretically centered about $V = 0$. Provided that an ideal 3-dB splitter/combiner is employed in the interferometer, it can produce perfect on/off states under electrical control. The reversed $\Delta\beta$ electrode—with its added complexity—is needed to achieve this important characteristic in the directional coupler.

The main difference between the two is quite fundamental: In the interferometer the phase shift or mismatch is induced between independent, uncoupled modes and the effect of this total accumulated shift is sensed by mixing or beating the two beams in the combiner. The periodic response is a natural consequence. Another consequence is that the response depends only on the integrated phase shift, not on the details of its spatial evolution. Therefore, nonuniformities in the electro-optically induced index change that may occur along the interferometer arms do not affect, for example, the extinction ratio.

An important difference between the two would appear to be the switching condition. To achieve complete modulation, all other factors being equal, the directional coupler requires a voltage-length product $\sqrt{3}$ times larger than that of an interferometer. However, for fast traveling-wave devices, the bandwidth per unit drive voltage is essentially identical for the two types of devices for a fixed electrode length [68]. In fact, it can be shown that the Mach-Zehnder and directional coupler switch/modulators are opposite limiting cases of an entire class of modulators that have a fixed bandwidth per unit drive voltage [69].

6.3.3 Electrodes and High-Speed Response

A. Crystal Orientation and Electrode Geometry

The electrode geometry employed depends on the device structure and crystal orientation. To ensure the minimum drive voltage, devices are designed so that the largest electro-optic coefficient—r_{33}— is used. To use r_{33}, the applied electric field and polarization direction of the incident light are oriented along the z axis. On a Z-cut crystal, the electrode is placed directly over the waveguide as shown in Fig. 6.10b, while for X- or Y-cut crystals, the electrodes

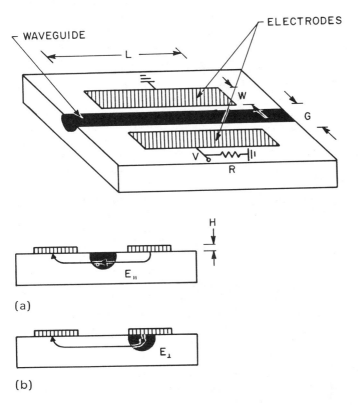

(a)

(b)

FIG. 6.10 Typical electrode/waveguide configuration.

are alongside the waveguide (Fig. 6.10a). In either case the applied electric field is a fringing, nonuniform field.

B. Electro-Optic Efficiency

General Form. Because the optical as well as the electrid field is spatially nonuniform, the effective applied field seen by the optical signal depends on the overlap between the two. Indeed, it is convenient to write the induced index change as

$$\Delta N_{eo} = \frac{-n^3}{2} r \frac{V}{d} \Gamma \tag{6.18}$$

where V/d is the ideal uniform electric field of a parallel-plate capacitor with a separation of d, r the relevant electro-optic coefficient, and Γ the optical-electrical field overlap parameter. From a device design viewpoint one wishes to maximize ΔN_{eo} for a given V. This condition may be constrained, however, by the implicit correlation of the fabrication parameters to other important device characteristics, such as the fiber-coupled insertion loss and speed. For the moment, we consider how ΔN_{eo} can be maximized; in Sec. 6.5 we discuss the overall optimization of the device performance.

Theoretical and Empirical Overlap Estimation. At first glance it is clear that a small gap is desirable and, indeed, it is. However, it is intuitively apparent that as the electrode gap is reduced, the electric field penetrates less deeply into the crystal. For a fixed optical mode depth, the overlap Γ will therefore decrease. The characteristic length, therefore, with which to compare the electrode gap is the optical mode width or depth. The calculated normalized electro-optic index change and overlap integral versus electrode gap is shown in Fig. 6.11 for both electrode configurations. As indicated, the induced index change in either case approaches a maximum as the gap is reduced to zero. However, there is very little increase as the electrode gap decreases below the characteristic mode size. Thus the device voltage can be reduced substantially by a narrow-gap electrode only if one fabricates correspondingly small dimension waveguides. Unfortunately, there are limits to the mode size achievable for $Ti:LiNbO_3$ waveguides. In addition, for effective fiber-waveguide coupling, larger modes compatible with typical optical fiber are required as discussed below. Since RF propagation loss also increases as the electrode dimensions are reduced, the waveguides should not be made too small.

The results in Fig. 6.11 indicate that the efficiency using a horizontally applied field is typically larger than for a vertical field. Empirically, the maximum ratio is approximately 1.5:1, but the

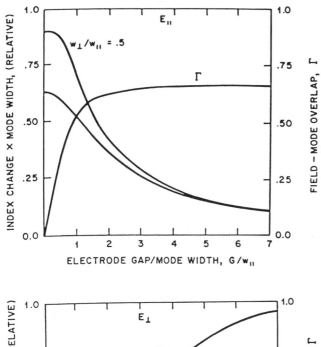

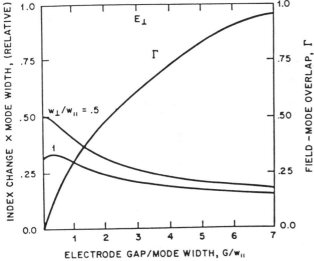

FIG. 6.11 Calculated electro-optic overlap.

difference is small when the waveguide modes are matched to typical fibers. Consequently, the choice of orientation is dictated more
by the function of the device. For directional couplers using the
push-pull electrode configuration on Z-cut, the voltage-length product
for switching is well approximated by 3.3 V-cm $\times$ λ^2 (μm). For typical devices the electrode gap is 4 to 5λ, indicating an overlap
efficiency of $\Gamma \sim 0.3$.

   Electrical/Optical Buffering. When using a vertically applied
field and the strongest electro-optic coefficient, it is necessary to
provide an intermediate insulating buffer layer—typically silicon
dioxide—between the electrode and waveguide to avoid strong optical attenuation. The buffer layer, however, also reduces the electric field strength in the waveguide. Fortunately, for a buffer
thickness sufficient to eliminate electrode loading loss, the increase
in drive voltage is only in the range of 20% [71]. Note that part
of the 1.5:1 ratio mentioned above is attributable to the use of a
buffer layer for the Z-cut orientation but not for the X-cut
orientation.

C.   Control Electrodes for High-Speed Performance

   Traveling-Wave Versus Lumped Operation. With respect to the
consideration of switching speed, two types of control electrodes are
used for Ti:LiNbO$_3$ switch/modulators. The simpler and slower is
the lumped electrode (Fig. 6.12a), in which the device is driven as
a capacitor terminated in a parallel resistor matched to the impedance
of the source line. In this situation, the modulation speed depends
primarily on the RC time constant determined by the electrode capacitance and the terminating resistance. To a smaller extent the speed
also depends on the resistivity of the electrode itself. The other
type of drive electrode is the traveling-wave electrode (Fig. 6.12b),
which is itself a miniature transmission line. Ideally, the impedance
of this coplanar line is matched to the electrical drive line and is
terminated in its characteristic impedance. In this case the switch/
modulator bandwidth is determined not by an RC time constant, but
by the difference in velocity of the optical and electrical signals
(velocity mismatch or walk-off) and electrical propagation loss. In
either case the electrode capacitance per unit length is a critical
design parameter.

   Electrode Geometries. The three most common electrode geometries for switch/modulators are shown in Fig. 6.13. The two-
electrode geometries, either symmetric or asymmetric, can be used
with the optical directional coupler on Z-cut lithium niobate to
achieve a push/pull index change using the large r_{33} coefficient.
This electrode is not suitable for the interferometric modulator
because unless a modified waveguide structure is used to allow the

LUMPED

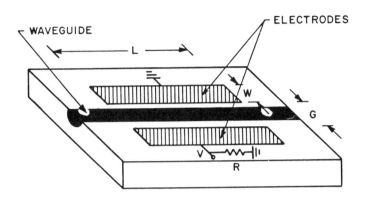

TRAVELING WAVE

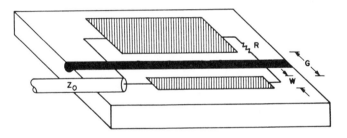

FIG. 6.12 Typical lumped and traveling-wave electrodes.

two interferometer arms to be brought very close together without introducing optical coupling between them [69,72], the electrode gap is unacceptably large to optimize for low-voltage operation. Instead, the three-electrode structure is typically used.

The electrode capacitance/length (C/L), which depends on the material dielectric constant, electrode gap-to-width ratio, and the electrode type, can be calculated using conformal mapping techniques. A generalized expression for C/L for the three different structures in Fig. 6.13 can be written as [73,74]

$$\frac{C}{L} = 2^{p-9} \, \varepsilon_{eff} \left[\frac{K'(r)}{K(r)} \right]^9 \tag{6.19}$$

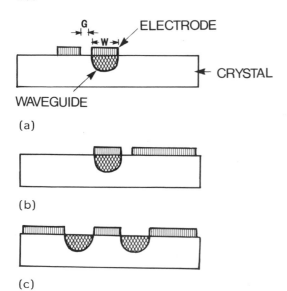

(a)

(b)

(c)

FIG. 6.13 Typical electrode geometries.

where

$$r = \left[1 + \left(\frac{2}{p} \left(\frac{W}{G} \right)^9 \right) \right]^{-1/p}$$

Here W and G are the electrode width and gap(s) as indicated in the figure and ε_{eff} is the effective dielectric microwave dielectric constant, which is given by $\varepsilon_{eff} = (\varepsilon_0/2)(1 + \varepsilon_s/\varepsilon_0)$ with ε_s the substrate permitivity [$\varepsilon_s = (\varepsilon_x \varepsilon_y)^{1/2} = 38.5$] and ε_0 that of the assumed air cladding. The function $K(r)$ is the complete elliptic integral of the first kind and $K'(r) = K(\sqrt{1 - r^2})$. Furthermore, $p = 1$ for the two-electrode symmetric and three-electrode structures, while $p = 2$ for the two-electrode asymmetric structure; $q = -1$ for the three-electrode and $+1$ for the other two structures.

Lumped Operation. The results of Eq. (6.19) for the capacitance/length for the symmetric two-electrode type is shown in Fig. 6.14. Also shown is the calculated optical bandwidth ($\Delta f = 1/\pi RC$) × length product, for a capacitance-limited lumped electrode modulator. This is the most useful figure of merit, assuming a matched termination (50 Ω), negligible electrode resistance, and no stray inductance or capacitance. The capacitance/length decreases and (Δf)L increases essentially logarithmically with increasing G/W.

However, to reduce the modulator drive voltage a small gap is desired; while to reduce the electrode resistance, to achieve the ideal capacitance limited results in Fig. 6.14, the electrode width should be increased. Thus there is a trade-off in lumped electrode design between a small G/W ratio to reduce the capacitance and a large G/W to reduce the drive voltage and electrode resistance.

Another limitation to the bandwidth of lumped-electrode modulators, electrical transit time, has a bandwidth × length product of about 2.2 GHz-cm. Consequently, there is no real advantage to use an electrode with a capacitance/length below about 3.2 PF/cm [see, e.g., Ref. 75). However, center-feed lumped electrodes can be used to reduce transit-time effects [76,77]. Lumped-electrode modulators with bandwidth × length products approaching this limit have been achieved [76-78].

D. Traveling-Wave Impedance and Design

Dependence on Geometry. A critical parameter for traveling-wave electrode design is the characteristic impedance, Z, which can be expressed as

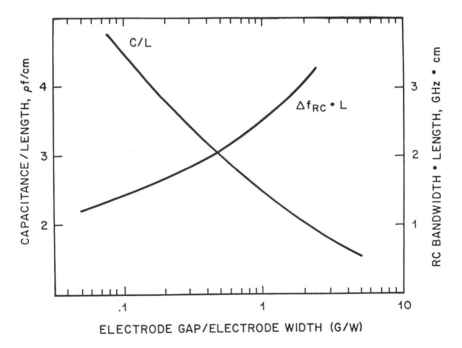

FIG. 6.14 Capacitance and bandwidth for the symmetric two-element electrode.

$$\frac{1}{Z} = \frac{c}{\sqrt{\varepsilon_{eff}}} \frac{C}{L} \tag{6.20}$$

where c is the speed of light. The characteristic impedance versus G/W is shown in Fig. 6.15, for the three common electrode structures.

Issues and Trade-Offs. The impedance calculations demonstrate an electrode design problem for Ti:LiNbO$_3$ modulators because of the relatively large microwave dielectric constant of the crystal—to make a 50-Ω coplanar transmission line a G/W value greater than 0.6 (two-electrode, symmetric) is required. For the other two electrode types the situation is even worse; for the three-electrode geometry a value of G/W $\cong$ 3 is necessary. The goal of a small gap to reduce drive voltage, together with a relatively wide electrode width to reduce electrode RF losses, argue for a lower than 50-Ω impedance coplanar

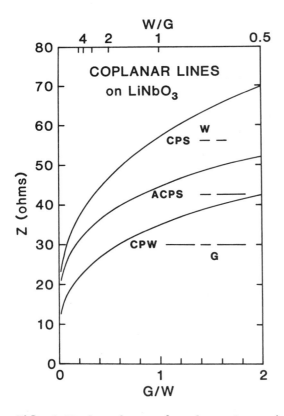

FIG. 6.15 Impedance of coplanar transmission lines on Ti:LiNbO$_3$.

electrode line. However, since most electrical sources are 50 Ω, this mismatch results in reduction of the effective drive voltage for a given power. In addition, the reflections at the impedance transition can adversely affect the drive electronics. Even if a lower-impedance source matched to the electrode impedance (Z) is used, the source power (compared to a 50-Ω source) must be increased by a ratio of 50/Z to achieve the same modulation depth. Therefore, as with lumped-electrode design, subtle trade-offs between bandwidth and voltage—in addition to the obvious inverse length dependence—must be considered in designing traveling-wave devices.

Optical/Electrical Walk-Off. Ignoring electrode loss, the bandwidth of traveling-wave devices depends on the length, velocity-mismatch, and—unlike lumped modulators—the device type. Consider first interferometric-type modulators. We assume that the electrode impedance is matched to the source and the termination resistor. A sinusoidal electrical drive signal launched on the drive electrode can be represented by

$$V(z,t) = V_0 \sin\left(\frac{2\pi N_m}{\lambda_m} z - 2\pi ft\right) \tag{6.21}$$

where $N_m = \sqrt{\varepsilon_{eff}/\varepsilon_0}$, λ_m, and f are the microwave refractive index, free-space wavelength, and frequency, respectively. The coordinate z is the position along the electrode. The voltage seen at any point along the electrode by a photon that enters the waveguide at a time t_0 can then be written as

$$V(z,t_0) = V_0 \sin\left[\frac{2\pi N_m f}{c}\left(1 - \frac{N_0}{N_m}\right)z - 2\pi ft_0\right] \tag{6.22}$$

where N_0 is the effective index of the guided optical mode. Now, for interferometric modulators the optical modulation is determined by the total induced phase shift introduced along the interaction length, L. The walk-off results in a frequency-dependent reduction in the integrated phase shift given by [79]

$$\frac{\int_0^L \Delta\beta(f)\ dz}{\overline{\Delta\beta}} = \frac{\sin(\pi T_w f)}{\pi T_w f} \sin[\pi(2t_0 - T_w)f] \tag{6.23}$$

where $\overline{\Delta\beta} = \pi n^3 r \Gamma V_0 L/\lambda G$ is the induced phase shift for dc operation and $T_w = (L/c)(N_m - N_0)$ is a transit time resulting from the velocity mismatch between the copropagating optical and microwave signals. For $N_0 = N_m$ the optical wave travels down the waveguide at the same speed as the microwave drive signal moves along the electrode, and consequently it experiences the same voltage over the entire electrode length. In this case the integrated value of $\Delta\beta$ is proportional to $V_0 L$, and arbitrarily long electrodes can be used to reduce the required drive voltage with no frequency limitation. However, for $N_0 \neq N_m$ there is a walk-off between the optical wave and microwave drive signal which results in a reduction or, for sufficiently large L or f, a complete cancellation of $\Delta\beta$. The latter case is the situation for $LiNbO_3$, where $N_m \cong 2N_0$. Because only the integrated e/o-induced phase shift is important for the interferometric modulator, its small-signal frequency response is the (sin f)/f functional dependence as given by Eq. (6.7.3).

The traveling-wave directional coupler (DC) frequency response is different from that of interferometric modulators [68]. The difference results from the fact that the behavior of a directional coupler depends on the distribution of the induced index change along the interaction length, not simply on its integral. The modulation response, which must be calculated numerically in the frequency domain, is shown in Fig. 6.15 together with the result of interferometric modulators. The DC response is braoder than that of the interferometric for the same electrode length and also has much lower sidelobes. Both of these features are consistent with a physically intuitive argument that goes as follows. In a one-coupling-length directional coupler the effect of phase mismatch is less significant at the ends of the coupler, where most of the light is in only one waveguide, and is most important in the center, where in the absence of applied voltage, the incident light is equally split between the two waveguides. As a result, the electrical/optical interaction is effectively symmetrically weighted along the coupler length. This weighting or apodizing results in the reduced frequency response sidelobes and in the increased bandwidth because the effective length overwhich the walk-off occurs is shorter. This shorter effective length is also responsible for the fact that the directional coupler switch voltage is larger by the $\sqrt{3}$ than that of an interferometric of the same electrode length. The bandwidth (50% optical modulation depth) under large-signal conditions can again be described by a bandwidth × length product which is about 15 GHz-cm for the directional coupler and about 9.6 GHz-cm for the interferometer.

RF Attenuation. To approach the velocity-mismatch limited results described above, it is essential that microwave loss in the electrode be minimized. A highly conductive, thick electrode is required. Electroplated gold has provided the best overall performance [80]. A scanning electron microscope photograph is shown in Fig. 6.16. Because of skin depth effects [81], the electrode loss varies

→| |← 15 μm

FIG. 6.16 Photomicrograph of gold-electroplated coplanar line.

approximately as $\sqrt{f}$. The RF power loss coefficient for a 3-μm-thick 15-μm-wide electroplated gold electrode with a 5 μm gap is about 1.5 dB/cm-$\sqrt{GHz}$ [16]. The effect of such a microwave loss on the optical modulator frequency response is shown in Fig. 6.17 for a 1-cm-long directional coupler modulator. The bandwidth reduction of about 20% is approximately the same for the interferometric type of modulator. Because of the $\sqrt{f}$ dependence the effect of microwave loss becomes much less critical, compared to velocity mismatch limits, as the device length is decreased.

E. Experimental Results

To a large extent the difficulty in attaining a flat and broadband response is the careful design of the coplanar line and RF launching and termination. If the drive signal is provided using coaxial

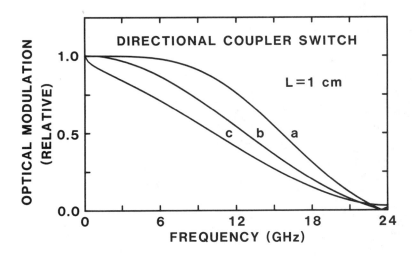

FIG. 6.17 Effect of RF propagation loss on modulator bandwidth: (a) large signal response, no RF attenuation; (b) small signal response, no RF attenuation; (c) small signal response including RF attenuation.

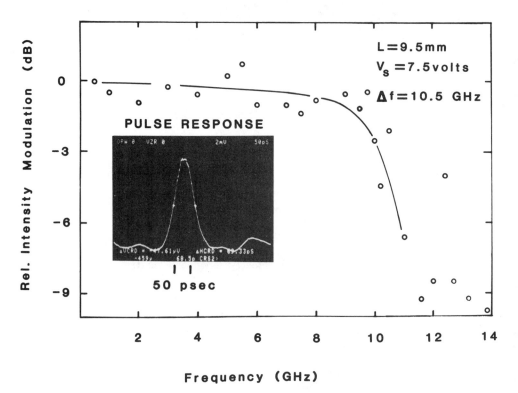

FIG. 6.18 Traveling-wave modulator frequency response.

cable, the transition from the coaxial to coplanar transmission line can be a source of difficulties. In general, it is best to keep the size and lengths of the coplanar line as small as possible. Using the technologies described above and careful design, broadband Ti:LiNbO₃ traveling-wave switches and modulators have now been demonstrated in several laboratories around the world. The first such devices were reported by Sueta and co-workers [82]. More recently, both interferometric and directional coupler traveling-wave modulators operating in the range of 20 GHz and above have been fabricated [83–85]. The optical frequency response and pulse response of a 1-cm directional coupler traveling-wave switch are shown in Fig. 6.18.

6.4 DEVICE INSERTION LOSS

Low optical insertion loss is essential for Ti:LiNbO₃ waveguide devices to find practical application. For nearly all applications, particularly optical fiber communication, permanently attached fiber pigtails are required. In this case, the total optical insertion loss includes fiber-to-waveguide coupling loss, waveguide propagation and bend loss, and possibly electrode loading loss.

6.4.1 Coupling Loss

A. Factors

Assuming perfect fiber-to-Ti:LiNbO₃ waveguide alignment, the two contributions to fiber-waveguide coupling loss are reflection, or Fresnel, loss and the loss caused by mismatch between the fiber and waveguide modes. The former can be reduced if not completely eliminated by using either index matching fluid or antireflection coatings on the lithium niobate end faces. The loss resulting from mismatch between the fiber and waveguide modes is then the principal source of fiber-waveguide coupling. Indeed, in early devices the total device insertion loss was dominated by fiber-waveguide coupling loss due to mode mismatch.

B. Mode Mismatch

To achieve low coupling loss it is necessary to fabricate titanium-diffused waveguides with a spot size well matched to that of the fiber. Because of the large index difference between lithium niobate and air, the waveguide mode shape in depth is typically quite asymmetric, while the fiber is circularly symmetric. Furthermore, for operation at λ = 1.3 to 1.5 μm and for fiber core sizes typical for lightwave communication applications, the fiber mode is relatively large. Thus relatively deep diffusion of the titanium into the lithium niobate is required. Fortunately, a deep diffusion also helps to reduce propagation loss due to surface scattering.

As discussed earlier the diffusion depth is determined by the diffusion time t, temperature T, and activation temperature T_A. The titanium metal thickness τ should satisfy two requirements. First, for the diffusion parameters required to achieve the desired depth, the titanium metal must be thin enough to allow complete diffusion. However, it must be sufficiently thick to produce a waveguide/substrate index difference, $\Delta n = n_g - n_s$, to provide strong guiding. Within these general constraints one should choose t, T, τ and the waveguide strip width W to produce a waveguide mode that matches the fiber mode as closely as possible. This selection can be accomplished approximately with the models described earlier. The final selection of diffusion parameters to ensure that the waveguides are optimized must, of course, be determined empirically.

The waveguide and fiber modes can be measured using the arrangement in Fig. 6.19. The near-field patterns are imaged onto an infrared vidicon; the output from the vidicon is displayed on a signal-averaging oscilloscope, and waveguide mode intensity profiles recorded in the width and depth directions. The mode size depends on wavelength and polarization.

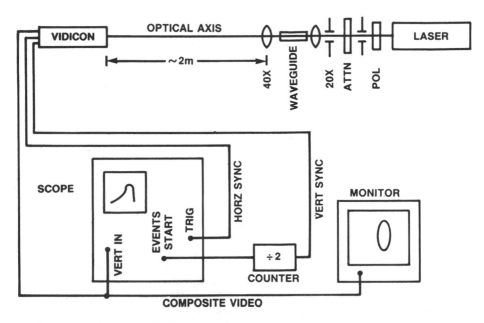

FIG. 6.19 Mode profiling arrangement.

An example of the measured width and depth waveguide mode profile with a 9-µm metal strip width and diffusion conditions chosen using the model described above is shown in Fig. 6.20. Also shown is the measured mode profile of a standard single-mode fiber. The waveguide mode in the crystal plane is essentially Gaussian and is an excellent match to the fiber. The mode profile in depth is asymmetric, as expected, because of the large waveguide–cladding index difference. The mode shape in depth is approximated by the Hermite-Gaussian function (dashed curve) over most of the profile. The intensity roll-off both near the surface and into the substrate is slower, however, than for the Hermite-Gaussian [86].

The longer tail on the air side of the waveguide is especially significant because of its better compatibility with the symmetric fiber mode. The measured 1/e intensity full width w_x and depth w_y for these waveguides, all of which were single mode, versus metal strip width is shown in Fig. 6.21 for both the TE and TM polarization. Also shown is the geometric mean mode diameter $\sqrt{w_x w_y}$. The mode dimensions are relatively constant over the range of strip widths between about 6 and 10 µm which correspond to well-guided modes. The rapid increase in mode size for small strip width results from poor confinement as these waveguides approach cutoff. The ratio of the mode width to depth, w_x/w_y, is approximately 1.5 for the well-confined modes and approaches unity for waveguides near cutoff. For these diffusion conditions there is a very good match between the fiber mode diameter and the waveguide mean mode diameter for strip widths from approximately 7 to 10 µm. This good match occurs for a slightly greater range of strip widths for the TE than for the TM mode.

The expected coupling efficiency can be determined by calculating the overlap between the fiber and waveguide modes. The power coupling coefficient can be written conveniently as [86, 87]

$$k = 0.93 \left\{ \frac{4(w/a)^2}{[(w/a)^2 + \varepsilon][(w/a)^2 + 1/\varepsilon]} \right\} \qquad (6.24)$$

where a is the fiber mode 1/e intensity diameter, w the geometric mean, $w = \sqrt{w_x w_y}$, and ε is the ratio of the waveguide mode width and depth, $\varepsilon = w_x/w_y$.

The constant factor of 0.93 is determined by modeling the mismatch between the symmetric fiber mode and the diffused channel waveguide mode assuming identical 1/e dimensions for the two [87]. Experimentally, this coefficient is found to be somewhat closer to unity [86]. The second term includes the further reduction in coupling efficiency due to any difference in the dimensions of the fiber and waveguide modes. Equation (6.24) indicates that the

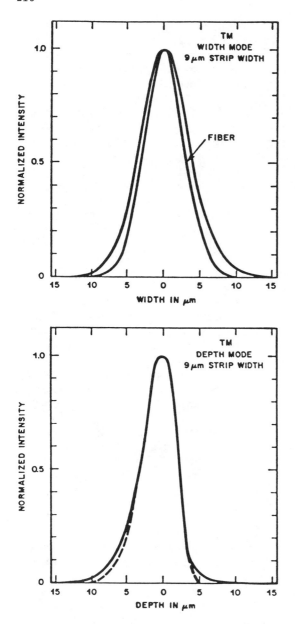

FIG. 6.20 Measured Ti:LiNbO$_3$ mode profiles.

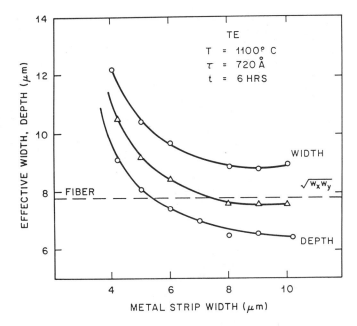

FIG. 6.21 Measured mode sizes for Ti:LiNbO₃ waveguides.

coupling efficiency can be specified simply by the ratio w/a and the waveguide mode eccentricity.

The dimensional dependence of the expected coupling loss [the term in parentheses in Eq. (6.24)] is shown in Fig. 6.22. The results in the figure indicate that regardless of the mode eccentricity, the optimum coupling is achieved for w = a. The dependence of the coupling loss on mode mismatch is weak for w near a. A mismatch in the mean waveguide mode size of 10% causes an increase in coupling loss of only about 0.04 dB relative to its value for w/a = 1 for each of the values of ε shown. The coupling loss is less sensitive to nonunity eccentricity. A 10% error from the optimum condition $\varepsilon = 1$ increases the coupling loss by only about 0.01 dB for w/a = 1.

To demonstrate the importance of mode matching for good fiber-waveguide coupling, measured fiber-waveguide-fiber total insertion loss results for the waveguides with mode data plotted in Fig. 6.21 are shown in Fig. 6.23. Results are shown for both the TE and TM modes. For the waveguides with minimum total insertion

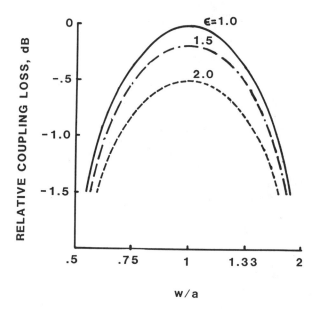

FIG. 6.22 Coupling loss versus mode size and eccentricity.

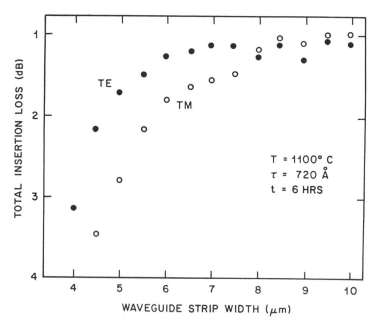

FIG. 6.23 Measured fiber-coupled insertion losses.

loss, the coupling loss per face is about 0.35 dB and the propagation loss is about 0.3 dB/cm. This measured coupling loss includes a calculated residual Fresnel loss of about 0.12 and 0.09 dB per face for the TE and TM modes, respectively, assuming a matching index of 1.65 [86]. This loss component can be reduced by better index match or by antireflection coatings. The rapid increase in total insertion loss for small metal strip widths is principally due to a large propagation loss resulting from poor mode confinement as waveguide cutoff is approached.

Calculated coupling losses using Eq. (6.24) and the measured mode sizes correlate accurately with the measured coupling loss displayed in Fig. 6.23. The insensitivity of the mode overlap to small changes in w/a together with the fact that for well-confined modes the mode width and depth are relatively constant over a large range of channel waveguide strip widths results in a coupling loss that is very insensitive to channel width. Very good coupling efficiency has been achieved for both TE and TM modes because for the diffusion conditions used, the mean mode sizes for well-guided modes are roughly equal for both polarizations. This is a consequence of the Δn for the extraordinary and ordinary indices not being greatly different for diffusion conditions employing modest values of Δn. However, for large values of titanium concentration, the index changes for the ordinary and extraordinary indices may be very different [45] and mode matching of the TE and TM modes simultaneously is more difficult. Fortunately, for the typically large fiber mode sizes, only modest values of Δn are necessary.

C. Waveguide Fiber Misalignment

Lateral Displacements. The sensitivity of the fiber-waveguide coupling efficiency to lateral and vertical (depth) translational misalignment is an important issue in considering manufacturable fiber attachment techniques. Example results for 1.3 μm are shown in Fig. 6.24. A slight asymmetry with respect to the vertical offset, which originates from the vertical waveguide mode shape, is evident. Because the mode size is relatively large to match to typical fibers, the coupling efficiency is not very sensitive to translational alignment errors. For both vertical and lateral translation, an offset of ±2 μm increases the coupling loss by only about 0.25 dB relative to the best alignment. This insensitivity to alignment accuracy simplifies the task of mechanical fiber-waveguide connectors.

Angular Alignment. In addition to the above contributions to the coupling loss noted above, one must also consider the consequences of misalignments of the waveguide and fiber axes. Experiments at 1.56 μm wavelength have shown that the angular misalignment must be maintained below 0.5° in order to attain an excess loss below

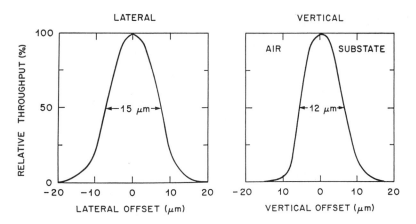

FIG. 6.24 Measured losses versus lateral offset.

0.25 dB [88]. The measured results are in good agreement with theoretical model calculations, assuming two nonidentical Gaussian-shaped mode profiles for the waveguides, as shown in Fig. 6.25. In general, since the sensitivity to angular misalignment is greater for larger mode sizes, the angular tolerances for $Ti:LiNbO_3$ wave-guide-to-fiber coupling is slightly more stringent for angles in the waveguide plane than out of the plane. This is a consequence of the slight asymmetry of the $Ti:LiNbO_3$ waveguide mode.

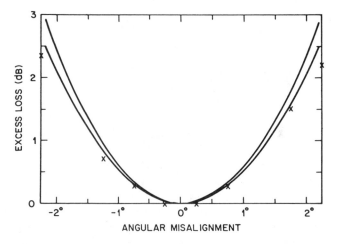

FIG. 6.25 Angular dependence of fiber-waveguide coupling loss.

D. Design

It is important to note that not all diffusion conditions that yield low fiber coupling loss necessarily also yield low propagation and bending loss, and consequently the lowest total insertion loss. This is especially important when using single-mode fibers with a relatively large mode size. In that situation, low coupling loss can be attained for weakly confined modes of a waveguide with small width and depth. The propagation and bending loss of such weakly confined guides can be quite large, however. Thus only modes that are first well confined should be considered when exploring trends in wave-guide mode sizes. The indicators provided by a waveguide design aid, as discussed in the preceding section, serve in identifying those diffusion conditions that yield good mode confinement. In Sec. 6.4.2 we consider propagation and bending loss in more detail.

6.4.2 Propagation Loss

The loss of light propagating through a crystal that is proportional to the length traversed can be divided into the categories of ab-sorptive and scattering losses. Within each of these catetories, the losses can be further classified as to whether or not they are either intrinsic or technical in nature. Ti:LiNbO₃ waveguide propagation losses in the spectral region 1.3 to 1.5 μm are below the level of 0.2 to 0.3 dB/cm [86,88], which is comparable to the experimental uncertainty. A lower upper limit of 0.05 dB/cm has been set by Suche and collaborators by measuring the finesse of resonators fabri-cated from Ti:LiNbO₃ waveguides [90]. Empirically, the propagation loss of Ti:LiNbO₃ waveguides is observed to decrease with increasing mode confinement. Below is a brief discussion of the sources of op-tical propagation loss.

A. Absorption

Electronic and Lattice Excitations. One component of the propa-gation loss is caused by the absorption of light. As mentioned be-fore, LiNbO₃ is nominally transparent in the visible and near-infrared region of the optical spectrum. Figure 6.26 shows the measured transmittance of bulk LiNbO₃ uncorrected for reflection loss at the input and output surfaces [30]. As with other crystals and glasses, the absorbance in the near UV is dominated by electronic absorption [91]. In the infrared beginning at about 4 μm the absorption is via the excitation of lattice vibrations. In the region between 0.5 and 4 μm, the tails of both the electronic and lattice states form a local mini-mum in the intrinsic absorption loss. Consequently, in this region the loss is dominated by scattering and impurity absorption.

Impurity Absorption. Typical impurities found in LiNbO₃ are the transition metals, notably Fe^{2+}, and hydrogen in the form of the

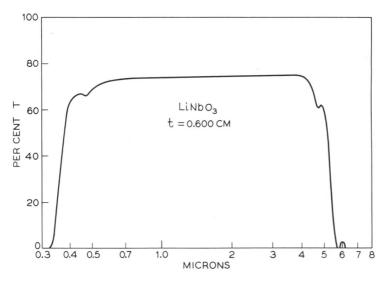

FIG. 6.26   Bulk optical transmission of $LiNbO_3$ (courtesy of K. Nassau).

hydroxyl ion, OH^-. The absorption band of the former is relatively broad, whereas the latter exhibits sharper absorption peaks—the most significant of which are near 2.8 μm and the vibrator's second harmonic at 1.4 μm. The impurity absorption is not a significant constribution to the total propagation loss in comparison to scattering losses. However, the Fe^{2+} center is closely linked to the phenomenon of photorefractive optical damage. This photo-induced refractive index change is quite evident at visible wavelengths but is virtually nonexistent at 1.3 μm [92].

Metallic Loading. If a metallic overlay is deposited on the surface of the waveguiding region, an optical wave propagating in the waveguide experiences an absorption referred to as metallic loading [93]. The absorption occurs through the excitation of eddy currents—more technically, surface plasmons—at the metallic surface and is strongest for the polarization normal to the interface. The absorption is efficiently reduced using a dielectric buffer layer of index lower than the substrate index, as indicated by theoretical analyses for $Ti:LiNbO_3$ waveguides [94]. At 1.3 μm, experiment has shown that an SiO_2 layer 1800 A thick is an appropriate buffer [95]. The metallic loading is increased for metals that do not exhibit ideal metallic behavior, that is, those that have a nonzero value of the real component of the index of refraction. Metallic loading can also be deliberately employed to fabricate polarizers with large extinction ratios [96].

B. Scattering

Volume Scattering. All materials, crystals and glasses alike, exhibit intrinsic scattering losses because of local fluctuations in the
index of refraction. This scattering is referred to as Rayleigh
scattering and has a $1/\lambda^4$ dependence. In glasses, the index fluctuations arise from density variations that are frozen in as the glass
is solidified. In single-domain crystals, scattering centers may arise
from random dislocations, faults, impurities, or compositional inhomogeneities. In addition, waveguides formed by the incorporation of
dopants may experience an increased scattering loss caused by random variations in the dopant density or stresses built up in the
crystal as a consequence of the incorporated material. These sources
of loss appear to dominate the transmittance of LiNbO$_3$ in the region
of relative transparency. Presumably, the light lost to volume scattering decreases with increasing effective index difference, ΔN, and
numerical aperture of the waveguide. Note, too, that in principle,
such random scattering in an anisotropic medium, such as LiNbO$_3$
can give rise to depolarization as well as contributing to the loss.

Surface Scattering. Waveguides may also exhibit scattering
losses because of surface roughness. Such losses can be large
because of the large index difference between the substrate and air.
Theoretically, these losses are predicted to increase as the mode
confinement is increased [97]. Inasmuch as this is opposite to experimental observation for Ti:LiNbO$_3$ waveguides, it is not considered a significant contribution to these diffused waveguides.

6.4.3 Bending Loss

A. Purpose of Bends

Waveguide bends provide a means of redirecting the guided optical
signal and of interconnecting devices to form optical circuits. For
example, waveguide bends are usually used at the inputs and outputs of directional couplers to separate the waveguides so that they
no longer interact and to permit connection to fibers, which have
cladding diameters of about 15 times the core diameter. A typical
waveguide bend geometry is illustrated in Fig. 6.27.
    The main concern in designing waveguide bends is optical loss.
In the bend region, the waveguide is no longer strictly a guided
mode, and optical power leaks from the waveguide as radiation. Con-
ceptually, the light is lost to radiation because the bend would require photons in the outer tail of the mode to travel at a speed
greater than the speed of light in the material in order to maintain
a front of constant phase directed orthogonal to the local radius
vector. From another viewpoint, the waveguide bend may be thought
of as locally modifying the index of refraction by the addition of an

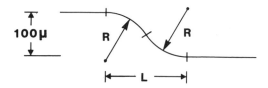

FIG. 6.27 Waveguide S-bend geometry.

index that increases in direct proportion to the distance from the
center of curvature. These ideas are illustrated in Fig. 6.28. In
the latter interpretation, light within the waveguide region penetrates
the index barrier and escapes as radiation.

B. Form of Loss Coefficient

The rate at which light is radiated in the bend depends on several
factors: the radius of curvature, the effective index of the mode,
the size of the mode, and the optical wavelength. For a planar
waveguide bend, the attenuation coefficient takes the form

$$C_1 e^{-C_2 R}$$

(6.25)

where C_1 and C_2 are independent of R but depend on the waveguide characteristics and wavelength [98]. This same exponential

STRAIGHT WAVEGUIDE   WAVEGUIDE BEND

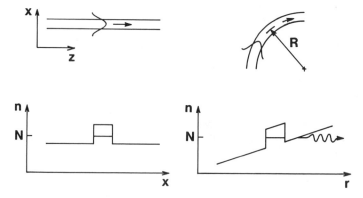

FIG. 6.28 Propagation of light in a waveguide bend.

dependence provides a good description of experimental measurements of Ti:LiNbO$_3$ channel waveguide bend loss [88,99,100]. By careful control of the diffusion parameters, experimental results of $C_1 \sim 15$ mm^{-1} and $C_2 \sim 0.4$ mm^{-1} have been achieved for Ti:LiNbO$_3$ waveguide bends in the region 1.3 to 1.5 µm. From these values it can be concluded that the excess loss for a 5-mm-long section of constant 20-mm radius is approximately 0.1 dB, whereas a 1-mm-long section of constant radius of 5 mm has nearly 10 dB of excess loss.

C. Role of ΔN

According to theory, the coefficient C_2 is proportional to the 3/2 power of the effective index, ΔN. The coefficient C_1 is a more complicated function of ΔN and also the mode size of the waveguide. Like C_2, C_1 increases for increasing ΔN, but it varies much less than C_2 over the range of fabrication parameters that yield single-mode waveguides for a given wavelength. Because C_2 appears in the exponent of the loss coefficient, the mode confinement represented by ΔN is the dominant factor determining the bend loss and should be made large.

D. Design

To minimize the bending loss of a single-mode waveguide, it is desirable to use as large an initial titanium thickness as possible for a given initial strip width and diffusion depth, without the waveguide becoming multimoded. Fortunately, this is consistent with the requirements to achieve low propagation loss and is compatible with the constraints imposed by fiber/waveguide coupling considerations. If waveguide bends with radii smaller than typical values of 20 to 30 mm are necessary, more sophisticated techniques may be employed [100].

6.5 AN APPLICATION TO OPTICAL COMMUNICATION

6.5.1 Optimized Modulator Performance

For application of high-speed waveguide modulators as external optical signal encoders in long-distance communication systems, broad bandwidth, low insertion loss, and low drive power are important. However, as discussed previously, these are generally competing requirements. Bandwidth and drive voltage both increase with decreasing device length. The trade-off between insertion loss and drive voltage results primarily from the need to make waveguides with a relatively large mode size (ca. 8 µm) to provide good mode match to a typical single-mode fiber. These relatively large modes result in a slightly reduced efficiency in the electro-optic interaction. By varying the waveguide diffusion parameters the trade-off between fiber-coupled insertion loss and voltage-length product for devices, such as the directional switch, can be explored. Results are shown in Fig. 6.29. There is nearly a factor of 2 increase in voltage required for the lowest-insertion-loss device compared to one designed only to minimize drive voltage. Fortunately, a reasonable trade-off between insertion loss and drive voltage can be achieved for a diffusion of 1025° and 6 h [101]. To optimize the drive voltage and the insertion loss separately, tapered waveguides to transform the relatively large mode size required for fiber coupling to the smaller modes need to minimize drive voltage may be required [102]. However, it will still be necessary that the smaller modes exhibit the low propagation loss that has been observed in waveguides made to achieve efficient fiber-to-waveguide coupling.

6.5.2 Systems Results

As a representative example illustrating the capabilities of the Ti: LiNbO$_3$ integrated optic waveguide technology, we consider its application to high-speed optical modulation and switching for communication. When used for the function of encoding information in optical form, Ti:LiNbO$_3$ waveguide modulators have several advantages over the direct current modulation of injection lasers. First and

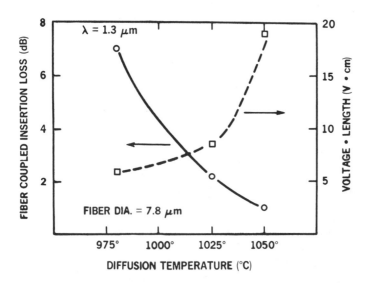

FIG. 6.29 Insertion loss/switching voltage trade-off.

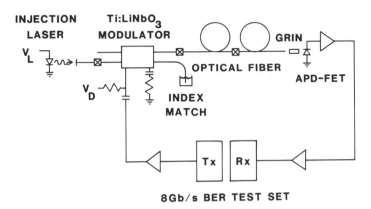

FIG. 6.30 Optical fiber transmission experiment.

foremost is the spectral quality of the transmitted optical signal.
Even when using single-frequency injection lasers, the output wavelength during a bit time is significantly chirped because of the index
change associated with changes in the current density, which is coupled to the photo density. This chirp is avoided with modulators external to the laser cavity that are based on parametric changes of the real
part of the index of refraction. In that case, the transmitted spectral
width of the optical signal is essentially that of the electrical drive
signal representing the data.

A second advantage is the ability to achieve high modulation extinction without sacrificing the purity of the transmitted signal. The
high spectral purity of the transmitted signal also permits the maximum utility of the available fiber bandwidth to multiplex other channels. Finally, the use of the external modulator allows the laser to
be optimized for spectral quality and power output without trade-
offs to attain high modulation bandwidth as well. The disadvantages
of the external modulator are its nonzero insertion loss and the complexity of an additional component in the system.

In the system experiments to explore and extend the state of the
art of optical communications, the advantages of the Ti:LiNbO$_3$ modulators have been shown to outweigh the disadvantages. A typical
optical fiber transmission experiment consists of a transmitter, fiber
span, receiver, and bit-error-rate test set as illustrated in Fig.
6.30. The goal of the experiments is to attain the highest possible
data rate for the transmitter and receiver, and the longest transmission distance before regeneration is necessary. That distance is
reached, by definition, when the BER becomes greater than 1×10^{-9}. At present, the optical fiber transmission systems that have
attained the longest repeaterless distance (117 and 68 km) at the

highest modulation data rates (4 and 8 Gbit/s, respectively) incorporate Ti:LiNbO$_3$ waveguide intensity modulators [103,104]. The characteristics of these Ti:LiNbO$_3$ devices designed for a 1.55-μm wavelength, include 1-cm active length, >4-GHz electrical bandwidth, 8-V switching voltage, 3-dB total insertion loss, −35-dB optical reflectivity, and a >25-dB extinction ratio at dc.

6.6 THE FUTURE

In this chapter we have discussed why and how Ti:LiNbO$_3$ devices are useful elements of advanced lightwave systems. Predicting where and when the technology will become viable in the marketplace is a much more difficult question to answer. Clearly it will be tied to the growth of the fiber communication network. At present, prospects for the Ti:LiNbO$_3$ technology are very promising. Issues such as reproducibility, aging, and packaging are being pursued aggressively. The next few years are certain to hold significant events in the history of the Ti:LiNbO$_3$ evolution.

REFERENCES

1. S. E. Miller, *Bell Syst. Tech. J.*, *48*:2059 (1969).

2. L. McCaughan and G. A. Bogert, Technical Digest of the Conference on Optical Fiber Communications, Paper TUQ20, San Diego (1985).

3. P. Granestrand, L. Thylen, B. Stoltz, K. Bergvall, W. Doldissen, H. Heidrich, and D. Hoffmann, Technical Digest of the Conference on Integrated and Guided-Wave Optics, Paper WAA3, Atlanta, Ga. (1986).

4. T. Tamir, ed., *Integrated Optics*, 2nd ed., Springer-Verlag, New York (1979).

5. R. G. Hunsperger, *Integrated Optics: Theory and Technology*, Springer-Verlag, New York (1982).

6. R. C. Alferness, *IEEE J. Quantum Electron.*, QE-17:946 (1981).

7. R. V. Schmidt and I. P. Kaminow, *Appl. Phys. Lett.*, 25:458 (1974).

8. R. V. Schmidt and I. P. Kaminow, *IEEE J. Quantum Electron.*, QE-11:57 (1975).

9. J. L. Jackel, V. Ramaswamy, and S. Lyman, *Appl. Phys. Lett.*, *38*:509 (1981).

10. R. J. Esdile, *Appl. Phys. Lett.*, *33*:733 (1977).

11. B. Chen and A. C. Pastor, *Appl. Phys. Lett.*, *30*:570 (1977).

12. O. Eknoyan, A. S. Greenblatt, W. K. Burns, and C. H. Bulmer, *Appl. Opt.*, *25*:737 (1981).

13. S. Yamada and M. Minakata, *Jpn. J. Appl. Phys.*, *20*:733 (1981).

14. R. C. Alferness, *Appl. Phys. Lett.*, *35*:748 (1979).

15. G. Eisenstein, S. K. Korotky, L. W. Stulz, J. J. Veselka, R. M. Jopson, and K. L. Hall, *Electron. Lett.*, *21*:363 (1985).

16. S. K. Korotky, G. Eisenstein, R. C. Alferness, J. J. Veselka, L. L. Buhl, G. T. Harvey, and P. H. Read, *J. Lightwave Technol.*, *LT-3*:1 (1985).

17. S. K. Korotky, G. Eisenstein, A. H. Gnauck, B. L. Kasper, J. J. Veselka, R. C. Alferness, L. L. Buhl, C. A. Burrus, T. C. D. Huo, L. W. Stulz, K. Ciemiecki Nelson, L. G. Cohen, R. W. Dawson, and J. C. Campbell, *J. Lightwave Technol.*, *LT-3*:1027 (1985).

18. F. J. Leonberger, C. E. Woodward, and R. A. Becker, Technical Digest of the Conference on Integrated and Guided-Wave Optics, Paper WA3, Pacific Grove, Calf. (1982).

19. H. J. Arditty, J. P. Bettini, Y. Bourbin, Ph. Graindorge, H. C. LeFevre, M. Papuchon, and S. Vatoux, in Proceedings of the Conference and Optical Fiber Sensors, Vol. II, Stuttgart, pp. 321–325 (1984).

20. K. Nassau, Proceedings of Processing of Guided Wave Optoelectronic Materials, *Proc. SPIE*, *460*:2 (1984).

21. A. Rauber, Chemistry and Physics of LiNbO$_3$, *Current Topics in Material Science*, Vol. 1, (E. Kaldis, ed.), North-Holland pp. 481–601 (1978).

22. R. S. Weiss and T. K. Gailord, *Appl. Phys. A*, *A37*:191 (1985).

23. M. E. Lines and A. M. Glass, *Principles and Applications of Ferroelectrics and Related Materials*, Oxford University Press, Oxford (1977).

24. S. C. Abrahams, J. M. Reddy, and J. L. Berstein, *J. Phys. Chem. Solids*, *27*:997 (1966).

25. S. C. Abrahams, W. C. Hamilton, and J. M. Reddy, *J. Phys. Chem. Solids*, *27*:1013 (1966).

26. P. Lerner, C. Legras, and J. P. Damas, *J. Cryst. Growth*, *3*:231 (1968).

27. J. R. Caruthers, G. E. Peterson, M. Grasso, and P. M. Bridenbaugh, *J. Appl. Phys.*, *42*:1846 (1971).

28. A. A. Ballman, *J. Am. Ceram. Soc.*, *48*:112 (1965).

29. K. Nassau, H. J. Levinstein, and G. M. Loiacono, *J. Phys. Chem. Solids*, *27*:983 (1966).

30. K. Nassau, H. J. Levinstein, and G. M. Loiacono, *J. Phys. Chem. Solids*, *27*:989 (1966).

31. B. T. Matthias and J. P. Remeika, *Phys. Rev.*, *76*:1886 (1949).

32. G. E. Peterson, A. A. Ballman, P. V. Lenzo, and P. M. Bridenbaugh, *Appl. Phys. Lett.*, *5*:62 (1964).

33. L. P. Kaminow and W. D. Johnston, Jr., *Phys. Rev.*, *160*:519 (1967).

34. I. Fujimoto, *Acta Cryst.*, *A38*:337 (1982).

35. I. P. Kaminow, *An Introduction to Electro-optic Devices*, Academic Press, New York (1974).

36. D. F. Nelson and R. M. Mikulyak, *J. Appl. Phys.* *45*:3688 (1974).

37. G. D. Boyd, R. C. Miller, K. Nassau, W. L. Bond, and A. Savage, *Appl. Phys. Lett.*, *5*:234 (1964).

38. G. D. Boyd, W. L. Bond, and H. L. Carter, *J. Appl. Phys.*, *38*:1941 (1967).

39. M. N. Armenise, C. Canali, and M. De Sario, *J. Appl. Phys.*, *54*:62 (1983).

40. C. E. Rice and R. J. Holmes, *J. Appl. Phys.*, in press (1986).

41. R. J. Holmes, Y. S. Kim, D. M. Smyth, and C. D. Brandle, Jr., *Ferroelectrics*, *51*:41 (1983).

42. R. J. Holmes and D. M. Smyth, *J. Appl. Phys*, *55*:3531 (1984).

43. M. Fujuma, J. Noda, and H. Iwasaki, *J. Appl. Phys.*, *49*:3693 (1978).

44. H. Naitoh, M. Nunoshita, and T. Makayama, *Appl. Opt.*, *16*:2546 (1977).

45. M. Minakata, S. Saito, M. Shibata, and S. Miyazawa, *J. Appl. Phys.*, *49*:4677 (1978).

46. K. Sugii, M. Fukuma, and H. Iwasaki, *J. Mater. Sci.*, *13*:523 (1978).

47. W. K. Burns, P. H. Klein, E. J. West, and L. E. Plew, *J. Appl. Phys.*, *50*:6175 (1979).

48. H. Ludtke, W. Sohler, and H. Suche, Digest of Workshop on Integrated Optics, Berlin, p. 122 (1980).

49. J. Crank, *The Mathematics of Diffusion*, 2nd Ed., Oxford University Press, Oxford (1975).

50. G. B. Hocker and W. K. Burns, *Appl. Opt.*, *16*:113 (1977).

51. J. Noda, N. Uchida, S. Saito, T. Saku, and M. Minakata, *Appl. Phys. Lett.*, *27*:19 (1975).

52. J. Ctyroky, M. Hofman, J. Janta, and J. Schrofel, *IEEE J. Quantum Electron.*, *QE-20*:400 (1984).

53. E. A. J. Marcatili, *Bell Syst. Tech. J.*, *48*:2071 (1969).

54. C. Yeh, L. Casperson, and W. P. Brown, *Appl. Phys. Lett.*, *34*:460 (1979).

55. R. M. Knox and P. P. Toulios, *Proceedings of the MRI Symposium on Submillimeter Waves* (J. Fox, ed.), Polytechnic Press, New York, pp. 497–516 (1970).

56. G. B. Hocker and W. K. Burns, *IEEE Quantum Electron.*, *QE-11*:270 (1975).

57. H. F. Taylor, *IEEE J. Quantum Electron.*, *QE-12*:748 (1976).

58. S. K. Korotky, W. J. Minford, L. L. Buhl, M. D. Divino, and R. C. Alferness, *IEEE J. Quantum Electron.*, *QE-18*:1796 (1982). [*Note*: The 9 in the expression of the expectation of ΔN should be a 3.]

59. U. Jain, A. Sharma, K. Thyagarajan, and A. K. Ghatak, *J. Opt. Soc. Am.*, *72*:1545 (1982).

60. M. D. Feit, J. A. Fleck, Jr., and L. McCaughan, *J. Opt. Soc.*, *73*:1296 (1983).

61. R. C. Alferness and L. L. Buhl, *Appl. Phys. Lett.*, *38*:655 (1981).

62. F. Heismann and R. Ulrich, *Appl. Phys. Lett.*, *45*:490 (1984).

63. R. C. Alferness and L. L. Buhl, *Appl. Phys. Lett.*, *40*:861 (1982).

64. O. Mikami, *Appl. Phys. Lett.*, *36*:491 (1980).

65. R. C. Alferness and J. J. Veselka, *Electron. Lett.*, *21*:466 (1985).

66. S. E. Miller, *Bell Syst. Tech. J.*, *33*:661 (1954).

67. H. Kogelnik and R. V. Schmidt, *IEEE J. Quantum Electron.*, *QE-12*:396 (1976).

68. S. K. Korotky and R. C. Alferness, *J. Lightwave Technol.*, *LT-1*:244 (1983).

69. S. K. Korotky, *IEEE J. Quantum Electron.*, *QE-22*:952 (1986).

70. D. Marcuse, *IEEE J. Quantum Electron.*, *QE-18*:393 (1982).

71. L. Thylen and P. Granestrand, *Opt. Commun.*, *7*:11 (1986).

72. M. Minakata, *Appl. Phys. Lett.*, *35*:145 (1978).

73. K. Kobota, J. Noda, and O. Mikami, *IEEE J. Quantum Electron.*, *QE-16*:754 (1980).

74. K. C. Gupta, R. Garg, and L. J. Bahl, *Microstrip Lines and Slotline*, Artech, Dedham, Mass. (1979).

75. R. C. Alferness, *IEEE Trans. Microwave Theory Tech.*, *MTT-30*:1121 (1982).

76. P. S. Cross and R. V. Schmidt, *IEEE J. Quantum Electron.*, *QE-15*:1415 (1978).

77. R. A. Becker, *IEEE J. Quantum Electron.*, *QE-20*:723 (1984).

78. P. Thioulouse, A. Carenco, and R. Guglielmi, *IEEE J. Quantum Electron.*, *QE-17*:535 (1981).

79. R. C. Alferness, S. K. Korotky, and E. A. J. Marcatili, *IEEE J. Quantum Electron.*, *QE-20*:301 (1984).

80. R. C. Alferness, C. H. Joyner, L. L. Buhl, and S. K. Korotky, *IEEE J. Quantum Electron.*, *QE-19*:1339 (1983).

81. J. D. Jackson, *Classical Electrodynamics*, 2nd ed., Wiley, New York, pp. 334–350 (1975).

82. M. Izutsu, Y. Yamane, and T. Sueta, *IEEE J. Quantum Electron.*, *QE-13*:287 (1977).

83. R. A. Becker, *Appl. Phys. Lett.*, *45*:1168 (1984).

84. C. M. Gee, G. D. Thurmond, and H. W. Yen, *Appl. Phys. Lett.*, *43*:998 (1983).

85. S. K. Korotky, G. Eisenstein, R. S. Tucker, J. J. Veselka, and G. Raybon, to be published.

86. R. C. Alferness, V. R. Ramaswamy, S. K. Korotky, M. D. Divino, and L. L. Buhl, *IEEE J. Quantum Electron.*, *QE-18*:1807 (1982).

87. W. K. Burns and G. B. Hocker, *Appl. Opt.*, *16*:2048 (1977).

88. J. J. Veselka, and S. K. Korotky, *IEEE J. Quantum Electron.*, *QE-22*:933 (1986).

89. S. Nemoto and T. Makimoto, *Opt. Quantum Electron*, *11*:447 (1979).

90. H. Suche, B. Hampel, H. Seibert, and W. Sohler, Proceedings of the Conference on Integrated Optical Circuit Engineering II, Boston, *SPIE Proc.*, *578*, 156 (1985).

91. D. Redfield and W. J. Burk, *J. Appl. Phys.*, *45*:4566 (1974).

92. G. T. Harvey, G. Astfalk, A. Y. Feldblum, and B. Kassahun, *IEEE J. Quantum Electron.*, *QE-22*:939 (1986).

93. A. Otto and W. Sohler, *Opt. Commun.*, *3*:254 (1971).

94. M. Masuda and J. Koyama, *Appl. Opt.*, *16*:2994 (1977).

95. L. L. Buhl, *Electron. Lett.*, *19*:659 (1983).

96. D. Eberhard and H. Bulow, *Proceedings of the 3rd European Conference on Integrated Optics*, (H.-P. Nolting and R. Ulrich, eds.), Springer-Verlag, Berlin, pp. 202–206 (1985).

97. P. K. Tien, *Appl. Opt.*, *10*:2395 (1971).

98. E. A. J. Marcatili and S. E. Miller, *Bell Syst. Tech. J.*, *48*: 2161 (1969).

99. W. J. Minford, S. K. Korotky, and R. C. Alferness, *IEEE J. Quantum Electron.*, *QE-18*:1802 (1982).

100. S. K. Korotky, E. A. J. Marcatili, J. J. Veselka, and R. H. Bosworth, *Appl. Phys. Lett.*, *48*:92 (1986).

101. R. C. Alferness, S. K. Korotky, L. L. Buhl, and M. D. Divino, *Electron. Lett.*, *20*:354 (1984).

102. M. Kondo, K. Komatsu, and Y. Ohta, Technical Digest of the Topical Meeting on Integrated and Guided-Wave Optics, Paper TuA5, Kissimmee, Fla (1984).

103. S. K. Korotky, G. Eisenstein, A. H. Gnauck, G. L. Kasper, J. J. Veselka, R. C. Alferness, L. L. Buhl, C. A. Burrus, T. C. D. Huo, L. W. Stulz, K. Ciemiecki Nelson, L. G. Cohen, R. W. Dawson, and J. C. Campbell, Technical Digest of the Conference on Optical Fiber Communications, Paper PDP1, San Diego (1985).

104. A. H. Gnauck, S. K. Korotky, B. L. Kasper, J. C. Campbell, J. R. Talman, J. J. Veselka, and A. R. McCormick, Technical Digest of the Conference on Optical Fiber Communications, Paper PDP9, Atlanta, Ga. (1986).

7

GaAs-Based Integrated Optoelectronic Circuits
Design, Development, and Applications

JAMES K. CARNEY   *Honeywell Physical Sciences Center, Bloomington, Minnesota*

LYNN D. HUTCHESON   *APA Optics, Inc., Blaine, Minnesota*

I. INTRODUCTION

The throughput of data and signal processors is being pushed to ever-increasing limits. The developments of faster, more complex silicon integrated circuits (ICs) and the use of parallel processing are largely responsible for this improved performance. At the same time, it has been necessary to improve the electrical packaging and interconnect technology in order not to compromise the speed of the IC.

In the past few years, the clock speeds of silicon VLSI circuits (very large scale integrated circuits) have been approaching 100 MHz for integration levels of a few thousand gates. The VHSIC (very high speed integrated circuit) program is attempting to raise the clock speed over 100 MHz for integration levels of tens of thousands of gates.

In an attempt to develop even faster circuits, major research programs have been started in gallium arsenide (GaAs) electronics, and to a lesser extent indium gallium arsenide phosphide (InGaAsP). One of the goals of these programs is to develop circuits having gigahertz clock rates. A few MSI-level GaAs parts have already been demonstrated with clocks above 1 GHz.

One of the problems that must be faced when designing a processor to operate at these higher speeds is the extreme difficulty of transmitting data at gigabit/second (Gbit/s) rates. The performance of electrical interconnects is adversely affected by increases in capacitance and reflections due to impedance mismatches. Multilevel board technology is being developed to address this probelm for chip-to-chip interconnects at hundreds of megahertz. The interlevel vias, however, are electrical discontinuities which become increasingly more troublesome as the frequency goes up.

One solution may be the use of optical interconnects to transmit the data. The fiber itself is a nearly ideal transmission medium, and optical sources and detectors have been demonstrated at operating frequencies above 5 GHz [1]. The advantages of using optics for long-line transmission or in a local area network are obvious. A strong case can also be made for using optical interconnects to send data between boards within a processor or even between chips on a board.

One major advantage of GaAs and InP over silicon in high-speed circuits is that III-V materials can emit light. It therefore becomes possible to integrate optical emitters on the IC to perform the I/O (input/output) functions. Integrated optoelectronic circuits (IOCs) are now being developed in a number of laboratories around the world. Interested parties in the United States include major computer manufacturers, defense contractors, telecommunications companies, and a number of universities. It has also been announced that Japan is establishing a research and development company for OICs to be staffed by 13 different companies.

In this chapter IOCs are examined from the standpoint of what it takes to fabricate the device and what performance can be expected. Special emphasis is placed on the use of IOCs in high-speed processors. For this reason, the examples will be from the GaAs/AlGaAs materials system. The slow rate of development of InP/InGaAsP electronics will prevent the use of IOCs from that system in signal processors for many years. Small InP/InGaAsP circuits may, however, find their way into other applications, such as repeaters for long-line optical transmission. In that case, the optical wavelength, not the integration level, is the determining factor.

When designing an optical interconnect, it must be remembered that the performance is dependent on the optical components as well as the electronics. Therefore, in order to see a significant improvement in the performance of an IOC over a hybrid circuit, the processing sequence must not compromise either the electronics or the optics, such as lasers, detectors, or optical modulators. This means that the circuit must accommodate the materials and processes needed for the electronics as well as those of the optics. This is where the similarities in the development of integrated optical components differ greatly from those in the development of silicon integrated circuits. The drive to increase the level of complexity of silicon ICs is based on considerations of cost, performance, density, and packaging ease. All parts of the silicon ICs are made with the same materials and processing sequence. For IOCs, however, the materials and processes for the optics and electronics are vastly different and the process for integration may be more complex still.

The obvious question, then, is: Why bother integrating the devices? There is no single answer to this question. It is actually application dependent. First, it should be stated that optical interconnects allow for higher performance than electrical interconnects in some cases. The major reasons for integrating the devices, then, fall into three areas: lower parasitics means higher performance, optical integration increases density, and fewer parts ease the packaging task. Not much emphasis should be placed on the reduction of cost for producing an integrated optical chip over an IC connected to a discrete optical device. This is because the integration of the optical components adds processing steps to the electronics, thereby increasing the cost per circuit and decreasing circuit yield. In addition, the crystal growth on a substrate for an integrated optical circuit costs just as much as one for a wafer to be made into discrete parts, but the integrated wafer will produce relatively few optical components, making each component more expensive. So instead of trying to justify the components from a cost standpoint, they should be examined on the basis of what can be done with this technology that cannot be done any other way.

In this chapter the components that make up an IOC—materials, electronics, and optoelectronics—are presented. The parameters that are important to a designer of interconnects (e.g., bandwidth, power, density, and bit error rate) are described. Other operating characteristics, such as temperature sensitivity, are discussed to provide an appreciation of what must be considered when the optical interconnect is taken out of the laboratory and designed into a system. The present status of IOCs is described for both optical transmitters and optical receivers. Finally, a few examples are given of the expected performance of IOCs and their impact on the system.

7.2 ELECTRONICS FOR INTEGRATION

Signal processors and supercomputers are continually pushing against the upper limits of performance of ultrahigh-speed electronics. Even though faster is always better in these applications, the ICs must also meet a few basic requirements before a designer will choose to insert them in a system. The ICs must be manufacturable and have demonstrated a reasonable level of reliability.

The ICs must also meet any application-dependent requirements. These may include wide temperature extremes, minimum delay through the circuit, or tolerance to high radiation.

For integrated optoelectronics ICs to meet the needs of high-speed processors, it is imperative that the IOCs be fabricated in a way that guarantees an adequate supply of chips. This will occur only if the IOCs are fabricated using a standard GaAs production process to which the extra steps for the optoelectronics have been added. It is very doubtful that a new GaAs electronics process will be developed solely for the purpose of allowing optoelectronic components to be integrated. In this section a description is given of the major GaAs technologies: their materials requirements, fabrication sequence, and demonstrated performance.

7.2.1 GaAs Processing

The cross sections of transistors fabricated using depletion-mode MESFETs (metal-semiconductor field-effect transistors), enhancement-mode MESFETs, and JFETs (junction FETs) are shown in Fig. 7.1. Each of these technologies is fabricated using undoped, semi-insulating (10^8 Ω-cm) GaAs as a starting substrate. The channel

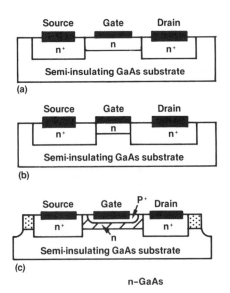

FIG. 7.1 Schematic cross-sections of GaAs field effect transistors: (a) Depletion-mode MESFET; (b) Self-aligned-gate enhancement-mode MESFET; (c) Junction FET.

and contact regions are formed by selective ion implantations which
are activated using a high-temperature annealing step.

Until now, nearly all GaAs IOCs have been fabricated using
depletion-mode MESFETs. Very few demonstrations, however, have
used the process as it exists on a production line. The researchers
have chosen instead to use a fabrication sequence which is more con-
ducive to experimentation and which does not require the same restrictions on substrate size and uniformity.

A schematic representation of the processing sequence for a
depletion-mode MESFET as it is found on a number of production facili-
ties is shown in Fig. 7.2. The starting material is typically a 3-in.
round, unintentionally doped LEC (liquid-encapsulated Czochralsky)
semi-insulating substrate having a (100) surface orientation. The
channel and ohmic contact regions are implanted with an n-type
dopant which is activated with an anneal at approximately 800°C.
The ohmic metal is next deposited and sintered at about 450°C for
30 s. The basic MESFET structure is completed by defining a
1-μm-wide gate on the channel region between the ohmics using a
lift-off technique. The gate metal forms a Schottky contact with the
lightly doped channel region. In the D-mode FETS, the channel
conducts current between the source and drain at a gate voltage of
0 V. If a negative bias is applied to the gate-source, the channel
is cut off and current does not flow. The voltage at which this
cutoff occurs is called the threshold voltage and is one of the most
important parameters in the electronics. The threshold voltage for
D-mode circuits is normally specified to be between −1.0 and −1.5 V.

D-mode MESFET GaAs [2] was one of the first technologies developed and is now in production at a number of foundries. D-mode
MESFET is thought of as being relatively fast. For example, serial-
to-parallel and parallel-to-serial converters have been fabricated
that operate at 3.0-GHz clock rates [3]. However, D-mode is also
power hungry at high speeds. These circuits require 20 mW per
gate.

The greatest amount of current research is in more advanced
GaAs technologies. The fabrication sequence for self-aligned gate
enhancement-mode (E-mode) MESFET [4] is shown in Fig. 7.3. The
channel is first formed with selective ion implantation. The Schottky
metal, a refractory silicide or nitride, is next deposited over the
entire substrate and the 1-μm-long gates are formed by etching.
The n+ ohmic regions for the source and drain are formed by implanting the region around the gate. The gate acts as an implant
mask, forming self-aligned sources and drains. The implants are
activated in a high-temperature (800°C) annealing step. Because
the Schottky metal for the gate is on the channel during the anneal,
it must be engineered to withstand the anneal temperatures while
maintaining excellent diode characteristics. The MESFET is completed by depositing the ohmic metal.

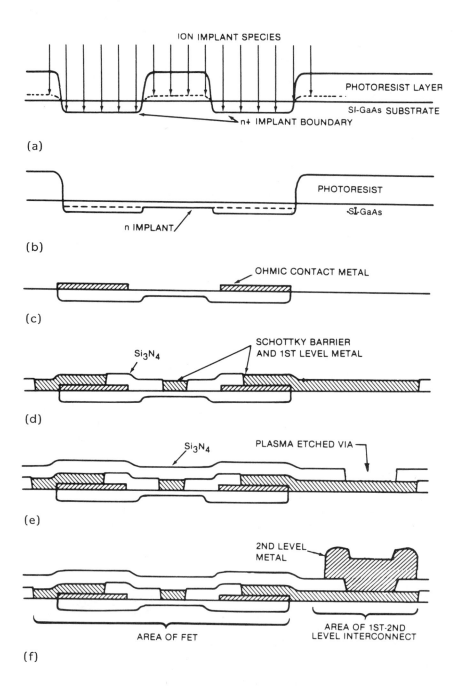

FIG. 7.2 IC processing sequence for depletion-mode MESFET: (a) Selective ion implantation of n-type dopant for n+ contact; (b) Selective ion implant for channel region; (c) ohmic contact deposition; (d) Schottky metal deposition using dielectric assisted lift-off; (e) Interlevel dielectric deposition and via etch; (f) Second-level metal deposition.

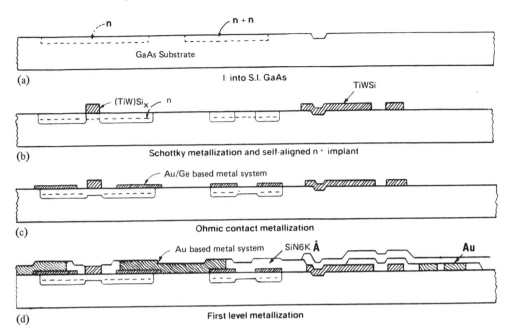

FIG. 7.3 IC processing sequence for self-aligned-gate enhancement-
mode MESFET: (a) Selective ion implantation of n-type dopant for
channel region; (b) Schottky metal deposition and etch followed by
n^+ ion implantation; (c) ohmic contact deposition; (d) first level metal
deposition using dielectric assisted lift-off.

The self-aligned gate E-mode MESFET has a great advantage
over the D-mode MESFET in both size and power. 16:1 serial-to-
parallel and parallel-to-serial converters have been demonstrated
that require only 110 mW to operate at a 800-MHz clock frequency
[5]. 1k-bit and 4K-bit RAMs have been fabricated with 3 ns access
times at 50 μW/bit [6].

The greatest disadvantage of the E-mode MESFET is that the
Schottky gate has a diode turn-on voltage of approximately 0.7 V.
The clamping effect of the diode limits the voltage swing of the
E-mode FET and can, therefore, limit the noise margin of the circuit. This means that the processing variations must be held to a
minimum. The standard deviation of the E-mode threshold voltage
should be 30 mV or less if a reasonable yield is to be obtained.

The fabrication of the JFET [7] is similar to that of the D-mode
MESFET, requiring ion implants into a semi-insulating substrate.
However, prior to depositing the gate metal, a p^+ layer is implanted
into the channel region. A p-type contact, instead of a Schottky

metal, is then deposited to form the gate. Although the JFET is somewhat more complicated to fabricate than the E-mode MESFET, the higher turn-on voltage of a p-n junction diode leads to improved noise margins. The JFET is therefore more tolerant of process variations and less sensitive to temperature than the E-mode MESFET.

The GaAs circuit families described above rely on selective ion implantation into semi-insulating substrates to form the transistors. The modulation doped FET (MODFET) and the heterojunction bipolar transistor (HBT) rely on the special properties of epitaxially grown GaAs/AlGaAs heterostructure to achieve high-performance devices. The fabrication sequence of an enhancement-mode MODFET [8] is shown in Fig. 7.4. This structure is obtained by using epitaxial layers rather than implants to form the channel region. The layers consist of a very pure GaAs layer grown on the substrate followed by an undoped AlGaAs spacer layer (less than 10 nm thick) followed by an AlGaAs layer doped n-type. The key to the high performance of the MODFET transistor is that the conducting channel exists in very pure, undoped GaAs. Therefore, there are no ionized doping sites to cause scattering and decrease the electron mobility. The electrons that are ionized in the n-AlGaAs layer will tend to fall into the lower-energy conduction band of the GaAs. The electrons will form a two-dimensional sheet charge at the GaAs/AlGaAs interface, where they are held at the discontinuity in the conduction bands by the attraction to the ionized donors. The sheet charge of electrons acts as a channel in the GaAs. Because the GaAs layer is grown undoped, the electrons have extremely high mobilities. This leads to very fast switching speeds. Ring oscillators have been fabricated with 1-μm gate lengths [9] that operate with gate delays as low as 11.6 ps at room temperature and 8.5 ps at 77 K.

The remainder of the MODFET fabrication is nearly identical to the E/D-MESFET. The gate metal is deposited and etched. Ion implantation is used to self-align the n^+ source and drain contacts. Because the doped and undoped AlGaAs layers are only a few tens of nanometers thick, the surface potential of the gate metal depletes the channel region. A positive gate-source voltage causes the channel to conduct.

A schematic representation for the heterojunction bipolar transistor from Texas Instruments [10] at various points in its process is shown in Fig. 7.5. The initial material is a heavily doped n-type substrate on which a multilevel heterostructure is grown. The heterostructure consists of:

1. A 1-μm-thick n-type $Al_{0.2}Ga_{0.8}As$ emitter
2. A 300 to 500-Å n-type graded AlGaAs layer
3. An undoped GaAs base layer about 0.2 μm thick which will be doped p-type by ion implantation
4. An n-type GaAs collector layer about 0.3 μm thick

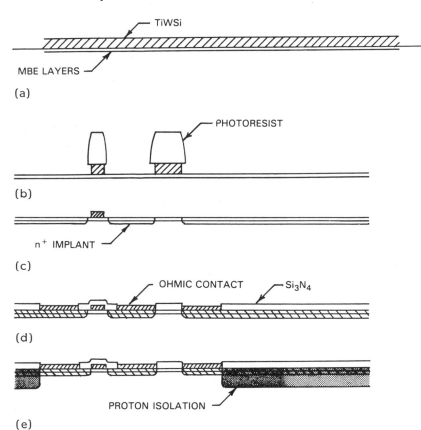

FIG. 7.4 IC processing sequence for self-aligned-gate enhancement-MODFET: (a) Schottky metal deposition on epitaxially-grown MODFET layers; (b) Schottky metal etch; (c) self-aligned gate n^+ ion implantation; (d) ohmic contact deposition; (e) FET isolation using proton implants.

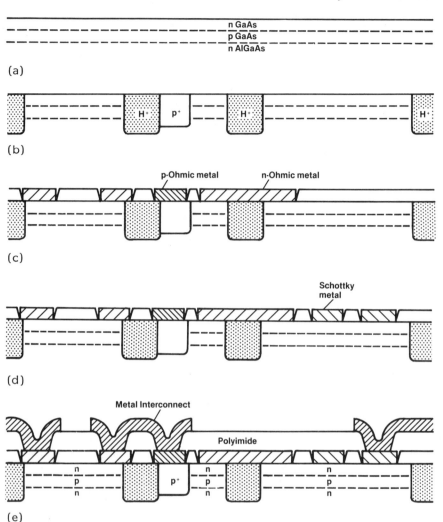

FIG. 7.5 IC processing sequence for Heterojunction Bipolar Transistors: (a) Growth of HBT layers by MOCVD or MBE; (b) Selective ion implantation of p^+-base contact and isolation region; (c) Deposition of p-ohmic and n-ohmic metals; (d) Deposition of Schottky metal; (e) Deposition of metal interconnects on polyimide interlevel dielectric.

TABLE 7.1 GaAs IC Demonstrations

Technology	Circuit	Speed	Power	Comment	Available from
D-MESFET	8:1 multiplexer	1.5 GHz	1 W	SCFL Logic	Gigabit Logic
	8:1 multiplexer	3 GHz	20 mW/gate	Buffered FET Logic	Tektronix
	8 × 8 multiplier	5.25 ns	2.2 W	SDFL Logic	Rockwell
E/D MESFET	32:1 multiplexer	400 MHz	110 mW	Direct coupled	Honeywell
	5 × 5 multiplier	4 ns	0.13-mW/gate	Direct coupled	Honeywell
	16 × 16 multiplier	10.5 ns	953-mW	Gate Array	Fujitsu
	4k RAM	3-ns access	720 mW		Fujitsu
	16k RAM	4.1-ns access	2.5 W		NTT
JFET	256-bit RAM	5-ns access	9 mW	Complementary	MacDonald Douglas
MODFET	5 × 5 multiplier	1.8 ns	0.4-mW/gate	Direct coupled	Honeywell
	1k RAM	3.4-ns access	290 mW		Fujitsu

First, the base layer is doped p-type by an unmarked sheet implant of beryllium (Be). A second, selective Be implant is then performed to fabricate heavily doped (ca. 10^{19} cm^{-3}) p$^+$ regions to contact the base layer. Both implants are activated using a short anneal to minimize excessive Be diffusion and maintain a high p-type doping concentration at the surface for ohmic contact formation.

Planar isolation of the various devices on a wafer is accomplished by a boron implant, which converts a 1-μm-thick surface layer into semi-insulating material. After the isolation implant, ohmic contacts are deposited. The Ti/Pt/Au Schottky metallization is then deposited. The schottky metal is also used as a first-level interconnect. The remainder of the process consists of completing the interconnect metallization.

In this process described above, the HBT is configured as integrated injection logic. Gate arrays with up to about 1000 gates have been designed in this logic [11]. The internal gates could be configured to provide either a gate delay of 1.5 ns at 0.2 mW per gate or 400 ps at 1 mW per gate.

A second type of HBT configuration which is under development at Rockwell is ECL [12]. This technology is still relatively new, but gate delays of as little as 20 ps per gate have been reported.

Table 7.1 contains a list of some of the most significant GaAs IC demonstrations. This table not only compares the speed and power of the different technologies, but also gives an indication of the level of maturity of the technologies.

7.3 INTEGRATED OPTOELECTRONIC DETECTORS AND RECEIVERS

Semiconductor optical detectors are two-terminal devices that convert optical inputs into electrical carriers. By connecting the detector to an appropriate circuit, the electrical carriers are collected and the signal is amplified to levels adequate to drive a digital IC.

The detector and amplifier must be designed as a unit. The integration of the detector with the amplifier provides a significant improvement in bandwidth and sensitivity over a hybrid circuit. The reason for the improvement is that the capacitance at the connection of the detector to the amplifier can be made as low as 0.2 pF for a monolithic circuit as opposed to 1.0 pF for a discrete detector/amplifier pair.

To obtain this improved performance, monolithic integration of the detector with electronic circuits must be achieved without compromising either the detector or the electronics. In this section the requirements for integration are described from the viewpoints of both fabrication and operation. A number of reported detector/

amplifier structures are described and prospects for future development are given.

7.3.1 Requirements for Optical Detectors

In the III-V system, the absorption of optical energy is across a direct bandgap and is, therefore, extremely efficient. More than 90% of the light at a wavelength of 830 nm is absorbed in 2.0 μm. Thus GaAs detectors require very little thickness to achieve high absorption efficiency.

In this section, four types of GaAs detectors that may be suitable for monolithic integration will be described. Each requires approximately 2 μm of GaAs, two electrical contacts, and a voltage supply. They also require an amplifier to raise the detected signal to digital levels. The four structures—(1) p-i-n photodiode, (2) avalanche photodiode, (3) photoconductor, and (4) back-to-back Schottky photodiode—are shown in Fig. 6.

The p-i-n photodiode requires the growth of epitaxial layers on the underlying substrate. The first layer grown is a heavily doped n^+ layer which is used for making electrical contact and reducing series resistance. The second layer is a very lightly doped region in which the majority of the absorption takes place. The top p-region can either be grown p-type or diffused. This device is operated with 20 V or less reverse bias. The photogenerated carriers are collected across the p-n junction and the resulting current is used to generate a voltage signal.

The avalanche photodiode (APD) is a very similar structure except for doping profiles and the addition of processes to prevent edge effects. APDs are operated at reverse-bias levels in excess of 60 V. The avalanching of photogenerated carriers provides gain. Therefore, the signal generated by the avalanche detector can be many times that from a p-i-n detector.

The photoconductor is fabricated by making ohmic contact to a lightly doped resistive region. This detector can be fabricated either in epitaxial or implanted material. The signal is generated by placing this device in series with a load and applying a voltage. The carriers generated by the light hitting the resistive region will lower the resistance of the detector, lowering the voltage across the device.

The back-to-back Schottky photodiode [13] is similar to the photoconductor. However, instead of ohmic contacts to the underlying GaAs, it has Schottky barriers. This device can be fabricated on a low-doped epilayer or directly on the semi-insulating substrate. In operation, a voltage is applied across the device in series with a load. Since both contacts form Schottky barriers to the substrate, one diode is always reverse biased, inhibiting current flow. In the presence of light, however, current can flow over the Schottky barrier and be detected. These detectors have been used to measure pulse widths less than 100 ps [14].

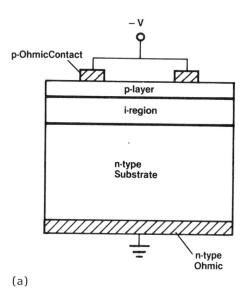

(a)

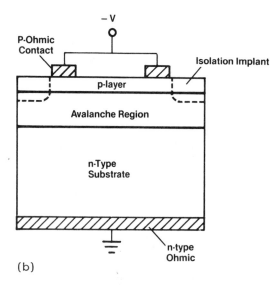

(b)

FIG. 7.6 Schematic Cross-sections of GaAs photodetectors. (a) p-i-n detector; (b) Avalanche detector; (c) Photoconductor; (d) Back-to-back Schottky detector.

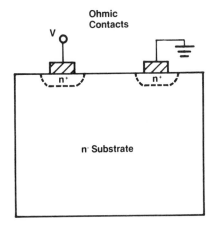

(c)

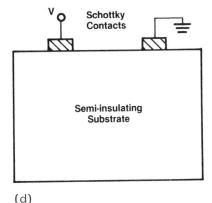

(d)

FIG. 7.6   (Continued)

A.   Receiver Operation

The receiver's function is to convert the optical signal to an electrical signal compatible with the digital electronics.   This requires
coupling the input fiber to the detector, designing and fabricating
a detector with sufficient bandwidth and sensitivity, interfacing the
detector with a preamplifier, and converting the analog signal from
the preamplifier to a digital signal.   The detector and first-stage
amplifier determine the overall performance of the receiver.   Hence
the electrical compatibility (bias voltage, resistance, capacitance,
etc.) of the detector and amplifier must be carefully analyzed.

A simple electrical representation of the detector connected to a FET amplifier is shown in Fig. 7.7. The bandwidth of the detector has two components. One is an RC time constant. In the example given, C is about 0.2 pF and R is 300 Ω. The bandwidth is then

$$\frac{1}{2\pi RC} = 2.6 \text{ GHz} \tag{7.1}$$

The second component of the detector bandwidth is associated with carrier transit time. The transit time of the detector (using n-type epitaxial material) for low reverse-bias voltage can be calculated from

$$t_t = \frac{d}{v} = \frac{d}{\mu_p E} = \frac{d^2}{\mu_p V} \tag{7.2}$$

where d is the distance between electrodes, μ_p the hole mobility (250 cm^2/V) and V the applied bias. For d = 2 μm and V = 2.5 V, t_t = 64 ps. For large reverse bias, the transit time is as short as

$$t_t = \frac{d}{v_{sat}} = \frac{2 \times 10^{-4}}{1 \times 10^7} = 20 \text{ ps} \tag{7.3}$$

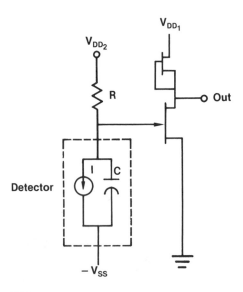

FIG. 7.7 Schematic circuit diagram of a p-i-n FET.

where v_{sat} is the hole saturation velocity:

$$v_{psat} \sim v_{nsat} \sim 1 \times 10^7 \text{ cm/s}$$

Therefore, the device in this example is RC-time-constant limited.

To understand the full effect of monolithic integration on the performance of the detector amplifier, the receiver sensitivity must be examined. The receiver sensitivity, which is the minimum time-averaged mean optical power ($\bar{P}$) that can be detected at or below a specified error rate, is [15]:

$$\eta \bar{P} = Q \frac{hc}{\lambda} <i^2>^{1/2} \tag{7.5}$$

where η is the coupling efficiency of the detector to the optical transmission medium, h is Planck's constant, c is the speed of light, and λ is the optical wavelength. Q is determined by the bit error rate (BER) from the integral.

$$\text{BER} = \int_Q^\infty \exp\left(-\frac{x^2}{2}\right) dx \tag{7.6}$$

For a bit error rate of 10^{-9}, Q is 6. This is the bit error rate of a telecommunications channel. Within a high-speed processor, the bit error rate will be more like 10^{-12} to 10^{-15} for error-free data transmissions. In this case, then, the value of Q will increase to be between 7 and 8. The additional operating margin can be obtained by increasing the amount of power from the optical source, increasing the coupling efficiency to the detector, or decreasing the mean square noise current of the detector/amplifier, $<i^2>$. For a detector/amplifier with a FET preamplifier operating at high bit rates, the noise current is given [16] approximately by

$$<i^2> = \frac{16kT\Gamma C_t^2 I_3 B^3}{g_m} \tag{7.7}$$

where B is the bit rate; kT the Boltzmann energy at temperature T; q the electronic charge; Γ the FET noise factor; C_t the total front-end capacitance, which is composed of the detector and the FET capacitances; I_3 an integral determined by the pulse shape and bandwidth of the circuit; and g_m the transconductance of the FET.

Therefore,

$$\eta P = AQ \left(\frac{C_t^2}{g_m}\right)^{1/2} B^{3/2} \tag{7.8}$$

A plot of ηP versus bit rate for various values of node capacitance is shown in Fig. 7.8. The values used to generate this figure are $g_m = 40$ mS, $\Gamma = 1.5$, $I_3 = 0.08$ (NRZ format and a raised cosine receiver output pulse function), and BER = 10^{-9}.

The sensitivity of the receiver depends strongly on the node capacitance, with the improvement being most significant for a reduction in capacitance at the highest bit rates. For example, a reduction in mean detectable optical power of nearly 5 dB is possible if the front-end capacitance is reduced from a relatively good hybrid receiver value of 1 pF to a value of 0.2 pF at B = 1 Gbit/s. This additional margin might then be used to permit a less expensive coupling scheme or to power split the signal from a laser to multiple detectors in an optical bus.

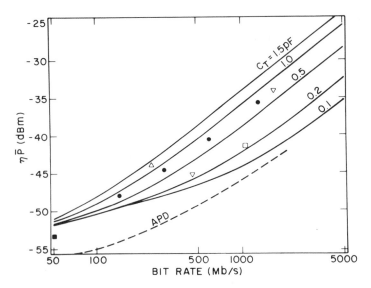

FIG. 7.8 Plot of sensitivity at 10^{-9} BER versus bit rate as a function node capacitance. The plot also contains a number of experimentally measured values of receiver performance. (From Ref. 18.)

Alternatively, the fivefold decrease in capacitance would permit the detector/amplifier to be operated at nearly five times the bit rate with no degradation in accuracy. This is a strong argument for integration.

B. Requirements for Integration

If a photodetector is to be integrated on a circuit with more than a few electronic components, it must meet a number of criteria. First, the detector must be compatible with the electronics processing. Therefore, it must be processed on a production line and be compatible with the substrates used for the electronics. For GaAs, this means a 3-in.-round semi-insulating substrate. Second, the material for the detector cannot interfere with the electronics. Thus, if epitaxial material is required, it must be excluded from the regions in which the electronics will be fabricated and the transition to the epitaxial region must be smooth enough to permit fine-line photolithography. Third, any process needed to fabricate the detector cannot degrade the performance of the electronics. For example, a very high temperature step may cause unacceptable surface damage. Fourth, the detector and electronics must be adequately electrically isolated from each other on the substrate. Finally, the detector must meet specifications after integration.

Although most of these sound obvious, they are not easily accomplished. If the devices are fabricated on semi-insulating substrates, a method must be found to make electrical contact to both sides of the detector other than via the substrate. The photoconductor and back-to-back Schottky have both contacts at the surface. One contact for the p-i-n and avalanche detector, however, is below the surface. A process step must be added to make contact to this area. Such a device has been fabricated [1] and is sold commercially by Ortel. The cross section of the device is shown in Fig. 7.9. Tests have shown the detector to have a cutoff frequency above 10 GHz.

To avoid extra processing, Honeywell researchers [17] grew the layers for the p-i-n detector in a well as one processing step toward a monolithic receiver. After the epitaxial layers were removed from the area in which the electronics were to be fabricated, the n^+ region was exposed at the edge of the well, as shown in Fig. 7.10, permitting easy access for contacting.

Although APDs provide an improvement in sensitivity over p-i-n detectors, they have a number of disadvantages. The high voltage required is not compatible with many applications. Not only is a high voltage supply undesirable in most systems, but it would be extremely difficult to electrically isolate the high voltage from the electronics. In addition, the avalanche gain is temperature dependent and would require extra expense to stabilize. For these reasons,

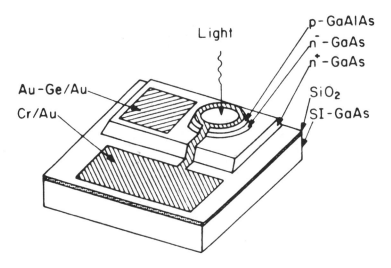

FIG. 7.9 Schematic of p-i-n "window structure" photodetector fabricated on a semi-insulating GaAs substrate. (From Ref. 1.)

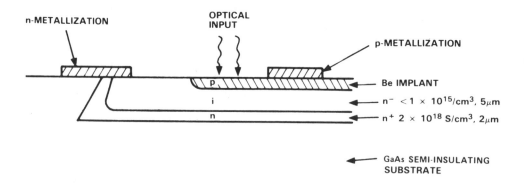

FIG. 7.10 Schematic cross-section of a p-i-n detector grown in a
well on a semi-insulating substrate. The ohmic contact to the back
of the detector is facilitated through the growth of the n^+ layer up
the side of the well.

APDs will not be used to any great extent in monolithic optoelectronic circuits.

Photoconductors are extremely easy to fabricate. On analysis, however, it can be shown that they will be slower than p-i-n detectors [18]. Therefore, photoconductors will be used only where cost is more important than performance.

C. Detector/Amplifier Demonstrations

The first and simplest detector/amplifiers reported were p-i-n/FETs fabricated at Bell Labs in InGaAs/InP [19]. Both the detector and FET were fabricated in epitaxial layers grown by LPE. One such example, a p-i-n/MISFET [20] is shown in Fig. 7.11. These devices were operated at 100 mb/s but did not perform as well as a hybrid p-i-n/FET.

The next level of sophistication is shown in the structure in Fig. 7.12 which was fabricated by Hitachi researchers in GaAs [21]. The schematic, shown in Fig. 7.13, is a transimpedance amplifier connected to the p-i-n diode. The layers for the detector and amplifier were selectively grown in the two regions by a two-step MOCVD process. As in the structure by Kolbas [17] the three p-i-n layers were grown first in a preetched (7-μm-deep) well. Next, the FET layers were grown, forming a nearly planar surface for photolithography. The p-n junction was formed by a Zn diffusion into the lightly doped GaAs absorption region. The technique of selectively growing the p-i-n and FET layers allows for the independent optimization of both circuit segments. For example, high-transconductance FETs require thin (<0.5-μm) n-type channels with high impurity concentration.

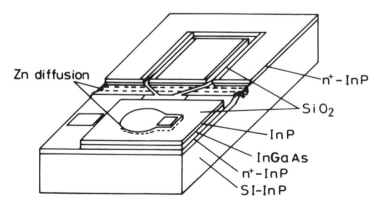

FIG. 7.11 Monolithic p-i-n/FET structure fabricated in InP/InGaAs. (From Ref. 20.)

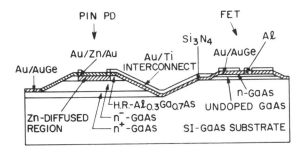

FIG. 7.12 Cross-section of a monolithic optical receiver fabricated in GaAs/AlGaAs. The p-i-n detector is fabricated in layers grown at a different time than the layers for the electronics. (From Ref. 21.)

The transimpedance amplifier consists of six GaAs MESFETs. The amplifier consists of two stages with a transimpedance of 1 kΩ. The output impedance is 50 Ω, and the circuit had a 400-ps rise and fall time.

A third example fabricated by Honeywell researchers [14] is shown in the photograph in Fig. 7.14. This GaAs circuit is an optical receiver consisting of a detector, an amplifier, and a 1:4 GaAs demultiplexer and operates at 1 GHz clock rate. The detector is an interdigitated back-to-back Schottky diode with 1-μm lines and 3-μm spaces fabricated directly on the semi-insulating substrate. The

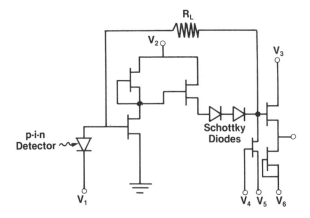

FIG. 7.13 Schematic diagram of the optical receiver IC in Figure 12. The p-i-n detector is fed into a transimpedance amplifier.

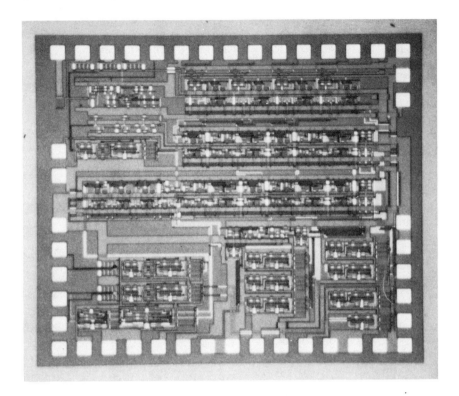

FIG. 7.14 Photograph of an monolithic optical receiver IC consisting of a back-to-back Schottky detector, an amplifier, and a 1:4
demultiplexer fabricated on a depletion-mode GaAs line. The detector is connected to the wide strip of metal along the upper left edge
of the chip.

circuit consists of depletion-mode MESFETs fabricated using selective
ion implantation on a 3-in. processing line. This is truly a produc-
tion-compatible part.

D. Expectations

Up until this point, most of the work in integrating detectors and
amplifiers has been directed solely at the processing issues, the
only goal being the demonstration of one part. There has been no
attention paid to the needs of a particular application.

In most applications, power is an issue. The demonstrations to
date are nearly all in depletion-mode GaAs. This is a fast technology,
but the chips require a large amount of power. The 1.0 to 1.5-V

swing also causes problems. First, a p-i-n detector produces a relatively small voltage swing, on the order of 1 to 10 mV. Therefore, a gain of a factor of 100 to 1000 is required in the amplifier. Because GaAs FETs have a relatively low gain, many stages of amplification are required to produce a GaAs level signal. Thus a lot of power is burned in the amplifier. It also leads to excessive latency (time delay from input to output).

The large voltage swing at the output of the amplifier may also couple back into the detector, causing excessive noise. This coupling can occur either through the power and ground lines or directly through the semi-insulating substrate. This coupling may be the reason that Makinchi et al. [21] measured an unexpectedly low sensitivity for the receiver in Fig. 7.12.

Many of the problems noted above may be alleviated by changing technologies. For example, E/D MESFETS have a voltage swing of approximately 0.5 V and require far less power than the D-mode versions. At the time of this writing, however, no E/D MESFET optical receivers have been reported.

Further advantage may be gained by switching to a MODFET technology. MODFET circuits can be designed to be much faster than E/D MESFETS. In addition, it may be possible to make better back-to-back Schottky detectors on a MODFET. The only processing difference is that the AlGaAs top layer must be etched away from beneath the Schottky fingers. This provides a "window" structure [1]. The light passes through the AlGaAs and is absorbed in the pure GaAs. The carriers, however, are prevented from recombining at the surface between the fingers by the remaining AlGaAs. This provides an increase in sensitivity and a possible improvement in the noise characteristics.

7.4 INTEGRATED OPTOELECTRONIC TRANSMITTERS

The fabrication of semiconductor laser diodes requires crystal growth and processing steps similar to IC manufacturing in order to define the electrical and optical cavity in the two dimensions perpendicular to the direction of light propagation. The length of the cavity is defined by partial mirrors which are formed by cleaving the semiconductor along parallel crystal planes. In contrast to the familiar gas laser, however, laser diodes are only 200 μm long.

The operation of the laser diode is achieved using current instead of voltage as in a gas laser. The gain medium is achieved by driving a current across a p-n junction, inverting the carrier population in the active region. Therefore, the diode requires electrical connections to two contacts. The inversion is aided by the growth of heterostructures near the p-n junction which provide barriers to

the injected minority carriers, confining them to very narrow regions.
The heterostructure also produce an optical waveguide due to a
change in the index of refraction between the two materials.

While the integration of the detector with the amplifier may prove
to be the limiting factor on the speed of the optical link due to circuit considerations, the integration of the laser with the associated
electronics will be more difficult from a materials and processing compatibility standpoint. There are three major reasons for this: (1)
lasers require a multilayered heterostructure up to 7 μm thick;
(2) they need two parallel mirrors separated by on the order of
200 μm; and (3) a method is needed that can achieve electrical and
optical confinement in the lateral dimension.

In this section the requirements for laser diodes are described.
These requirements translate into operational characteristics. The
types of lasers that can meet these characteristics are described.
Some examples of integrated structures are given and expectations
for further development discussed.

7.4.1 Requirements for the Laser

Semiconductor lasers require, at a minimum, a p-n junction in material of extremely high optical quality, two parallel mirrors perpendicular to the junction which define the laser's length, and two electrical contacts. To make a laser that requires low power to operate,
it is necessary to confine the electrical current through the device
and the optical field propagating in the device in both vertical and
lateral dimensions. These dimensions will be on the order of 1 μm.
This confinement is achieved in the vertical dimensions by the use
of a heterostructure to change the band-gap and the index of refraction. The problem is a little more difficult in the horizontal dimension. A number of different laser types have been used to
achieve this end.

A cross section of a ridge guide laser is shown in Fig. 7.15a.
In this structure the electrical confinement is achieved by restricting
the current to flow through a narrow stripe in an insulator on the
top surface of the device. Although the current does spread laterally, this can be controlled to some extent by the resistivity of the
epitaxial layers. The optical confinement in the horizontal dimension is provided by etching away some of the material on the exter-
ior of the waveguide. This will effectively lower the index of refraction of the etched region, creating an optical waveguide.

The buried heterostructure laser is shown in Fig. 7.15b. The
lateral confinement is achieved by etching away the heterostructure
on either side of the laser's active region and regrowing material
with a higher band-gap and lower index of refraction. The current
is then restricted to flow through the hourglass-shaped region, and

**Active Region
and High Index Guide**

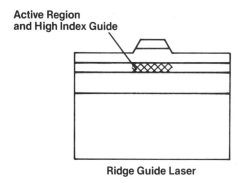

Ridge Guide Laser

(a)

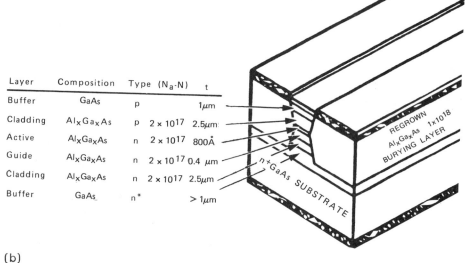

Layer	Composition	Type	(N_a-N)	t
Buffer	GaAs	p		$1\mu m$
Cladding	Al_xGa_xAs	p	2×10^{17}	$2.5\mu m$
Active	Al_xGa_xAs	n	2×10^{17}	800Å
Guide	Al_xGa_xAs	n	2×10^{17}	$0.4 \ \mu m$
Cladding	Al_xGa_xAs	n	2×10^{17}	$2.5\mu m$
Buffer	GaAs	n*		$> 1\mu m$

(b)

FIG. 7.15 Schematic cross-sections of laser diode structures. (a)
Ridge Guide; (b) Buried Heterostructure; (c) Transverse Junction
Stripe.

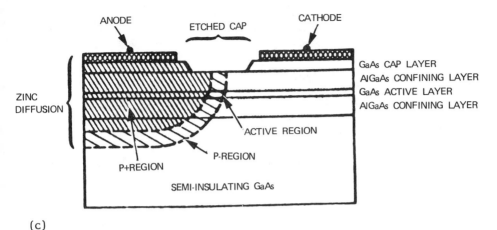

(c)

FIG. 7.15 (Continued)

the highest current density is achieved within the active region itself. These devices are commercially available from Ortel at very
low threshold currents.

In the two structures described above, the epitaxial layers are
grown on a conducting substrate and the current flows in the contact on the top layer and out a contact on the back of the substrate.
In the transverse junction stripe laser (TJS, Fig. 15c), the situation is
different. The epitaxial layers are grown on a semi-insulating substrate and both electrical contacts are on the top surface. The device is fabricated by first growing the double heterostructure using
all n-type layers. A selective zinc diffusion follows which forms
p-n junctions in all the layers. If a bias is applied to the structure, current will only flow transversely through the p-n junction
in the active layer because it has the lowest bandgap. TJS lasers
are available from Mitsubishi with threshold currents under 20 mA.

A high-quality reflector is also a necessity for a low-threshold
laser. In production, the two end mirrors are formed by cleaving
the processed wafer along a crystal plane perpendicular to the length
of the device. The change in the index of refraction between the
semiconductor and the air provides about 30% reflection. In a monolithically integrated IC, this cleaving process is a major restriction
because one dimension of the electronics must then be less than the
length of the laser. To avoid this problem, it is necessary to fabricate at least one on-chip mirror.

Etched laser facets using either dry or wet chemicals have been
demonstrated. This usually results in a significant increase in the

laser threshold. One of the most promising approaches is the micro-
cleaved mirror [22]. In this technique, chemical etching is used to
form a cantilever of epitaxial material at one end of the laser. This
cantilever is then broken off, leaving a cleaved surface. The threshold is not significantly affected because the mirror is as smooth as
one formed in the normal cleaving process.

    Other methods of forming noncleaved reflectors, such as dis-
tributed feedback and distributed Bragg reflectors have been demonstrated. The processing for such devices is very difficult and will
probably not lend itself to a production integrated optoelectronic
circuit.

7.4.2 Operation of the Laser

The semiconductor laser can be considered a current-sensitive device. Figure 7.16 shows the optical power emitted from one facet
of a discrete laser as a function of the current through the device.
At low values of current, the optical output consists of spontaneous
emission. Above the threshold current, the optical power increases

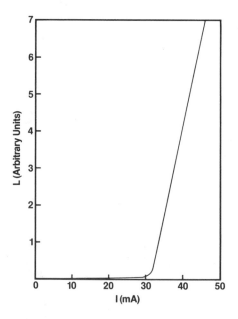

FIG. 7.16 Plot of the optical output power from a laser diode versus the input current. This TJS laser was grown by 3 in. MOCVD
and had a 32 mA threshold current, T = 300 K pulsed.

rapidly with current. The slope of the plot above threshold gives a value for the differential quantum efficiency (change in photons out versus electrons in). The differential quantum efficiency, η, is calculated by

$$\eta = \frac{q\lambda}{hc} \frac{\Delta P}{\Delta I} \qquad (7.9)$$

Therefore, to a zeroeth-order approximation, the amount of optical power above threshold is

$$P = \eta \frac{hc}{q\lambda} (I - I_{th}) \qquad (7.10)$$

The gain within a semiconductor laser is very temperature dependent. This characteristic causes the lasing threshold current to increase exponentially as the temperature increases. This is approximated by the formula

$$I_{th}(T) = I_{th}(25°C) \exp\left(\frac{T - 25°C}{T_0}\right) \qquad (7.11)$$

where $I_{th}(25°C)$ is the threshold current at 25°C and T_0 is the characteristic temperature of the laser. The value of T_0 is between 100° and 300°C, depending on the laser structure and materials. Therefore, a device with a room-temperature threshold of 50 mA and a T_0 of 100°C will have a threshold of 70 mA at 70°C and 106 mA at 100°C.

When discrete lasers are manufactured, they are normally mounted on a copper heat sink, which holds the operating temperature near that of the ambient. If the laser is integrated with GaAs electronics, the device will unavoidably be operating at temperatures above the ambient due to the heat generated by the circuit and with compromise IC packaging requirements. Therefore, the laser that is chosen for the integration must be capable of operating not only at the maximum ambient temperature of the IC, but at the increased temperature due to the heating of the IC. In general, this points to a very low threshold laser.

Two methods exist for modulating a laser diode: direct and indirect modulation. The major differences between the two approaches are requirements on the electrical drivers, spectral purity of the optical pulses, and the device size. These differences become more or less important depending on the application. Both direct and indirect optical modulation have been demonstrated with speeds in excess of 10 GHZ.

The modulation of the current through the laser diode is the simplest way to modulate its output. The modulation characteristics

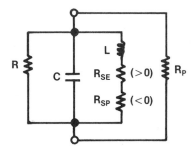

FIG. 7.17 Small signal circuit model for a laser diode. The value of the inductor is inversely proportional to the optical density in the laser cavity.

of laser diodes have a resonance peak in the range of 1 to 4 GHz due to the dynamic coupling of the electron and photon populations in the active region. Above this peak, the electrical and optical signals rapidly decouple. The position and height of the peak can be modified by changes in the design of the laser and by supplying the laser with a constant bias current. The reason for the biasing is to maintain a relatively current high level of optical power within the cavity. The effect of the optical density is illustrated by the small-signal model of a laser developed by Katz at Cal Tech [23]. The schematic representation is shown in Fig. 7.17. The junction capacitance for the laser is actually very large. However, it is shorted out by an inductor, the size of which is determined by the optical density in the cavity. Therefore, biasing of the laser above threshold is a necessity.

One of the detrimental effects of modulating the laser directly at gigahertz rates is that the optical spectrum of the resulting pulse is not stable. Many optical modes are generated, creating a rather broad spectrum. The broad spectral width causes an excessive amount of pulse dispersion in the fibers, limiting the distance over which a high-speed signal can be sent.

The requirement to bias the laser "on" leads to several additional problems. First, a feedback loop that determines what bias current to apply to the laser must be designed into the circuit. This is required because of the strong dependence of the threshold current on temperature as well as changes expected as the laser ages.

A common method for achieving this feedback is to place an optical detector near the rear facet of the laser to monitor the power. The signal generated in the monitor is then used to control the bias current to the laser. This technique therefore, requires a detector and associated control electronics.

Second, the biasing of the laser above threshold means that there is still background light when the signal goes to logical "0". Therefore, the choice of the decision level at the receiver becomes difficult. The "0" light level at one receiver a short distance from the transmitter may be higher than the "1" level at a more distant receiver.

Indirect modulation of a laser diode is accomplished by coupling the light from a laser-driver by a DC current into a high-speed optical switch. The light level of a logical "0" for an indirectly modulated laser is actually very nearly 0. Therefore, the decision point is much easier to determine and design into the receiver. The only requirement is that the light amplitude for a logical "1" must be greater than a certain level at the receiver.

The spectral width of a pulse generated by indirect modulation is extremely narrow. This is because the pulses are produced by "sampling" the mode-stabilized output of a laser operating with a continuous-wave (CW) current input. The high spectral quality of the pulses makes this approach ideal for communication over long distances of fiber optic cables at very high speed.

A number of integrated optic approaches have demonstrated indirect intensity modulation of laser diodes. Some of the types include polarization, mode cutoff, mode conversion, diffraction, deflection, interferometer, directional coupler, and so on. For high-speed applications, the modulators are dictated by a refractive index change via the electro-optic effect. The change in refractive index due to the electro-optic effect is

$$\Delta n = \frac{\alpha n_0^3 r_{41} E}{2} \tag{7.12}$$

where r_{41} is the electro-optic coefficient ($r_{41} = 1.5 \times 10^{-12}$ m/V for GaAs), n_0 the refractive index ($n_0 = 3.5$ for GaAs), E the applied electric field, and α an overlap figure having a value between 0 and 1, depending on the overlap of the electric and optical fields.

The interferometer and directional coupler modulators show the most promise from an efficiency and speed consideration. A discussion of implementations and results of these two modulators fabricated in GaAs-based materials follows.

A common interferometric technique is the Mach-Zender interferometer shown in Fig. 7.18, in which interference is produced by the coherent interaction of the light traveling over different optical path lengths. Light is coupled into a single-mode channel waveguide and is divided into two equal beams by a Y-junction beam splitter.

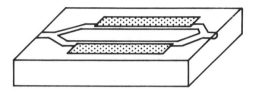

FIG. 7.18   Layout of a Mach-Zender interferometer.

The device is designated to have equal path lengths for each optical beam. By applying a voltage to the electrodes, the optical path lengths can be changed. A Mach-Zender interferometer waveguide modulator has the well-known transfer function given by

$$I = I_0 \cos^2\left(\frac{\pi V}{2V_\pi}\right) \tag{7.13}$$

where I is the intensity-modulated output, I_0 the intensity of the input light, and V_π is known as the half-wave voltage. The half-wave voltage is the amount of applied voltage such that a phase change of π radians between the two arms is produced. The recombination of the two beams at the output arm results in an optical field distribution corresponding to a higher-order mode. Since the output waveguide is single mode, this higher-order mode is cut off and the light radiates into the substrate, yielding zero optical energy in the waveguide.

The MOCVD-grown GaAs/AlGaAs rib-waveguide structure shown in Fig. 7.19 [24] has been developed for a Mach-Zender interferometer modulator. The rib waveguides were fabricated using the electrodes on top of the substrate as a self-aligning process to define the waveguide. The Y branch angle was less than 2° and the arms of the modulator was separated by 20 μm. TE polarized light from a 1.15-μm HeNe laser was end-fire coupled into the cleaved input facet. The output of the device was focused onto a Ge photodetector. Figure 7.20 shows a plot of the normalized output power as a function of applied dc voltage. The extinction ratio of 17 dB was achieved and the device had a half-wave voltage $V_\pi = 14.6$ V. Although this device was not tested for high speed, it is anticipated that this device would be capable of modulation to 3 GHz.

A directional coupler modulator shown in Fig. 7.21 operates on the basis of modulating the efficiency of coupling light from one waveguide to another waveguide. If two waveguides are in close proximity to each other, coupling of energy takes place through the overlapping evanescent fields of the two waveguides. The amount of coupling

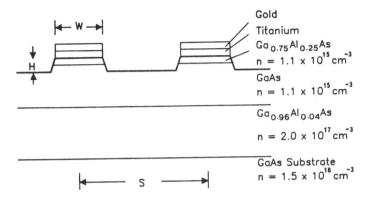

FIG. 7.19 Cross-section of GaAs/AlGaAs waveguide structure developed for a Mach-Zender interferometer modulator. (From Ref. 24.)

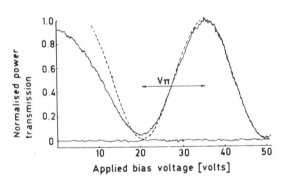

FIG. 7.20 Plot of the output power versus voltage for the modulator in Figure 19. (From Ref. 24.)

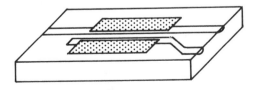

FIG. 7.21 Layout of an integrated optical directional coupler.

depends on the relative phases of the waveguide modes that are mutually coupled. When the phase constants of the coupled modes are equal to each other, a phase-matching condition is satisfied and a total transfer of power can take place. The coupling is a strong function of the waveguide parameters (separation, dimensions, index, etc.) and the mode propagation constants. One advantage of the directional coupler modulation is that it can be theoretically faster [25] by a factor of $\sqrt{3}$ than the interferometric device for the same drive voltage and device length. As shown in Fig. 7.22a [26], the device can have lumped electrodes in which the modulating electrodes act like a capacitor with its bandwidth limited by the RC time constant. The configuration shown in Fig. 7.22 is known as the traveling-wave directional coupler modulator. If the group velocity of the optical wave is matched to the phase velocity of the microwave signal, there are essentially no bandwidth limitations.

GaAs/AlGaAs has a strong advantage over other materials such as LiNbO$_3$ because the dielectric constant at microwave frequencies is only slightly larger (ca. 5%) than the dielectric constant at optical frequencies. This means that modulation speeds of several gigahertz can be achieved before velocity matching is required. One of the first demonstrations of the directional coupler modulator, fabricated in GaAs [27], exhibited a modulation depth of 13 dB and a 3-dB modulation bandwidth of 100 MHz.

There are two major disadvantages with the indirect modulators demonstrated to date. The first is that the devices are typically a few millimeters long, making them undesirable for integration with electronics. The second is the large voltage required to modulate the devices. Typical devices require swings of 10 V for complete

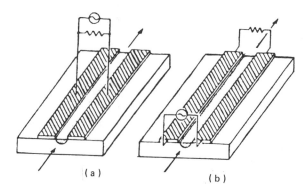

FIG. 7.22 Directional couplers for which the electrodes are analyzed as (a) a lumped capcitor; (b) microstrip line with termination.

modulation. This would be an extremely difficult specification for the drive electronics to meet.

The greatest problem with the large voltage, however, is that the power required to drive the device becomes enormous. The power to drive a capacitive load, C, is $P = 1/2 \ CV^2$, where C is the modulation frequency. If C is 10 pF and V = 10 volts, then P = 0.5 W/GHz. A directly modulated laser requires only 0.1 W and is not significantly dependent on frequency.

7.4.3 Requirements for Integration

The requirements for monolithically integrating a laser diode with high-speed electronics fall into three categories: compatibility, required components, and performance. The decisions that must be made to obtain a working component will be application dependent.

In general, the materials and processing for the laser and the electronics will be different. In addition, one of the major requirements for monolithic integration is that the epimaterial for the laser covers only specific regions of the substrate while the rest of the surface consists of high-quality planar material for the fabrication of the electronics. As a second requirement, the epitaxial layers for the laser must be grown on a semi-insulating substrate. It is also necessary that the resulting substrate after the laser processing is highly planar to permit the fine-line lithography for the electronics. Finally, it is necessary to isolate the laser from the electronics both electrically and optically.

The structure of the laser diode will place some requirements on the fabrication process. There must be a method for making contact to both terminals of the diode. This is straightforward in the TJS laser but may require extra etching steps for the ridge guide laser and the buried heterostructure. The structure must also contain an on-chip laser mirror in order not to restrict the size of the IC. Finally, it would be advantageous to integrate the power monitor on the same device.

The performance of the laser and electronics must not be degraded due to the monolithic integration. For example, the laser may be greatly affected if the electronics causes a large increase in the temperature of the integrated circuit. The correct operation of the electronics may be affected by the light generated by the laser. This must be considered during the design and layout of the monolithic integrated circuit.

The laser and its driver must be considered as an I/O port by the circuit designer. Therefore, the circuits must be designed to meet certain specifications such as bandwidth and power. These values can typically be obtained using relatively straightforward calculations of RC time constants and the current flow through the

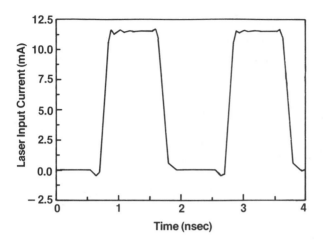

FIG. 7.23 Simulation of the current supplied to a laser diode by a GaAs MESFET laser driver. A voltage pulse is applied to the input of the driver at t = 0.5 nsec. The current pulse to the laser begins at approximately t = 0.7 nsec.

devices. Another specification which may have an impact on the performance of the system is the delay through the laser/driver. If the optical I/O is in a critical timing circuit, this must be considered in the same way as any gate delay. Figure 7.23 shows a SPICE simulation of a laser/driver being driven as 1 Gbit/s. The first pulse was applied at time t = 0.5 ns. The current to the laser starts to change at t = 0.7 ns. Therefore, there is a 200-ps delay through the optical I/O.

7.4.4 Optical Transmitter Demonstrations

The first monolithic laser/electronics demonstration was reported out of the laboratory of Amnon Yariv at Cal Tech [28]. The component consisted of an AlGaAs laser which was integrated with a GaAs Gunn oscillator. Since that time, there have been numerous demonstrations of different laser and electronic devices in both GaAs and InP materials systems. The early demonstrations were limited to single lasers, grown by LPE, integrated with a single transistor. The laser mirrors were formed by cleaving. These parts demonstrated the feasibility of integrating lasers with electronics, but they were far from being practical.

In more recent years, advances have been made in the areas of crystal growth, circuit design, and complexity which brings these components closer to production. One major concern is that the crystal growth process must be compatible with production processing of electronics. This eliminates LPE. Both MBE and MOCVD are capable of growing on the 3-in.-round substrates which are the standard for current-production GaAs ICs.

Demonstrations have also been reported which show an increase in the complexity of the electronics, an on-chip mirror, and power monitor. Figure 7.24 [29] contains an optoelectronic IC which uses a differential laser driver, an on-chip mirror fabricated by wet chemical etching, and a photodiode with a transimpedance amplifier.

A more sophisticated demonstration structure is shown in Fig. 7.25 [30]. The laser is grown in a well etched in the semi-insulating substrate and has a multi-quantum-well active region. The back facet was formed by reactive ion etching and resulted in a threshold

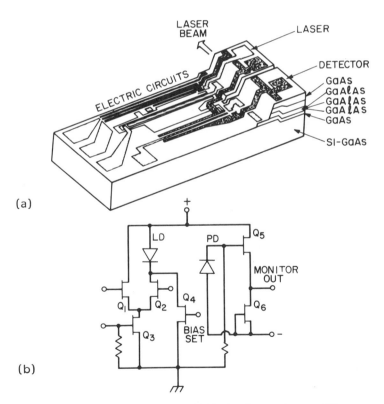

(a)

(b)

FIG. 7.24 Monolithic GaAs/AlGaAs laser transmitter consisting of a laser with on-chip mirror, differential laser driver, and a monitor photodiode with a transimpedance amplifier. (From Ref. 29.)

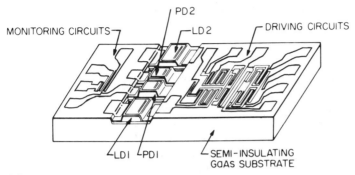

(a)

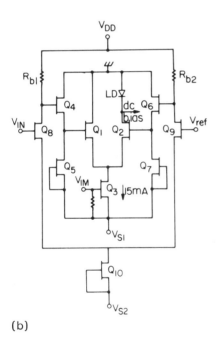

(b)

FIG. 7.25 Integrated optoelectronic laser transmitter fabricated with a quantum well laser and selective ion implantation. The transmitter was operated at rates up to 2 Gbit/sec. (From Ref. 30.)

of 40 mA. The electronics were formed by selective ion implantation into the semi-insulating substrate. The circuit design uses input buffers as well as a differential drive. It was demonstrated at modulation rates up to 2 Gbit/s using nonreturn-to-zero format.

The most complex demonstration to date has been from Honeywell researchers [31]. Figure 7.26 is a micrograph of a TJS laser integrated with a 4:1 multiplexer. The chip is approximately 1.8 × 1.8 mm^2. The 4:1 multiplexer (MUX) and its associated circuitry is contained within the area surrounded by bonding pads in the lower half of the chip. The MUX is formed by selective ion implantations and contains 36 NOR gates (approximately 150 D-mode MESFETs).

The TJS laser was grown by liquid-phase epitaxy in an etched well which shows up as the dark lines at the upper left of the chip. Resurfacing of the wafer resulted in as little as 1-μm steps between the laser structure and the semi-insulating substrate for the electronics. The rear facet was formed with an undercut mirror process. This chip was tested at speeds up to 160 MHz.

7.4.5 Expectations

Integrated transmitter structures are still in their infancy. The greatest immediate challenge is still one of materials and processing compatibility between the optoelectronic and electronic components. The problem, however, is solvable. Once this is accomplished, the question is still how to get these circuits into applications. This subsection contains a few thoughts.

First, the IOCs must be fabricated with one of the higher-speed, lower-power GaAs technologies such as E/D MESFET or MODFET. Depletion-mode MESFET is a good technology to use when fabricating demonstration units, but the power required for the circuit is too high to make it attractive for high-speed applications. Not only would the system power be high, but the heat generated would be detrimental to the laser threshold.

Second, the threshold of the laser diode must be brought down. The development of quantum-well lasers may be the answer to this problem. Figure 7.27 shows a multiple-quantum-well (MQW) laser having a ridge waveguide structure that is being developed for monolithic integration. The MQW laser structure is grown by MOCVD and consists of five 100-Å GaAs wells separated by four 40-Å $Al_{0.2}Ga_{0.8}As$ barriers. The ridge waveguide is ion milled, having a width of 5 μm. The milling is adjusted to stop approximately 0.2 μm above the active region. Room-temperature CW threshold currents as low as 12 mA and a differential quantum efficiency as high as 60% have been measured for devices with a cavity length of approximately 200 μm.

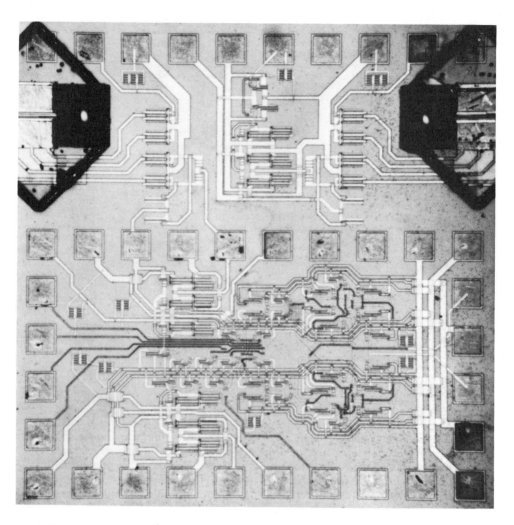

FIG. 7.26 Photomicrograph of an OIC consisting of a TJS laser with on-chip mirror, laser driver FETs, and a 4:1 multiplexer. The circuit contains approximately 150 depletion-mode MESFETS.

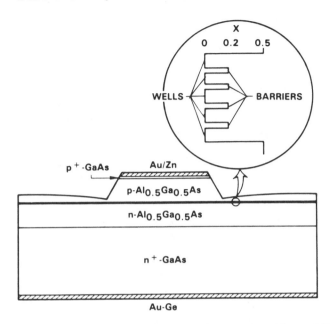

FIG. 7.27 Low-threshold, ridge waveguide multiple quantum well laser being developed for monolithic integration. (From Ref. 32.)

Third, electronic circuit designers must feel comfortable designing with IOCs. This may be the most difficult of all because the implications are that the circuits have been proven reliable, can meet all environmental specifications, are available in abundance, and are the obvious best solution to an I/O problem. This still requires a great deal of work.

7.5 MATERIALS FOR IOCs

An IOC begins with a GaAs substrate and requires the growth of epitaxial layers to fabricate the optoelectronic components. These materials must provide a method for the electronics to be compatible with the optoelectronics. In production, this means that the substrates must be at least 3 in. round and the epitaxial growth system must be able to handle the 3-in. substrate. Further requirements are placed on both the substrate and the epilayers by the devices that are fabricated. These are described below.

7.5.1 Substrates

The requirements of the starting GaAs substrate are dependent on the electronics and the optoelectronics. The electronic technologies which are fabricated using selective ion implantation require uniform, high-resistivity substrates which maintain their properties after the implant anneal steps. The circuits which require epitaxial layers are not as dependent on the electrical properties of the substrate, but do depend on the density and type of defects. Optoelectronic components, especially lasers, are very susceptible to defects in the substrate which propagate up through the active region. These defects have proved to be one of the major causes of short-lived lasers.

There are two basic methods for growing GaAs substrates: horizontal Bridgeman (HB) and liquid-encapsulated Czochralsky (LEC). HB material was used in the early stages of GaAs electronics. The boules were doped with chrome during growth to obtain semi-insulating properties. This led to problems during fabrication, however, because the chrome diffused during the implant anneal, causing nonuniform resistivity and even conversion from semi-insulating to conductive. Some of these problems have been worked out by cleaning up the crystal growth so that less chrome is required. It is also possible to grow and cut an HB boule in order to obtain 3-in.-round wafers [33]. However, the substrates of choice have been 3-in. LEC material.

Typical LEC wafers have resistivities in the range of 1×10^8 Ω-cm. These resistive properties are maintained fairly well during implant annealing. The major problem with the quality of the substrates is the defect density. Present 3-in.-round LEC wafers have an etch pit density on the order of 1×10^4 cm^{-2}. Nonuniformities in the threshold voltage of FETs and defects in the epitaxial growth have been attributed to these defects.

"Zero-defect" LEC material has been reported recently (available from Sumitomo). These wafers are fabricated by doping with indium, which ties up any defects and thus prevents them from propagating. Because the indium is isoelectronic, it does not cause the same electrical problems as chrome during implant anneals. The major drawback to this material is its cost. Each wafer is approximately three times as expensive as a standard LEC wafer. Only about 10 good In-doped wafers, as opposed to 80 standard wafers, can be obtained from a boule of GaAs. Therefore, the cost of characterizing the boule is proportionally higher. It is possible, however, that the success of integrated optoelectronics depends on the development of the In-doped material. The reason is that the laser lifetime may not be long enough on the high-defect standard LEC material. Therefore, low-defect density is a must.

7.5.2 Epitaxial Growth

The requirements for the growth of the epitaxial layers on the substrate are also dependent on the type of component to be fabricated. The electronics technologies requiring epilayers generally need low background carrier concentrations (around 1×10^{13} cm^{-3}) and the ability to control doping accurately. Lasers are not as dependent on electrical properties, but need high photoluminescence efficiency. Waveguide structures need low carrier concentration (low capacitance) and excellent morphology (low scattering). Quantum-well lasers and MODFET-type electronics also require extremely sharp interfaces (on the order of a few angstroms) between layers of GaAs and AlGaAs.

There are two epitaxial growth techniques that can be considered for use in the production of integrated optoelectronic circuits: molecular beam epitaxy (MBE) and metal-organic chemical vapor deposition (MOCVD). Liquid-phase epitaxy (LPE) has been used for years to manufacture high-quality lasers and light-emitting diodes. The substrates used in LPE, however, are limited in size to approximately 2 in.^{2}, which is not compatible with GaAs electronics.

MBE [34] has been the standard technique for the development of MODFET structures. The system can achieve the low background carrier concentrations, layer thicknesses, and sharp interfaces needed for these devices. MBE has also been used to demonstrate extremely low threshold lasers [35] and optoelectronic circuits [36]. There are two drawbacks to MBE as a production process for OICs. The first is the limited throughput of the machines. MBE can grow on only one 3-in. wafer at a time, and the growth rate is approximately 1 μm/h. Therefore, the growth of a laser structure would take up the better part of an 8-h shift. The second problem is the defect density of the grown layers. A better-than-average MBE will produce a 2-μm-thick layer with approximately 1000 defects per square centimeter. This is higher than can be tolerated for VLSI-level circuits or for waveguide structures. The number of defects also increases as the layers are grown thicker. This may lead to additional problems with lasers as well. Some pioneering work has been done to reduce the level of defects to less than 100 cm^{-2} [37], but this technique is far from production.

MOCVD [38] is now being used in production to fabricate the lasers sold by Spectra Diode Labs. MOCVD layers grown by Spire Corporation are also being used in the fabrication of HBT structures for Texas Instruments. MOCVD has the advantage that it can be used for multiple wafer growths and the growth rate is higher than MBE. The layers can be made with very low defect levels, and the photoluminescent efficiency is extremely high. The major drawback to MOCVD is its background carrier concentration. Typical MOCVD

layers have a background level of around 1×10^{15} cm^{-3}. This is
not acceptable for either MODFET electronics or waveguide structures.
The major limitation appears to be in the purity of the starting metal-
organic sources. This is bound to improve.

A second possible limitation of MOCVD in production is the sharp-
ness of the interfaces. Sharp interfaces are determined by the memory of the system. This is a measure of how long it takes the composition of the gas at the wafer to switch from one composition to
the next desired composition. A long memory will eliminate the possibility of growing quantum-well structures and will degrade the properties of a MODFET. The situation may be improved by going to
low-pressure growth of MOCVD or by slowing down the growth during transitions. The details, however, still need to be worked out.

7.6 APPLICATIONS OF INTEGRATED
OPTOELECTRONIC CIRCUITS

It is anticipated that future systems of communication, instrumentation, sensors, and data processing will require monolithic integrated
optoelectronic technology. In addition, it is expected to complement
the well-established technologies of microelectronics and fiber optics.
The needs and requirements for applying this technology to telecommunications have been explored extensively and are well documented [39]; they are covered in detail in Chapter 6 of this book.
The areas that have generated a lot of interest the past few years
include optical techniques for feeding and controlling GaAs monolithic microwave integrated circuits (MMIC), optical digital and analog
computing, and optical interconnect for improving VLSI/VHSIC
performance.

An example of an MMIC application [40] is shown in Fig. 7.28.
This particular MMIC is to be used in a phased-array antenna for
satellite communications above 20 GHz. Each MMIC module requires
several RF lines, bias lines, and digital lines to provide a combination of phase and gain control information. This presents an extremely complex signal distribution problem with topology and signal
interference being quite severe. One solution is to use a fiber optic
distribution network interconnecting monolithically integrated optical
components with GaAs MMIC array elements. Optical fibers can transmit both analog and digital signals as well as provide small size,
light weight, mechanical stability, decreased complexity, and large
bandwidth. As can be seen in the figure, RF transmission requires
that an identical signal be fed to all modules in parallel. Phase and
amplitude control of GaAs MMICs can be achieved via a single fiber
interconnect to an array processor rather than several electrical
connections.

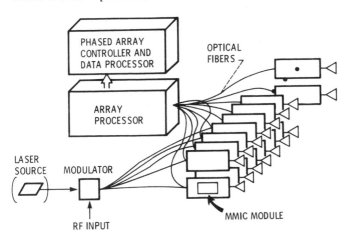

FIG. 7.28 Block diagram illustrating how optical interconnects can be used in a phased array antenna for satellite communications. Optics are used to carry both the 20 GHz signal as well as the control signals to the phased array elements.

Another application area that can benefit significantly from monolithic integration is optically interconnecting high-speed VLSI/VHSIC chips, boards, and computing systems [41]. There is a growing demand to increase the throughput of high-speed processors and computers. To meet this demand, denser, higher-speed ICs and new computing architectures are being developed. Electrical interconnects and switching have been identified as bottlenecks to the advancement of computer systems. Two trends brought on by the need for faster computing systems have pushed the requirements on various levels of interconnects to the edge of what is possible with conventional electrical interconnects. The first trend is the development of higher-speed, denser switching devices in silicon and GaAs. Switching speeds of logic devices are now exceeding speeds of 1 Gbit/s, and high-density integration has resulted in the need for interconnect technologies to handle hundreds of output pins. The second trend is the development of new architectures for increasing the parallelism, and hence the throughput, of a computing system.

A representation of processor and interconnect complexity for present and proposed computing architectures [42] is shown in Fig. 7.29. The dimension along the axis is the number of processors required for the architecture. On the left end of the axis is the von Neumann type of architecture, which has very few processors, but the processors tend to be very complex. Progressing to the right, the number of processors per system increases until it reaches a neural network requiring millions of processors, but the processors are much less complex than in the von Neumann case. Looking at Fig. 7.29, it becomes apparent that as the number of processors increases, the number of complexity of the interconnects within the system increase dramatically. In fact, at the far right of this scale, the interconnects become an integral part of the computing architecture, and the boundary between the processors and the interconnects becomes blurred.

Optical interconnects have been proposed for applications at all levels of complexity. At the course end of the scale a single electrical output from a GaAs IC operating at 2 GHz may more easily be transmitted optically. At the fine end of the scale the shear number of interconnects is overwhelming. These machines would be simplified by multiplexing of lines.

Before an optical interconnect can be considered for use in any system, the device must be designed to meet the needs of the application. The application determines the specifications for circuit performance, the input/output interface voltages, temperature range, size, and power budget. For example, in a satellite, power, size,

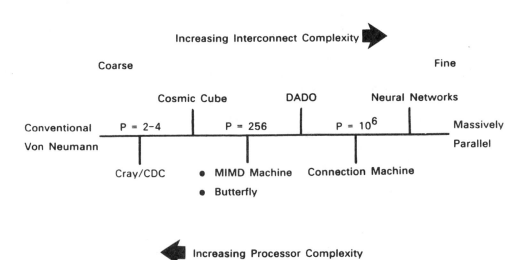

FIG. 7.29 Representation of the transfer of complexity from processor to interconnect for present and proposed architectures. (From Ref. 42.)

and weight are of greatest importance. The ICs must be capable of operating over a wide temperature range (-55 to $+125°$C) and in a high-radiation environment.

The shear speed of the ICs is of greatest importance in a super-computer. The chips will operate in a stable, temperature-controlled environment. Depending on the machine, the temperature may range from slightly above room temperature to $-20°$C or be cooled by liquid nitrogen to 77 K. For this case, radiation-hardened chips will probably not be needed.

Integrated optoelectronic circuits will also find applications in numerous areas of distributed processing. These include optical networks to connect multiple individual microprocessors, optical back-planes within each processor to connect boards, and chip-to-chip interconnects for CPU to memory on a board.

The remainder of this applications section will be devoted to a discussion of the expected performance of optical interconnects for these applications. First-order calculations will be used to identify the limiting factors in the operation of the device.

The first application for integrated optoelectronics that will be described is an optical network for processor-to-processor communications. In a distributed network, multiple processors communicate with each other on an electrical bus. An electrical bus is typically 32 to 40 bits wide. In a local area network, the processors can be separated by up to 1 km and the transmissions are generally asynchronous. This means that the bits are not transmitted at set times known a priori to both the sender and receiver. Instead, the receiver must synchronize itself with the incoming signal. Syncing is accomplished by coding the data in a way that guarantees a certain number of transitions in a given time period. The transitions are detected and generate a signal that controls the frequency and phase of a local oscillator.

One common type of local area network (LAN) is the ring network. In the ring, each node transmits to only one other node and receives from only one other node. Therefore, each line is a point-to-point interconnect. The nodes are typically connected by a parallel electrical bus consisting of perhaps 40 coax or twisted pairs of cables. Optical interconnects are being developed to greatly reduce the complexity and size of such a bus. A schematic of the components in an optical bus interface unit (BIU) for a ring bus are shown in Fig. 7.30. The information enters the BIU in parallel from the microprocessor. The parallel lines are converted to serial data for transmission by the multiplexer and coder. At the receive end, the signal is detected and amplified. The timing is recovered using a phase-locked loop and decoder.

The complexity of the node, excluding the media access controller, is approximately 2000 gates. The bandwidth of the optical signal

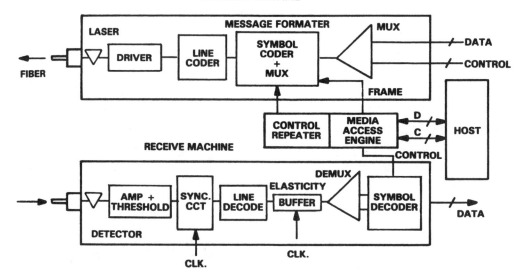

FIG. 7.30 Block diagram of the components within an optical bus
interface unit for a ring bus.

is determined by the number and bandwidth of the parallel lines and
the overhead due to the coding.

As an example, 40 lines at 50 MHz each are multiplexed using a
code that produces 20% overhead. The serial transmission rate would
be 2.4 GHz. A rough order of magnitude for the power required
for the optical node is as follows:

Components	Power requirement	Power
400 gates	0.5 mW/gate	200 mW
1600 gates	0.05 mW/gate	80 mW
Laser/driver	2 V at 50 mA	100 mW
Detector/amplifier	2 V at 30 mA	60 mW
PLL		100 mW
40 parallel I/O to the host	10 mW each	400 mW
		840 mW

The assumptions are that E/D MESFETs are used for the GaAs electronics and that the parallel outputs from the silicon IC to the GaAs are CMOS compatible, unmatched with 20 pF of capacitance, 50% duty cycle.

The power required for a comparable electrical node that transmits 40 individual 50-MHz lines is as follows:

40 gates (coding)	0.05 mW/gate	20 mW
40 I/O onto the network (3.3-V swing into 50 Ω 50% duty)	100 mW each	4000 mW
40 I/O to the host (3.3-V swing into 20 pF)	10 mW each	400 mW
		4.42 W

Therefore, the optical approach provides a significant power savings.

One measure of an optical communication system performance is bit error rate (BER). For the example above, a 2.4-GHz data rate will be used and requires that a BER of 10^{-12} will be required to meet a reasonable system performance. As shown in Fig. 7.31, the BER is strongly dependent on the power at the receiver. A plot calculated from Smith's data [43] of the received optical power versus log (BER) is shown. As can be seen from the figure, a small change in the received power changes the BER by orders of magnitude. This is true for all communication systems. Therefore, a design of at least −35-dBm optical power must strike the optical detector. For most optical interconnect applications other than telecommunications, the maximum fiber length will be approximately 1 km. Assuming a 2-dB/km transmission loss and a 10-dB laser-to-fiber coupling loss, the required optical power must exceed −23 dBm. Since most lasers emit powers of 10 dBm or more, the error rate specification can easily be met.

Because there is more optical power available than required for a point-to-point link, more complex architectures are possible by splitting the power to many points. The maximum fanout for an optical channel can be determined from the minimum power required by the detector (for a given BER), the optical source power, and the losses throughout the system. Figure 7.32 shows a comparison of maximum fanout for optical and electrical interconnects versus data rate [44] for a busing structure. The fanout limitations are plotted for various conditions: (1) maximum theoretical optical fanout; (2) practical limitation to optical interconnects assuming a 6-dB design margin; and (3)−(5) terminated electrical lines of length 2-,

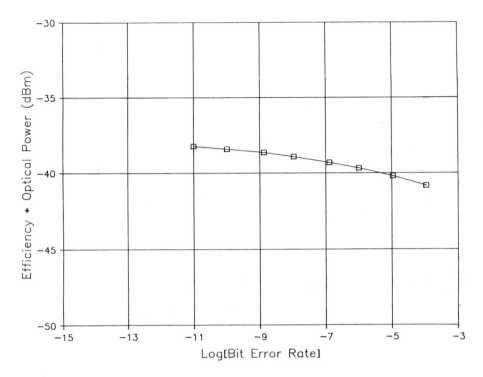

FIG. 7.31 Plot of Bit Error Rate versus optical power. (From
Ref. 43.)

10-, and 50-cm, respectively. From this plot of fanout, it is clear
that optical interconnects can provide a significant advantage over
electrical interconnects under the right circumstances.

Moving inside a processor, optical links are being investigated
for the computer backplane to reduce the number and volume of
cables. A schematic of how two boards in the processor would communicate is shown in Fig. 7.33. All the chips on the board are
silicon except for an I/O chip which is GaAs and contains the optical and high-speed electrical functions. Most processors are synchronous, meaning that a single clock controls all operations. Therefore, the interconnect has the time between two clock pulses to send
a bit, have it propagate to the receiver board, and be valid and
ready for latching into the input registers.

In a 32-bit machine, each electrical bus contains between 32 and
40 parallel lines. Each line will operate at up to 50 MHz for future
systems and would therefore require impedence matching.

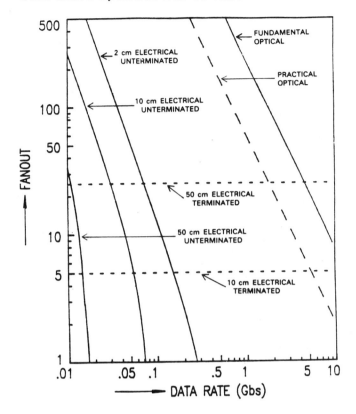

FIG. 7.32 Plot of the fanout limits under various conditions. (From Ref. 44.)

For a board-to-board interconnect, the parallel electrical lines are multiplexed into a single optical channel. The question is, then: How fast must the multiplexer and demultiplexer operate? This is determined by calculating the time available for multiplexing and dividing it into the number of parallel lines.

Assume 32 lines to be transmitted in 20 ns (50-MHz system clock) across 1 m of fiber. The clock period for the high-speed serial rate of the multiplexer is designated by T. The time delays associated with transmitting the data from board to board are approximately as follows:

Settling time of silicon output buffer   1 ns

Latching of parallel data in MUX   T

Time to transmit last bit	32 T
Latency of laser/driver	0.3 ns
Delay through fiber (2×10^8 m/s)	5 ns
Latency of detector/amplifier	0.7 ns
Latch in D-MUX	T
Settling time at silicon buffer input	0.5 ns

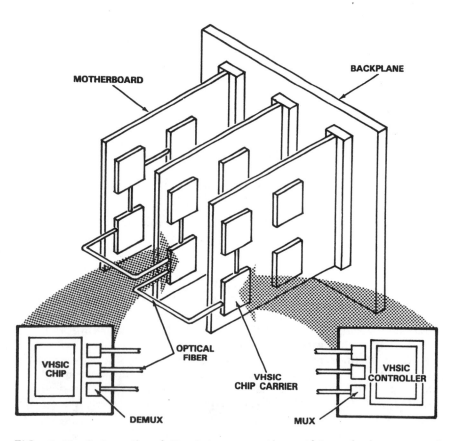

FIG. 7.33 Schematic of the interconnections of boards in a computer back-plane using optical interconnects.

Therefore,

$$20 \text{ ns} - 7.5 \text{ ns} = (32 + 2)T \qquad\qquad (7.14)$$

$$\frac{1}{T} = 3.7 \text{ GHz}$$

This is obviously a very fast multiplexer. The speed, however, is a result of the need to have the information valid in 20 ns. If the system could be designed with asynchronous interconnect timing, the speed of the multiplexer would be a more manageable

$$f = \frac{32}{20} \text{ ns} = 1.6 \text{ GHz}$$

Optical interconnects may also find an application at the chip-to-chip level. High-speed multiplexing and demultiplexing can be used to decrease greatly the interconnect complexity of VLSI-level parts. The advantages are twofold. First, the number of signal lines coming out of a package will be greatly reduced by multiplexing. Second, the elimination of the capacitive loading due to electrical routing on the circuit board should make it possible to achieve higher I/O rates from the VLSI chips.

One final example in which to consider integrated optical interconnects is in the CPU-to-memory link of an all-GaAs high-speed processor. Optical interconnects can provide a speed advantage in this chip-to-chip application. In this case, speed is the critical factor. In all processors, accessing a high-speed memory is one of the greatest bottlenecks to throughput. A schematic for an electrical implementation of the CPU to memory is shown in Fig. 7.34. Ideally, the CPU will request data from the RAM by transmitting the address of the bits to be accessed. At the RAM, the address bits are decoded and the accessed bits appear at the bit line outputs. The data are then transmitted back to the CPU, where they are latched at the input buffers.

The minimum amount of time from the clock pulse which starts the address on its way toward the memory until the returning data can be latched at the buffers determines the fastest clock speed at which the processor can operate. This time can be estimated for the electrical implementation as well as the optical by estimating the delays encountered at each stage during the retrieval of data from the memory.

In both cases it is assumed that the latch in the final stage of the GaAs electronics is capable of driving approximately 1.0 mA into either the laser driver or the output line drivers. For the sake of the comparison, assume that the line driver and the laser driver can

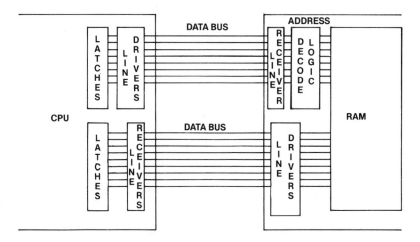

FIG. 7.34 Block representation of the connection of a CPU chip to memory. The memory address is generated and transmitted by the CPU and decoded at the RAM. The desired word appears at the RAM line drivers for transmission back to the CPU.

sink 5 mA of current. This would require an E-mode MESFET approximately 200 μm wide. Such a FET would have a gate capacitance of 0.4 pF. The delay encountered for the latch to raise the input of the driver to 0.5 V can be estimated from

$$t = C \frac{\Delta V}{I} = 0.4 \text{ pF} \frac{0.5 \text{ V}}{1.0 \text{ mA}} = 200 \text{ ps}$$

First, the electrical case will be analyzed. Assume that the CPU is surface mounted on a multichip package approximately 1 cm from a memory with 1-ns access time. The capacitance of the pads plus the interconnect lines is estimated to be 2.0 pF. Then the time required to cause a voltage swing of ΔV on the line can be estimated from

$$t = C \frac{\Delta V}{I} = 200 \text{ ps} \tag{7.15}$$

for ΔV = 0.5 V and I = 5 mA.

The total delay for the data request is estimated from this delay and the other delays in the following list.

Delay times

Buffer on output latch	0.2 ns
Rise time of line	0.2 ns
Media delay	0.05 ns
RAM access time	1.0 ns
Buffer delay	0.2 ns
Rise time	0.2 ns
Media	0.05 ns
Settling time	<u>0.1 ns</u>
	2.0 ns

Therefore, the electrical interconnect will limit the maximum clock rate to 500 MHz.

In an optical link, the output line drivers are replaced by a laser and laser driver, and the line receivers are replaced by a detector and amplifier.

For the laser, C is approximately 1 pF, but ΔV will only be determined by the series resistor, R_s. $\Delta V = IR_s = 5$ mA $\times$ 10 $\Omega =$ 50 mV. Thus

$$t = C \frac{\Delta V}{I} = 0.01 \text{ ns} \tag{7.16}$$

The 5-mA pulse into a prebiased laser will produce a change in the output power P.

$$P = \frac{h\nu}{q} \Delta I = 2.75 \text{ mW} \tag{7.17}$$

In a short point-to-point interconnect, the losses in the optical coupling would be only about 10 dB. Therefore, -7 dBm of optical power would fall on the detector. This is orders of magnitude more than required to meet the 10^{-12} sensitivity. The extra power, however, is needed to reduce the number of gain stages in the amplifier.

-7 dBm of optical power would provide a large enough electrical signal at the detector so that only two small stages of amplification would be required. The delay through the amplifier would be only about 0.15 ns. Therefore, the total data request delay for an integrated optoelectronic link is as follows:

Delay times

Buffer on latch output	0.2 ns
Laser driver	0.01 ns
Media	0.05 ns
Detector/amplifier	0.15 ns
RAM access time	1.0 ns
Output buffer	0.2 ns
Laser driver	0.01 ns
Media	0.05 ns
Detector/amplifier	0.15 ns
Settling time	0.1 ns
	1.72 ns

Therefore, as described, the optical interconnect would provide a cycle time that is shorter than the electrical interconnect.

A major advantage of the optical interconnect will be realized if the laser would require a pulse current of only 1 mA and the optical coupling is improved. The buffer delay would be reduced to zero but the improved coupling would prevent the detector/amplifier delay from degrading. The total time, therefore, would be reduced to 1.3 ns. The reduction of current into the driver would result in a power saving for the link. The electrical link, on the other hand, would get slower. The rise time would increase to approximately 1.0 ns, making the data request delay 2.8 ns.

In this section some of the applications for optical interconnects in signal processors have been highlighted. The following statements and areas of required research can be identified:

1. Optical interconnects outperform electrical interconnects in LANs.
2. Optical interconnects require less volume than electrical interconnects in a computer backplane and do not require impedance matching when multiple taps are used.
3. Optical interconnects are faster than electrical if the optoelectronic components are integrated.
4. Optical interconnects will benefit from the use of advanced GaAs technologies (E/D MESFET, MODFET) and a high-efficiency laser which requires only 1 mA of current to provide a signal above the spontaneous emission.
5. Optical interconnects will work extremely well at 77 K because the laser threshold will be very low, the efficiency increases, and the electronics can run faster.

REFERENCES

1. Lau, K. Y., Ury, I., and Bar-Chaim, N. (1984). Proceedings of SPIE, vol. *408*: 29.

2. VanTuyl, R. L. and Liechti, G. A., (1974). *IEEE J. Solid State Cir.*, *SC-9*: 269.

3. McCormack, G. D., Rode, A. G., and Strid, E. W., (1982) Tech. Digest of GaAs IC Symposium, New Orleans, p. 25.

4. Yokoyama, N., Mimura, T., Fukura, M., and Ishikawa, H., (1981). ISSCC Digest of Tech. Papers, p. 218.

5. Mactaggart, R. M., (1986) private communication.

6. Yokoyama, N., Onodera, H., Shinoki, T., Ohnishi, H., Nishi, H., and Shibatomi, A., (1984). Technical Digest of ISSCC, p. 44.

7. Zuleeg, R., Notthoff, J. K., and Lehovee, K., (1978). *IEEE Trans on Electron Dev.*, *ED-25*: 628.

8. Cirillo, N. C., Abrokwah, J. K., and Jamison, S. A., (1984) Tech. Digest of GaAs IC Symposium, Boston, p. 167.

9. Cirillo, N. C., Abrokwah, J. K., Fraasch, A. M., and Vold, P. J., (1985). *Electron. Lett.* *21*: 772.

10. Yuan, H. T., Doerbeck, F. D. and McLevige, W. V., (1980). *Electron. Lett.*, *16*: 637.

11. Yuan, H., McLevige, W. V., Shih, H. D., and Hearn, A. S. (1984). Digest of Tech Papers IEEE ISSCC, San Francisco, p. 42.

12. Asbeck, P., Miller, D., Anderson, R., and Eisen, F., (1983). Technical Digest of the GaAs IC Symposium, Phoenix, p. 170.

13. Sugeta, T., Uritsu, T., Sakata, S., and Mizushima, Y., (1980). *Japan. J. Appl. Phys.*, *19*: 459.

14. Ray, S. and Walton, M. B., (1986). Proceed. of IEEE Microwave and Millimeter IC Symp., Baltimore, p.

15. Smith, R. G. and Personick, S. D., (1979). *Topics in Applied Physics*, *39*: 89.

16. Personick, S. D., (1973). *Bell System Tech. Journal*, *52*: 843.

17. Kolbas, R. M., Abrokwah, J. K., Carney, J. K., Bradshaw, D. H., Elmer, B. R., and Biard, J. R., (1983). *Appl. Phys. Lett.*, *43*: 821.

18. Forrest, S. R., (1985), *J. Lightwave Technol.*, *LT-3*:347.

19. Leheny, R. F., Nahory, M. A., Pollack, M. A., Ballman, A. A., Beebe, E. D., DeWinter, H. C., and Martin, R. J., (1980). *Electron. Lett.*, *16*:353.

20. Kasahara, K., Hayashi, H., Makita, K., Taguchi, K., Suzuki, A., Nomura, H., and Matsushita, S., (1984). *Electron. Lett.*, *20*:314.

21 Makiuchi, M., Wada, O., Miura, S., Hamaguchi, H., Hachida, H., Nakai, K., Horimatsu, H., and Sakurai, T., (1984) Tech. Digest of IEDM, p. 862.

22. Blauvelt, H., Bar-Chaim, N., Fekete, D. Margalit, S., and Yariv, A., (1982). *Appl. Phys. Lett.*, *40*:289.

23. Katz, J., Margalit, S., Harder, C., Wilt, D., and Yariv, A., (1981). *IEEE J. Quant. Electron.*, *QE-17*:4.

24. Rodgers, P. M., (1985). Technical Digest Third European Conference on Integrated Optics, Berlin, May 6–8.

25. Korotky, S. K. and Alferness, R. C., *J. Light Tech.*, *LT-1*: 244.

26. Sueta, T. and Izutsu, M. (1982). *J. Opt. Comm.*, *3*:52.

27. Cambell, J. C., Blum, F. A., Shaw, D. W., and Pawley, K. L., *Appl. Phys. Lett.*, *27*: 202.

28. Lee, C. P., Margalit, S., Ury, I. and Yariv, A. (1978). *Appl. Phys. Lett.*, *32*:806.

29. Matsueda, H., Sasaki, S. I., Nakamura, M., (1983). *J. Lightwave Tech.*, *LT-1*:261.

30. Nakano, H., Yamashita, S., Tanaka, T., Hirao, N., and Naeda, N., (1986). *J. Light. Tech.*, *LT-4*:574.

31. Carney, J. K., Helix, M. J., Kolbas, R. M., (1983). Technical Digest of the GaAs IC Symposium, Phoenix, p. 48.

32. Kilcoyne, M. K., Kasemset, D., Asatourian, R., Beccue, S. (1986). SPIE Proceedings, Vol. 625, p. 127.

33. Nishine, S., Kito, N., Fujita, K., Sekinobu, M., Shikatani, O., and Akai, S., (1982) Tech. Digest of GaAs IC Symposium, New Orleans, p. 58.

34. Cho, A. Y., (1983). *Thin Solid Films*, *100*:291.

35. Tsang, W. T. (1982). *Appl. Phys. Lett.*, *40*:217.

36. Sanada, T., Yamakaski, J., Wada, O., Fujii, T., Sakurai, T., and Sasaki, N. (1984). *Appl. Phys. Lett.*, *44*:325.

37. Abrokwah, J. (1986). Private communication.

38. Dupuis, R. D., Moudy, L. A., and Dapkus, P. D., (1979). *Gallium Arsenide and Related Compounds*, p. 1.

39. Midwinter, J. E., (1985). *J. Light. Tech.*, *LT-3*:927.

40. Bhasin, K. B., Anzic, G., Dunath, R. R., Connolly, D. J., (1986), 11th Annual Communications Satellite Systems Conference, AIAA, San Diego, Ca., March 16–20.

41. Hutcheson, L. D., (1986), Technical Digest of Conference on Lasers and Electro-Optics, San Francisco, Ca., June 9–13.

42. Hutcheson, L. D., Haugen, P. R. and Husain, A. (1985). "Gigabit per Second Optical Chip-to-Chip Interconnects," *SPIE Proceedings*, Vol. 587.

43. Smith, R. G. and Personick, S. D., (1980), in *Semiconductor Devices for Optical Communications*, Springer-Verlag, p. 89.

44. Haugen, P. R., Rychnovsky, S., Husain, A., and Hutcheson, L. D., (1986). *Optical Eng.*, *25*:1076.

8

Integated Optical Logic Devices

BOR-UEI CHEN   *PCO, Inc., Chatsworth, California*

8.1 INTRODUCTION

Optical data transmission systems offer the potential of operating at extremely high data rates. The frequency of the optical carrier is on the order of 10^{14} Hz. Some of the components required for such an optical system operating at 100 GHz are available, at least in the laboratory. One such component is the high-speed semiconductor laser. GaAlAs lasers with optical pulse widths as short as 0.65 ps have been demonstrated by passive mode locking of a buried optical guide laser in an external cavity [1]. Another optical component is the high-speed optical detector; devices having a rise time as short as 35 ps are available commercially. Silicon-on-sapphire (SOS) films radiation-damaged by ion implantation have shown a photoconductivity with the relaxation time reduced by several orders of magnitude to 8 ps [2]. The transmission-medium, single-mode optical fiber, having an attenuation loss as low as 0.2 dB/km at 1.55 μm and zero dispersion at 1.3 μm, is used for long- and medium-haul telecommunications. At a wavelength around 1.3 μm, the pulse broadening of such fibers can be negligibly small. For the optical wavelength close to the minimum-dispersion wavelength region, a single-mode fiber transmission system can be operated at a 20 to 100-Gbit/s data rate over a distance over 100 km [3]. The latest single-mode dispersion-shifted optical fiber, where both zero dispersion and minimum attenuation

loss of 0.2 dB/km occur at 1.55 μm, is now available from Corning
Glass Works. With this new fiber, one can fully utilize the low-loss
and wide-bandwidth characteristics of optical glass fibers.

Using these extremely fast optoelectronic components, it is possible
to convert a high-speed electrical signal into an optical signal, transmit that optical signal over a long distance, and then reconvert the optical signal back into an electrical signal. Currently, however, the
high-speed signal processing is done in the electrical domain. The
electronic signal processing has its fundamental speed limitation at
about 1 GHz for silicon integrated circuits and at about 10 GHz for less
matured GaAs integrated circuits. On the other hand, integrated optical devices offer the potential of operating at higher speeds and at lower powers, which makes them very attractive for use as signal processors. In particular, because of their small size, integrated optical devices offer the potential of achieving picosecond switching time while
requiring only picojoules or less of switching energy. Other advantages
of optical circuitry include immunity to EMI, RFI, and no ground-loop
requirements. The ability of performing logic operations in the optics
domain will further enhance the system flexibility because many electrical-to-optical and optical-to-electrical conversions can be eliminated.
In this chapter we discuss integrated optical logic devices and logic devices that are capable of being implemented in an integrated optics

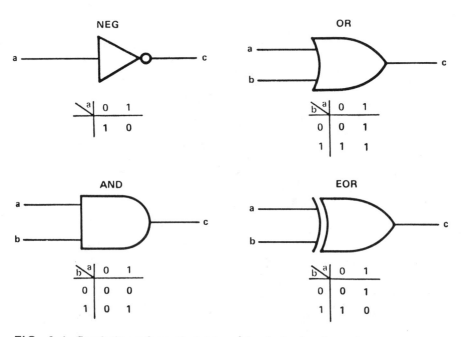

FIG. 8.1 Symbols and truth table of basic logic elements.

format. Some of the devices can be implemented in optical data systems at data rates higher than 1 Gbit/s without further technology development. Some of the devices having the potential of picosecond switch time and picojoule switching energy will still require many years of research efforts before they are realized. The optical logic devices discussed here include types for which imput and output data are a combination of electrical and optical formats.

Most optical devices are operated in a linear fashion; however, logic devices must perform nonlinear functions. There are several possible ways to introduce appropriate nonlinearities into an optical device. These possibilities include (1) electrically induced material effects, such as the electro-optic effect and the Franz-Keldysh effect; (2) conversion of optical fields into an electrical signal using a square-law detector such as a photodetector; and (3) optically induced material effects, such as saturable absorption and optical Kerr effect. We will use the types of nonlinearity employed as the basis for categorizing the various optical logic devices. The truth tables and symbols of basic logic devices are shown in Fig. 8.1.

8.2 ELECTRO-OPTIC-EFFECT DEVICES

For electro-optic-effect devices, the output optical intensity is a non-linear function of the change in index of refraction. The change in the index of refraction is induced by an electrical signal through the electro-optical effect. The most commonly used electro-optic material for integrated optic devices is lithium niobate ($LiNbO_3$). If the input data exists in an electrical format, they can be applied directly to the logic device. If the input data are in an optical format, part of optical signal has to be converted into an electrical signal through the use of a photodetector and an amplifier.

The first device configuration we will consider is the well-known $\Delta\beta$-reversal switch. Even though the device is usually considered as an optical switch, it is also possible to use it as an optical inverter or as an optical Exclusive-OR (EOR). The device is shown schematically in Fig. 8.2 together with its operating characteristics. The input signals are in electrical format and are applied to the electrodes in the interaction region. The interaction region is designed such that for zero potential difference between the electrodes, the optical power in the two channels is completely exchanged (the "cross" state). The presence of a potential difference between the electrodes makes the complete exchange of optical energy impossible. With the appropriate potential difference, all the energy will exit from the interaction region in the same channel that it entered. If we consider the case where only input port S_0 is excited and if we monitor the output of port R, the output intensity is nonzero (the "1" logic output state) when there is no electrical signal applied onto the electrode. However, if the appropriate potential is now applied, the output intensity becomes zero (the "0" state). In this manner, the device operates as an optical inverter.

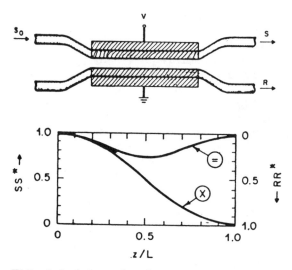

FIG. 8.2 Schematic of a Δβ switch and its operating characteristics.

If neither electrode is tied permanently to ground potential, the device can operate as an EOR. In this case, only input port S_0 is excited and output port S is monitored. If the potential difference is zero and either both electrical signals are present or both are absent, the optical output at port S is always zero. However, if only one of the electrical signals is present, which corresponds to the (0,1) or (1,0) input conditions, the output is nonzero.

The switching speed of these devices is primarily limited by the electrical RC time constant, where R is the impedance placed in parallel with the electrodes to match the driver impedance and C is the capacitance of the electrodes. For a typical LiNbO$_3$ device, the parameters are an electrode length of 5 mm, an electrode separation of 5 μm, and the capacitance is on the order of 2 pF. These parameters imply that the rise time of charging the capacitor is about 100 ps and the required switching voltage is on the order of 5 V. This corresponds to a capacitor charging energy of approximately 25 pJ. To date the fastest reported Δβ-reversal switch has a rise time of 110 ps [4]. If one replaces the lumped-electrode design with traveling-wave electrodes, the speed of the device can be improved dramatically. Experimentally, a switching speed up to 20 GHz has been demonstrated in many laboratories.

Another useful device configuration is that of the Mach-Zehnder waveguide interferometer, shown in Fig. 8.3. The optical output power, P_{out}, from the interferometer is given by the expression

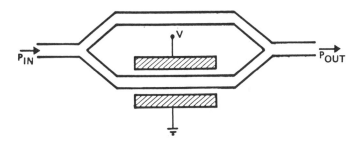

(a)

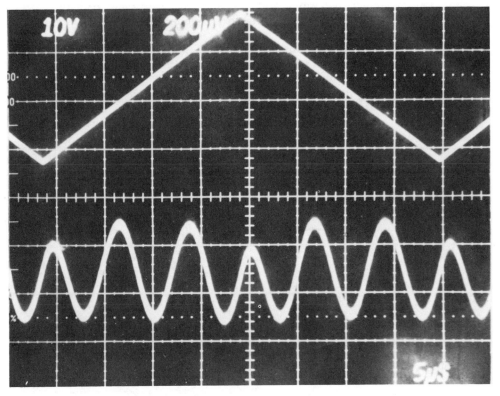

(b)

FIG. 8.3 Schematic of a waveguide interferometer and its operating characteristics, the upper trace is the applied electrical signal and the lower trace is the corresponding optical output. (From Ref. 6.)

$$p_{out} = \frac{\xi p_{in}(1 + \cos \phi)}{2} \tag{8.1}$$

where p_{in} is the optical input power, ϕ the electrically induced optical phase change on one arm of the interferometer, and ξ a constant
accounting for losses of the device. An example of the output from
such an interferometer is shown in Fig. 8.3. The lower trace is the
output optical power. This device has been proposed as the basic
unit for various logic operations [5]. One such configuration of EOR
logic is shown in Fig. 8.4. This device is a waveguide interferometer with three applied electrical signals. Two electrodes are connected to the electrical input and the third is always electrically
biased at the "1" state corresponds to an electrical signal that gen-
erates an 180° optical phase shift via the electro-optic effect. During operation, if signals a and b are equal, either both 0's or both
1's, the optical beams in the two arms of the interferometer are 180°
out of phase and the output from the interferometer is zero. If,
however, only one of the input signals is present, the two beams
are in phase and the output is nonzero. Thus we have the device
operating as an EOR logic. Depending on the electrode arrangement,
one can also design other types of logic operations, as shown on
Fig. 8.4. Using the waveguide interferometer, analog-to-digital (A/D)

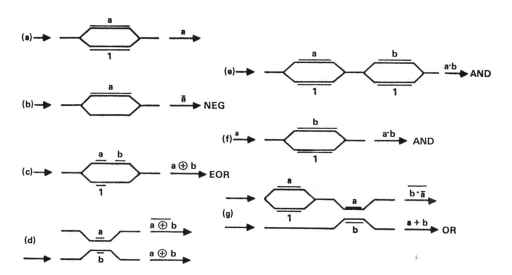

FIG. 8.4 Electro-optical EXCLUSIVE-OR based upon a waveguide
interferometer. (From Ref. 5.)

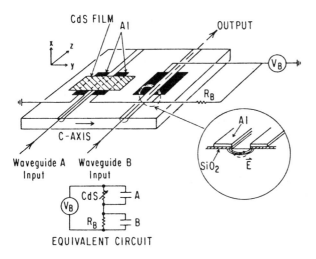

(a)

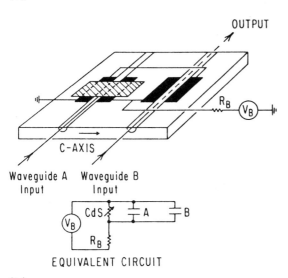

(b)

FIG. 8.5 Optical logic devices using photoconductor overlays. (From Ref. 8.)

converters [6] and digital-to-analog (D/A) converters [7] have been demonstrated.

The devices discussed up to this point are designed for use with electrically formatted input data. It is also possible to use the electro-optical effect with optically formatted input data. However, it is necessary first to convert the optical signal to an electrical signal. One conversion technique is the use of a photoconductor placed over an optical waveguide. The presence of an optical signal in the waveguide under the photoconductor will reduce the resistance of the photoconductor. The change in resistance is used to control the voltage applied across a pair of electrodes on a second waveguide. The voltage change induces a change in the phase or loss of the second optical signal through the electro-optic effect. Using this conversion technique, an AND gate and an inverter have been constructed using an evaporated layer of cadmium sulfide as the photoconductive overlay [8,9]. These particular devices are shown schematically in Fig. 8.5. The devices are designed such that if most of the bias voltage appears across the waveguide electrodes, the waveguide is driven into cutoff and any optical energy present in the waveguide will be scattered into the substrate. The use of CdS film as the photoconductor will result in relatively slow devices since the photocurrent decay time is approximately 50 μs. The use of a faster photoconductor would increase the usefulness of these devices.

From the preceding discussion, the following conclusions can be drawn concerning optical devices based on the linear electro-optic effect. First, the ultimate speed of these devices will be limited by the electrical rise time of the electrodes and other associated electronic circuits. Second, these devices require only small switching energy. The electrode charging energy is a few tens of picojoules for a 5-mm-long electrode pair. There is no threshold requirement for the optical power. Third, these devices can be optically cascaded to form a complicated logic circuit, performing several levels of optical logic operations.

8.3 OPTICAL-TO-ELECTRICAL CONVERTER DEVICES

Optical logic devices can also be based on the nonlinear conversion of optical fields into electrical signals, for example, by a square-law photodetector. When two optical fields are combined coherently, the output power (square of the sum of optical fields) will depend on the relative phase relationship of the two input fields. At optical frequencies, even the fastest photodetectors are sensitive only to the time-averaged optical power and not the optical fields. The logic devices discussed in this section are designed for use with optically formatted input data. Such devices are only capable of adding and subtracting coherent optical fields.

One configuration that can be used as an OR gate or as an EOR gate is shown in Fig. 8.6. This device consists of three coupled waveguides and a pair of electrodes used to introduce an additional phase shift into one of the input channels. The interaction regions of this device is designed such that equal amounts of power are coupled from the input channels into the output channel. If the coupling coefficient K and the coupling length L satisfy the equation

$$KL = \frac{\pi}{2\sqrt{2}} \tag{8.2}$$

the output optical field is described in the truth table of Fig. 8.6. The application of an 180° phase shift to the electrodes will enable the device to function as an EOR gate. Changing the phase shift to 120°, the device will function as an OR gate. However, it is

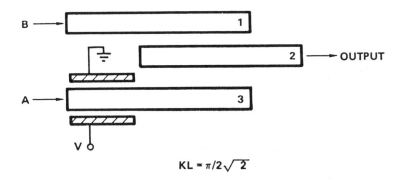

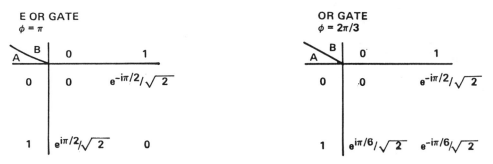

FIG. 8.6 Optical logic device utilizing three coupled waveguides.

important to note that the truth tables shown in Fig. 8.6 are not exactly the ones of EOR and OR gates. Using the EOR gate as an example, the difference occurs for the cases when only one of the optical inputs is present. In these instances, the phase of the output
depends on the input condition. This phase ambiguity prevents the
device from being optically cascaded with another device that is
sensitive to the phase of its optical inputs. The phase-ambiguity
problem of the output field of the EOR gate can be overcome by incorporating a photodetector overlayer over the top of channel 1, and
connecting it serially with the electrode to the voltage supply, as
shown in Fig. 8.7. A portion of the optical signal in channel 1 is
absorbed by the photoconductive overlay, thus reducing the shunt
resistance. As a result, the electro-optic π phase shifter is activated whenever channel 1 is excited. When only channel 2 is excited, the phase shifter is set at zero. The phase of the output
signal at channel 3 will be identical to that when only channel 1 is
excited. When both channels are excited simultaneously, the π
phase difference between the two inputs results in a zero output.
The modified circuit configuration allows the devices to perform multilevel logic functions. However, the incorporation of a photodetector

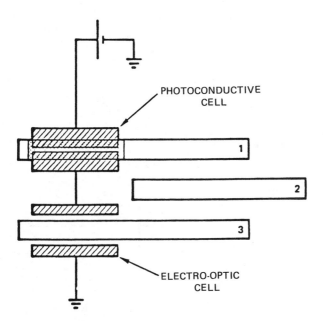

FIG. 8.7 Three coupled waveguide EOR gate incorporating a photo-
conductive-electro-optic phase reversal of one input branch.

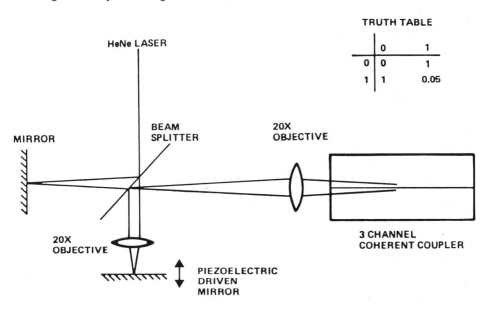

FIG. 8.8 Experimental testing of logic operation of 3 channel coupler and the measured truth labels for EOR function.

overlay converts the devices into the one based on the electro-
optic effect.

A logic device based on the coupled channel waveguide has
been constructed, and logic operations such as EOR, OR, and inverter have been demonstrated. The waveguide circuit was fabricated in a glass substrate by Na^+-K^+ ion-exchange process. The
experimental setup is shown in Fig. 8.8. A beam splitter was
used to divide the test He-Ne laser into two beams, which are coupled separately into the two outside channel waveguides. The required phase shift was generated externally through a change in
the optical path traveled by one of the beams obtained by placing
one of the reflecting mirrors on a piezoelectric-driven stage. Depending on the phase difference (determined by the voltage applied
to the PZT stage), the device operated as either an EOR or an OR
gate. The extinction ratio (i.e., the ratio of logic "1" and "0"
states) of 20:1 was measured. The experimentally measured truth
table is shown in Fig. 8.8.

It is also possible to use a two-coupled channel waveguide configuration to perform the same logic functions. However, the length
of the interaction region and the required phase shifts differ from
those of three-coupled channel waveguide devices. Figure 8.9 shows

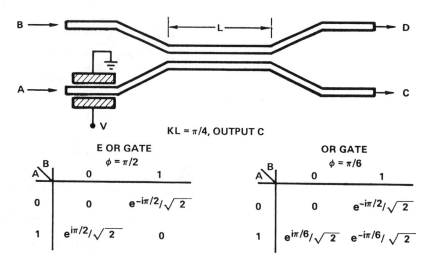

FIG. 8.9 Schematic and truth tables of two coupled channel waveguide logic devices.

the schematic and truth tables of two coupled-channel waveguide logic devices. If there is one input present, the device works as a 3-dB power splitter (i.e., the product of coupling coefficient and coupling length equals $\pi/4$, $KL = \pi/4$). To obtain EOR and OR operations, the phase shifter has to be biased at 90° and 60°, respectively.

The following conclusions can be drawn concerning optical logic devices based on addition of coherent fields and nonlinear conversion of the optical field into an electrical signal. First, these devices require coherent optical inputs originated from the same laser. Second, these devices cannot be optically cascaded because of the phase ambiguity. Third, these devices only require a dc bias voltage and do not require any high-speed switching voltage. Thus there is no energy consumed to charge the electrode capacitor. Fourth, the speed of these devices is limited by the speed of the detector.

8.4 OPTICALLY INDUCED EFFECT DEVICES

The third type of optical logic devices is based on the optically induced nonlinear material effects. These nonlinear effects result in the optical output depending nonlinearly on the optical input. This type of device is designed for operations involving either coherent or incoherent optical signals and no electronic circuit is needed.

The nonlinear optical effects usually take place at very high optical intensities; the index of refraction is then described by

$$n = n_0 + n_2 E^2 \qquad (8.3)$$

where n_0 is the linear refractive index and the second term, $n_2 E^2$, is the nonlinear contribution. The term E^2 is the mean square of the optical field inside the medium. For an isotropic material and a linearly polarized light, the term n_2 is expressed in terms of the third-order nonlinear susceptibility as

$$n_2 = \frac{12\pi}{n_0} \chi^{(3)} \qquad (8.4)$$

where $\chi^{(3)}$ represents the diagonal element of the nonlinear susceptibility tensor. Equations (8.3) and (8.4) are expressed in esu units; however, it is more convenient to rewrite Eq. (8.3) as

$$n = n_0 + \gamma I \qquad (8.5)$$

where γ, the nonlinear coefficient, and I, the optical intensity, are both in SI units. The nonlinear index coefficient γ is related to the nonlinear index n_2 by

$$\gamma (m^2/W) = \frac{40\pi}{cn_0} n_2 \text{ (esu units)}$$

$$= \frac{480\pi^2}{cn_0^2} \chi^{(3)} \qquad (8.6)$$

Table 8.1 lists the third-order nonlinear susceptibility of some materials that are potentially useful for integrated optics devices. In general, all the $\chi^{(3)}$ values are small and are independent of the third-order nonlinear phenomena involved except at certain characteristic wavelengths where resonance effects occur. The largest reported value for $\chi^{(3)}$ is approximately 10^{-2} esu, which was measured InSb at a wavelength of 5.3 µm and a temperature of 5 K [11]. Another material found to possess a large nonlinear coefficient is GaAs, the measured value of $\chi^{(3)}$ is approximately 10^{-6} to 10^{-7} esu at a wavelength of 0.82 µm and a temperature below 120 K [12]. The large nonlinearity seen in GaAs is attributed to the free exciton resonance. Far from the exciton resonance, the nonlinear coefficient is reduced to approximately 10^{-10} esu.

TABLE 8.1   Third-Order Nonlinear
Susceptibility χ

Material	$\chi^{(3)}(10^{-14}$ esu)
LiF[a]	0.3
BSC glass[a]	0.8
NaCl[a]	1.2
MgO[a]	2.4
CS_2	43
MBBA	3000
Fused SiO_2[a]	0.7
PTS polymer[b]	8.5×10^4
Ge[b]	4.0×10^4
Air[a]	4.1×10^{-5}
He[c]	4.0×10^{-25}/atom
Xe[c]	11.0×10^{-22}/atom
GaAs[d]	1.6×10^4
InSb[e]	10^{12}

[a]C. C. Wang and E. L. Baardsen, *Phys. Rev.*, *185*:1079 (1969); *Phys. Rev.*, *B1*: 2827 (1970).
[b]C. Santeret et al., *Opt. Commun.*, *18*: 55 (1976).
[c]J. F. Ward and G. H. C. New, *Phys. Rev.*, *185*:57 (1969).
[d]Far from exciton resonance.   At exciton resonance $\chi^{(3)} \cong 10^{-6}$ to 10^{-7} esu.
[e]D. A. B. Miller and S. D. Smith, *Opt. Commun.*, *31*:101 (1979).

Another nonlinear material effect that can be employed is saturable absorption. Saturable absorption is a term applied to the phenomenon whereby the absorption of a material decreases with the increasing incident optical intensity, and at some point the absorption
of the material saturates and therefore becomes only weakly dependent on the incident optical intensity. For a two-level energy system, the absorption coefficient α can be described by

$$\alpha = \alpha_B + \frac{\alpha_0}{1 + I/I_S} \tag{8.7}$$

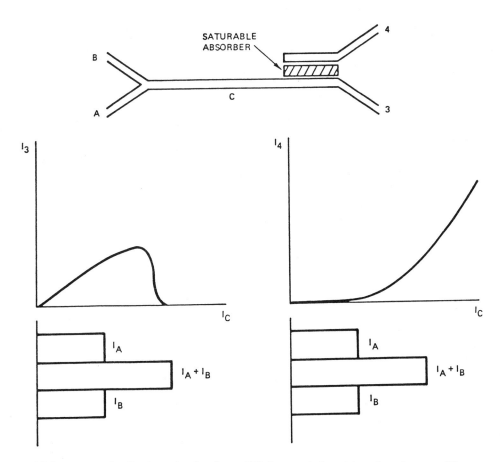

FIG. 8.10 Optical logic device utilizing a saturable absorber. (From
Ref. 13.)

where α_B is the background unsaturable absorption, α_0 the saturable absorption at low intensities, and I_S the saturation intensity.

A proposed logic device utilizing nonlinear material effects is shown in Fig. 8.10. This device is capable of performing both the EOR and the AND functions. The device consists of two sections. The first section is a power combiner used to combine the optical inputs into a single channel. The second section consists of two adjacent waveguides with a saturable absorber, and the interaction regions are designed such that if only one of the optical inputs is present, the absorber will not saturate and the optical power will not be coupled into the upper level. In this case, the optical energy will exit from port 3. If, however, both inputs are excited simultaneously, the combined optical power is sufficient to saturate the absorber. The resulting change in the absorption and the accompanying change in refractive index will couple the optical power from the lower channel to the upper one, exiting from port 4. If port 4 is monitored, the AND function is realized, and if port 3 is monitored, the EOR function is realized.

Other possible configurations based on optically induced nonlinear material effects are capable of performing logic functions. They all share the following characteristics. First, the speed of the devices will ultimately be limited by the response time of the nonlinear material effects. Response times in the range 10^{-14} to 10^{-12} s have been predicted. Second, the optical power required will depend on the strength of the nonlinarity, which can be very high. Significant power reduction can be obtained by the resonant enhancement effect, if it exists. Third, these devices do not necessarily require inputs that originate from the same laser pulse. Consequently, the devices can be optically cascaded.

8.5 BISTABLE OPTICAL DEVICES

The bistable optical devices have been receiving a lot of attention recently. Especially interesting have been predictions that such devices have the potential of achieving subpicosecond switching speeds and requiring only picojoules of switching energy [14]. A bistable optical device consists of an optical element whose optical output (or possibly the optical input) is sampled and fed back into the device in such a way that its operating characteristics are modified. It is the effect of the feedback that causes the astable operation of the device. Bistable optical devices are usually classified as being intrinsic device or hybrid device.

Intrinsic device utilizes an optically induced feedback effect, such as the optically induced effects described in Sec. 8.3. The most common intrinsic bistable optical device configuration consists of a

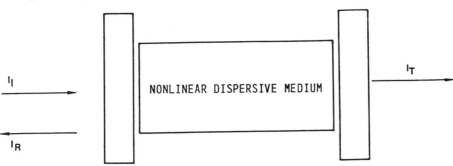

FIG. 8.11 Nonlinear Fabry-Perot Resonator.

nonlinear optical medium placed inside a Fabry-Perot resonator as shown in Fig. 8.11. The optical feedback is provided by the partially reflecting resonator mirrors. The smallest device reported was of the Fabry-Perot resonator type, a 5-μm-thick GaAs wafer with 90% reflectance coatings deposited on its surface to form the resonator structure [12]. The optical saturation of the exciton effect in the GaAs and the corresponding change in refractive index provided the necessary nonlinearity. The reflective coating results in a resonator with a finesse of 16 in regions far below the exciton resonance, and absorption is small. Optical bistability was observed from 5 to 120 K, with laser wavelengths 10 to 25 Å longer than the wavelength of free exciton peak. The bistability is primarily dispersive due to the light-induced change in exciton absorption. The holding optical intensity was about 10^5 W/cm^2. In the experiment, it was not possible to measure the switch-up time. However, rise times of less than 1 ps have been predicted for saturation of the exciton effect in GaAs [15]. The measured switch-down time was 40 ns, due primarily to the long lifetime of excited carriers. Schemes such as impurity doping have been proposed for reducing the carrier decay time.

The exciton absorption effect on optical bistability can be enhanced by using multiple quantum well structures. Figure 8.12 shows a bistable device consisting of 61 periodic layers of GaAs and GaAlAs with a total thickness of less than 4.9 μm. Free exciton resonance in pure GaAs decreases with increasing temperatures, resulting in a lower nonlinear refractive index. However, the excitons of the superlattice have a greater binding energy and therefore a greater nonlinear response at room temperature. Room-temperature bistable operation was observed and the holding optical intensity was reduced to several kW/cm^2 [16]. It should be noted that material physics plays a crucial role in the development of intrinsic bistable

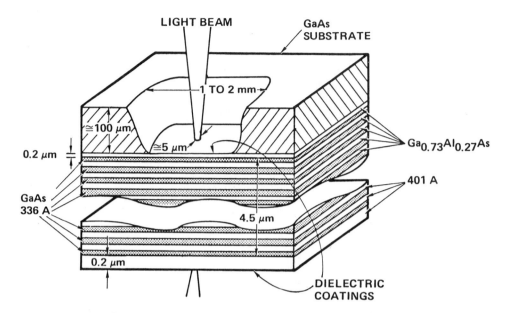

FIG. 8.12 Room temperature bistable optical device with a 61-layer GaAs/GaAlAs superlattice as the nonlinear medium. (From Ref. 17.)

optical devices. A lot of research effort is still needed to understand the bistable mechanism. One good example is the latest report of room-temperature operation of the bistability of bulk GaAs devices, with holding optical intensities comparable to those of quantum well devices [18].

Figure 8.13 illustrates the calculated operating characteristics of a Fabry-Perot resonator containing a nonlinear medium of zero absorption. The plot shows the output power versus input power for various values of the zero-field-cavity detuning parameter δ_0. The reflectivities of the resonator used in the calculation are 90%. It is clear from these plots that the operating characteristics can be adjusted by the initial resonator detuning. If the initial detuning is zero ($\delta_0 = 0$, i.e., the resonator is tuned to a transmission peak), the device operates as a power limiter. The power-limiting function ensures that only small variations of the transmitted intensity occur for even relatively large variations in the input intensity. If the detuning is increased to $\delta_0 = 0.18$ rad, the device can be operated as an optical amplifier. In this case, the nonlinear Fabry-Perot resonator exhibits a large gain around the point of $\beta I_{in} = 0.16$. In this operation mode, the device can also be used to perform the logic function of an AND gate. If the detuning is increased to $\delta = 0.32$ rad, the device becomes bistable. The bistability can be used as an optical memory, AND gate, OR gate, or EOR gate.

To perform the logic function of an EOR gate, it is necessary to use the reflected intensity from the nonlinear Fabry-Perot as the output channel. The transmitted intensity is again capable of operating as an AND gate. To physically separate the reflected beam from the incident beam, ring resonators with input and output taps has been proposed and analyzed [19]. Figure 8.13 shows that normalized incident switch-up intensity is approximately $\beta I_{in} = 0.55$; however, it will take an infinite amount of time to lengthen the switch-up time near the critical intensity. The lengthening of switch-up time near the critical intensity was predicted theoretically in an

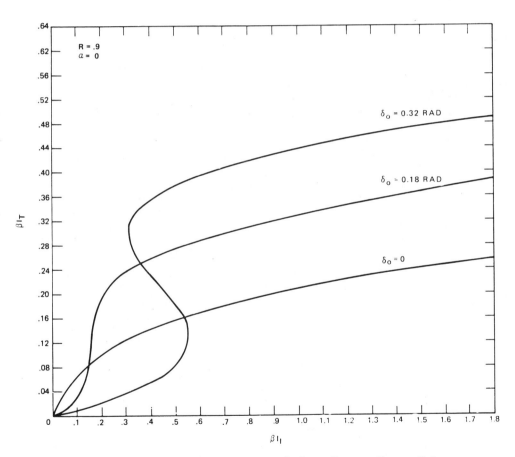

FIG. 8.13 Typical operating characteristics of a nonlinear Fabry-Perot resonator. β is a parameter related to the operating wavelength and nonlinear coefficient.

analysis of two-level atoms within a Fabry-Perot resonator [20]. The theoretical prediction of critical slowing down was confirmed in an experiment of hybrid bistable optical devices [21]. Figure 8.14 shows the experimental results; clearly shown is the increase in the device switching time as the input voltage approaches the critical switch-up voltage. Although the exact form of the curve is true only for the hybrid device used, it is expected that a similar curve will be applicable for all optical bistable devices. Numerical calculations have been employed to analyze the transient response of a nonlinear Fabry-Perot resonator. Some of the results are shown in Fig. 8.15. The critical switch-up intensity is 26.2 kW/cm^2. For an input intensity of less than 26.2 kW/cm^2, the device remains in the low-output state. The overshoot and ringing features of all output traces are characteristic of the linear properties of the Fabry-Perot cavity. For input intensities greater than 26.2 kW/cm^2, the device is switched to the high-output state. The switch-up delay is reduced at a higher input intensity. For sufficiently large inputs the switching speed approaches the cavity decay time. In this example, the nonlinear response time was assumed to be 10 ps. The device response time is limited by the cavity decay time (τ_c = 33 ps) and the effect of critical slowing down.

Bistable optical devices offer the potential of operating at high switching speeds and low switching energies. Figure 8.16 shows plots of calculated switching energy versus cavity decay time for a lossy nonlinear Fabry-Perot resonator. The magnitude of the nonlinearity was the same as that reported for the exciton resonance in GaAs. It was also assumed that the speed of the device was limited by the cavity decay time, not the decay time of the nonlinearity. The solid lines represent a cavity of fixed length and varying mirror reflectivities. The dashed lines represent a cavity with constant mirror reflectivity but varying length. Calculations of this type predict that the nonlinear Fabry-Perot type bistable optic devices should be capable of subpicosecond switching speeds and possibly require only femtojoules of switching energy. Thus, because of its versatility, high speed, and low power requirements, the bistable optical device appears to hold great promise for use as an optical logic device.

Intrinsic integrated optic bistable devices are difficult to fabricate using available materials. The nonlinear coefficients are usually small, thus requiring high input optical power. In addition, the waveguide technology does not exist to allow us to fabricate these devices with reasonable attenuation losses. Even though the intrinsic bistable devices are more attractive in ultimate device performance, the hybrid approach is still of great importance for devices operating at speed up to a few gigahertz. In the hybrid circuit, the feedback is provided by an electronic signal, for example, an electro-optic

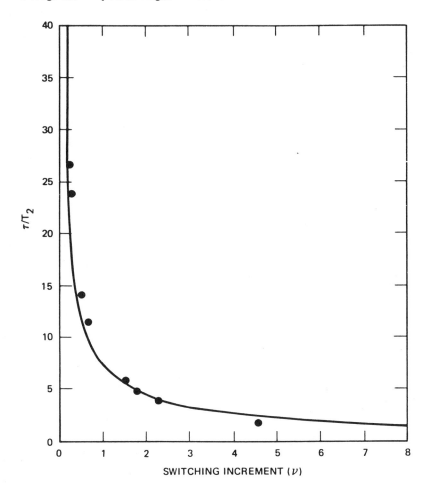

FIG. 8.14 Characteristic switching time versus increment of input switching signal beyond critical value. Where $\nu = V - V_c/V$, τ is the device response time, is the feedback time, and V_c is the critical switching voltage. (From Ref. 21.)

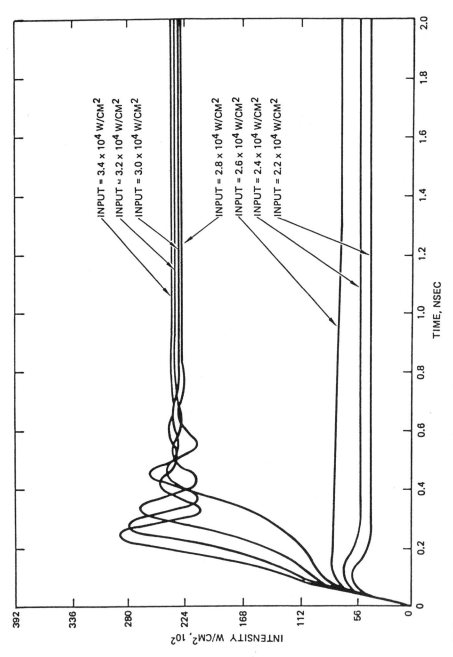

FIG. 8.15 Transient response of NLFP including a 10 psec nonlinear response time. Linear cavity decay time $\tau_c = 33$ psec. (From Ref. 22.)

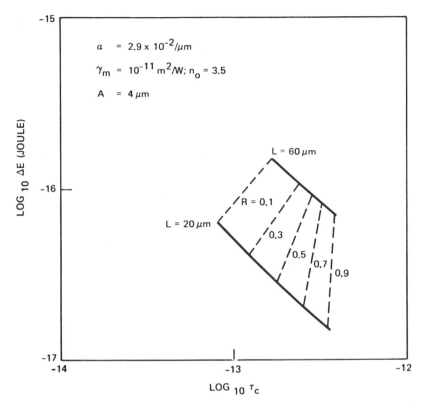

FIG. 8.16 Calculated switching energy versus cavity decay time for a lossy nonlinear Fabry-Perot resonator, where a is the absorption, γ_m is the nonlinear index coefficient, n_0 is the linear refractive index, A is the cross sectional area of the cavity, L is the cavity length and R is the reflectance.

phase modulator. The advantage of this approach is that the "artificial nonlinearity" is many orders of magnitude higher than the intrinsic nonlinearity of any known material. The characteristics of those devices have been discussed in previous sections.

The hybrid bistable optical device was first demonstrated using a Fabry-Perot resonator containing an electro-optic phase modulator [23]. A beam splitter and a detector at the output sample the output power and convert it into an electrical signal. The electrical signal is then applied to the electro-optic phase modulator. Various nonlinear characteristics have been demonstrated: devices such as the optical limiter, pulse shaper, optical triode, optical limitar, differential

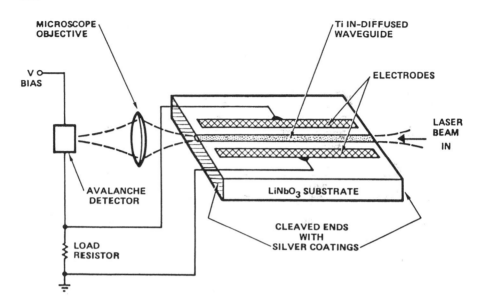

FIG. 8.17 Experimental setup for the characterization of hybrid integrated bistable optical device. (From Ref. 24.)

amplifier, optical switch, and logic elements. The integrated optics version of the hybrid optical bistable devices has been demonstrated using a Ti-diffused LiNbO$_3$ waveguide [24]. The schematic diagram of the experimental setup is shown in Fig. 8.17. The ends of X-cut LiNbO$_3$ were cleaved along the (10.2) plane, which intersects the X plane at an angle of 32.7° from the Z axis. To form the Fabry-Perot resonator, a silver coating was applied to the cleaved parallel ends to have a reflectivity of 50% on each end. He-Ne laser light was coupled into and out of the single-mode Ti-in-diffused waveguide with 20× microscope objectives. The transmitted light is focused to an avalanche photodiode; the detected signal in the form of electrical voltage is then applied to the electrodes of the phase modulator. Optical hysteresis was observed for optical power levels at the detector as low as 5 nW. The switching time for the configuration is limited by the RC time constant, which was about 200 µs. The optical switching energy was then approximately 1 pJ. In a similar experiment, a switching time of 50 ns was measured for an input power of 10 µW. The switching energy was less than 0.5 pJ.

    Table 8.2 summarizes the experimental results of various bistable optical devices reported to date.

TABLE 8.2 Experimental Results of Bistable Optical Devices

Type of device	Switching power (W)	Switching time (s)	Switching energy (J)
Intrinsic: Fabry-Perot resonators			
CS_2	3×10^5	5×10^{-10}	1.5×10^{-4}
Na vapor	1×10^{-2}	1×10^{-5}	1×10^{-7}
GaAs	2×10^{-1}	4×10^{-8}	8×10^{-9}
InSb	1×10^{-2}	$<5 \times 10^{-7}$	$<5 \times 10^{-9}$
Hybrid: Fabry-Perot resonator, $LiNbO_3$	1×10^{-5}	5×10^{-8}	5×10^{-13}
Liquid-crystal matrix	5×10^{-7}	4×10^{-2}	2×10^{-8}
Nonlinear interface glass, CS_2	2×10^5	2×10^{-12}	4×10^{-7}

Source: Ref. 14.

Optical bistability can also be demonstrated without using the Fabry-Perot resonator. The curcial criterium for observing bistability is to provide an adequate feedback to modify the transmission characteristics of the media. Incoherent mirrorless bistable devices have been demonstrated using a $LiNbO_3$ electro-optic modulator and a multimode laser [25].

REFERENCES

1. J. P. van der Ziel, W. T. Tsang, R. A. Logan, R. M. Mikulyak, and W. M. Augustyniak, Subpicosecond pulses from passively mode-locked GaAs buried optical guide semiconductor lasers, *Appl. Phys. Lett.*, *39*:525 (1981).

2. P. R. Smith, D. H. Auston, A. M. Johnson, and W. M. Augustyniak, Picosecond photoconductivity in radiation-damaged silicon-on-sapphire films, *Appl. Phys. Lett.*, *38*:47 (1981).

3. C. Lin and D. Marcuse, "Dispersion in Single Mode Fiber: The Question of Maximum Transmission Bandwidth," Paper TUC5, IOOC81, San Francisco (Apr. 1981).

4. R. C. Alferness, N. P. Economou, and L. L. Buhl, Fast Compact optical waveguide switch modulator, *Appl. Phys. Lett.*, *38*:214 (1981).

5. H. F. Taylor, Guided wave electro-optic device for logic and computation, *Appl. Opt.*, *17*:1493 (1978).

6. F. J. Leonberger, High-speed operation of $LiNbO_3$ electro-optic interferometric waveguide modulators, *Opt. Lett.*, *5*:312 (1980).

7. M. Papuchon, C. Puech, and A. Schanpper, "A Digitally Driven Integrated Amplitude Modulator," Paper TuE2, Digest of the Topical Meeting on Integrated and Guided Wave Optics, Incline Village, Nev. (Jan. 1980).

8. L. Goldberg and S. H. Lee, Optically activated switch/modulator using a photoconductor and two channel waveguides, *Radio Sci.* *12*:537 (1977).

9. L. Goldberg and S. H. Lee, Integrated optical half adder circuit, *Appl. Opt.*, *18*:2045 (1979).

10. W. D. Bomberger, unpublished results.

11. D. A. B. Miller and S. D. Smith, Two beam optical signal amplification and bistability in InSb, *Opt. Commun.*, *31*:101 (1979).

12. H. M. Gibbs, S. L. McCall, T. N. C. Venkatesan, A. C. Gossard, A. Passner, and W. Wiegmann, Optical Bistability in Semiconductors, *Appl. Phys. Lett.*, *35*:451 (1979).

13. W. D. Bomberger, T. K. Findakly, and B. Chen, unpublished results.

14. P. W. Smith and W. J. Tomlinson, Bistable optical devices promise subpicosecond switching, *IEEE Spectrum*, 26 (June 1981).

15. C. V. Shank, R. L. Fork, R. F. Leheny, and J. Shah, Dynamics of photoexcited GaAs bandedge absorption with subpicosecond resolution, *Phys. Rev. Lett.*, *42*:112 (1979).

16. J. L. Jewell et al., Technical Digest of the Topical Meeting on Optical Bistability, Paper ThAZ, Optical Society of America (1983).

17. H. M. Gibbs et al., *Appl. Phys. Lett.*, *41*:221 (1982).

18. J. L. Jewell et al., Technical Digest of the Conference on Lasers and Electro-Optics, Paper WOZ (1984).

19. D. Sarid, Analysis of bistability in a ring-channel waveguide, *Opt. Lett.*, *6*:552 (1981).

20. R. Bonifacio and A. L. Lugiato, Cooperative effects and bistability for resonance fluorescence, *Opt. Commun.*, *19*:1972 (1976).

21. E. Garmire, J. H. Marburger, S. D. Allen, and H. G. Winful, Transient response of hybrid bistable optical devices, *Appl. Phys. Lett.*, *34*:374 (1979).

22. S. M. Jensen, High speed optical logic devices, *SPIE Proc.*, *218*:33 (1980).

23. P. W. Smith, E. H. Turner, and P. J. Maloney, Electro-optic nonlinear Fabry-Perot devices, *IEEE J. Quantum Electron.*, *QE-14*:207 (1978).

24. P. W. Smith, I. P. Kaminow, P. J. Maloney, and L. W. Stulz, Integrated bistable optical devices, *Appl. Phys. Lett.*, *33*:24 (1978).

25. E. Garmire, J. H. Marburger, and S. D. Allen, Incoherent mirrorless bistable optical device, *Appl. Phys. Lett.*, *32*:320 (1978).

9
Nonlinear Integrated Optics

GEORGE I. STEGEMAN, JAMES J. BURKE, AND COLIN T. SEATON
Optical Sciences Center, University of Arizona, Tucson, Arizona

9.1 INTRODUCTION

Nonlinear optics was made possible by the invention of the laser.
It is based on the premise that an optical field can be of sufficient
strength to perturb the response of matter. Since it is the electrons associated with atoms and molecules that determine the optical
properties of matter, it is their response to an applied optical field
that determines the nonlinear properties. It is customary to discuss optical nonlinearities in two regimes. At wavelengths far from
any atomic or molecular resonances, the response becomes nonlinear
when the applied fields become comparable to those that bind the
electrons in their equilibrium configurations. Alternatively, near
a resonance, the absorption can be sufficiently strong to alter the
thermal distribution of electrons in the various energy levels and
hence affect the response to an incident field. Typically, the second mechanism, although it involves attenuation and hence limited
propagation distances, is the more efficient of the two, especially
in the emerging field of optical bistability. For example, it is the
source of very large cubic nonlinearities in semiconductors.

 There are a number of characteristics of nonlinear interactions
that lend themselves to the utilization of integrated optics waveguides. The strength of the interactions is proportional to some

power of the local electric field. The field is proportional to the square root of the intensity (i.e., power per unit area). In a thin-film or channel waveguide, one or both of the guiding dimensions can be limited to optical wavelengths. This reduction in beam cross-sectional area, which can be maintained for centimeter distances, reduces the power required to produce a high power density.

It is generally true that to obtain useful efficiencies, nonlinear interactions must be phase matched. That is, the vectorial sum of the incident light wave vectors must be equal to the wave vector of a radiation field at the sum frequency. This is not a trivial condition for plane-wave interactions because of the dispersion in refractive index in the near-infrared and visible regions of the spectrum. In the case of guided waves, dispersion also occurs with film thickness. Furthermore, for thick-enough films, a number of discrete modes can occur. These two factors make phase matching easier to guided waves than for their plane-wave counterparts.

There are highly nonlinear materials that can easily be manufactured in thin-film form, but not as bulk media. A classical example is that of the polydiacetylenes, which exhibit very large and fast third-order nonlinearities.

On the other hand, there are also problems associated with using guided waves for nonlinear optics. High-quality films are required both to maintain long propagation distances and to withstand the high power densities implicitly required for nonlinear optics. It has been shown [1,2] that waveguides can be fabricated which can handle power densities comparable to those of bulk media if careful fabrication techniques are employed. Thin-film fabrication techniques, however, are generally more complex and difficult than for bulk media. Techniques such as sputtering, CVD, MBE, LPE, and so on, are both expensive and time consuming to implement and master.

Despite the difficulties, impressive progress has been made in the field of nonlinear integrated optics. Second harmonic generation has been by far the most extensively pursued interaction. The goal is usually to generate radiation at a desired frequency: for example, to double laser diode radiation into the visible. The initial work in GaAs slabs of Anderson and Boyd [3] was followed by a flurry of both theoretical and experimental work in the early and mid-1970s. The most successful waveguide for efficient doubling of infrared radiation has been in-diffused $LiNbO_3$, despite its optical damage properties. In fact, it now appears that conversion efficiencies of $\simeq 1\%$ should be feasible with 1-mW input powers. Recently, highly nonlinear organic materials have been used in waveguide form to demonstrate very efficient harmonic generation. Efficient sum and difference frequency generation has also been demonstrated by a number of groups with the emphasis again on $LiNbO_3$ waveguides.

Initial results on parametric oscillators are very encouraging, and progress in this area can be anticipated in the near future.

One of the nonlinear interactions unique to integrated optics is the mixing of two counterpropagating waves to generate a sum-frequency-wave plane wave radiated normal to the waveguide surface. The radiated field corresponds to the convolution of the two input waves and can be used for signal processing on a picosecond time scale. This particular interaction can also be used as an optical picosecond transient digitizer.

An area of currently growing activity is the application of third-order nonlinearities to waveguide geometries. In particular, it has been predicted that degenerate four-wave mixing can result in efficient signal processing, and very recently the first experiment on degenerate four-wave mixing in planar waveguides has been reported. Numerical calculations have also shown that optical bistability can be achieved with very low power levels in an integrated optics context. The great interest in optical bistability ensures that this phenomenon will be pursued vigorously in the near future. Very recently, nonlinear directional couplers and CARS (coherent anti-stokes Raman scattering) have been reported in optical waveguides. In fact, it has been predicted that both second- and third-order nonlinear phenomena in thin-film waveguides have the potential for studying monolayers on surfaces, and one should expect experiments along these lines in the near future. Nonlinear directional couplers could prove very valuable for optical logic. So also could gratings on thin films fabricated from materials with intensity-dependent refractive indices. There are also third-order nonlinear phenomena unique to guided waves: namely, nonlinear couplers and nonlinear guided waves whose field distribution and wavevector vary with guided wave power. In both these cases, initial experiments have been reported. In summary, one can expect that phenomena based on third-order nonlinearities in waveguides will be an area of escalating activity in the next few years.

In this chapter we review the field of nonlinear integrated optics. We include here only planar and channel waveguides, and exclude nonlinear effects that occur in optical fibers. For a review of this subject area, the reader is directed to excellent review articles by Stolen [4–6]. The work discussed here probably does not include everything published to date on nonlinear integrated optics, but reflects our interests and prejudices.

This chapter is organized as follows. In Section 9.2 we deal with the formal aspects of the application of guided waves to nonlinear optics. There we discuss the basic concepts and the methods used to analyze nonlinear guided-wave phenomena. We attempt here to present all of the nonlinear interactions in a single notation. The

experiments reported to date on second-order nonlinear phenomena
are reviewed in Sec. 9.3. In Section 9.4 we deal with many of the
recent predictions of third-order processes in thin film waveguides,
as well as with the handful of experiments performed to date. Finally, in Sec. 9.5 we summarize progress to date and identify areas
which, in our opinion, will flourish in the near future.

9.2 NONLINEAR OPTICS OF GUIDED WAVES: BASIC
CONCEPTS AND ANALYSIS TECHNIQUES

In this section we discuss the theoretical basis for nonlinear integrated optics. All the individual aspects of this problem have been
investigated before and are well understood. The formalism of nonlinear optics in terms of the nonlinear polarization is well known.
The analysis of guided waves driven by nonlinear polarization fields
is also well documented, especially in the form of coupled-mode theory [7,8]. The calculation of nonlinearly generated waves will, as
we shall see, involve nothing more than standard electromagnetic
theory.

9.2.1 Bulk Nonlinear Polarization

The basic equation of nonlinear optics for the nonlinear polarization
induced inside a medium is very well known. For the most general
case we assume that there are a number of incident waves of frequency ω_a present in the medium, with each written in the form

$$\underline{E} = \frac{1}{2} \underline{E}(\underline{r}, \omega_a) \exp (i\omega_a t) a_a(x) + c.c. \tag{9.1}$$

where clearly the $\underline{E}(\underline{r}, \omega_\alpha)$ terms are the Fourier components of the
incident waves with $\underline{E}(\underline{r}, \omega_a)^* = \underline{E}(\underline{r}, -\omega_a)$. We shall define the amplitude function $a_a(x)$ later. The total polarization induced in the
medium can be written as a sum of terms of the form

$$P(\underline{r}, t) = \frac{1}{2} P^L(\underline{r}, \omega) \exp (i\omega t) + \frac{1}{2} P^{NL}(\underline{r}, \omega) \exp (i\omega t)$$

$$+ c.c. \tag{9.2}$$

where the frequencies ω are determined by the incident fields. For
every frequency ω present in the medium (which includes both the
incident and generated waves),

$$P_i^L{}_{(\underline{r},\omega)} \exp (i\omega t) = \varepsilon_0 \chi_{ij}^{(1)}(-\omega,\omega)E_j(\underline{r},\omega) \exp (i\omega t)a(x) \qquad (9.3)$$

where this term leads to a frequency-dependent refractive index. For every pair of fields present in the material (say, $\omega_1 \geqslant \omega_2$),

$$P_i^{NL}(\underline{r},\omega) \exp (i\omega t) = \varepsilon_0 \chi_{ijk}^{(2)}(-\omega,\pm\omega_2,\omega_1)E_j(\underline{r},\omega_1)$$
$$\times E_k(\underline{r},\pm\omega_2) \exp [i(\omega_1 \pm \omega_2)t]$$
$$\times a_1(x)a_2(x) \qquad (9.4)$$

This interaction clearly leads to sum and difference frequency generation at $\omega = \omega_1 \pm \omega_2$. For every three waves present (which can include any single wave taken three times, etc.), for example, $\omega_1 \pm \omega_2 \pm \omega_3 \geqslant 0$,

$$P_i^{NL}(\underline{r},\omega) \exp (i\omega t) = \varepsilon_0 \chi_{ijk}^{(3)}(\omega,\pm\omega_3,\pm\omega_2,\omega_1)E_j(\underline{r},\pm\omega_1)$$
$$\times E_k(\underline{r},\pm\omega_2)E_l(\underline{r},\pm\omega_3) \exp [i(\omega_1 \pm \omega_2$$
$$\pm \omega_3)t]a_1(x)a_2(x)a_3(x) \qquad (9.5)$$

In this case the signs must be chosen so that $\omega \geqslant 0$. Furthermore, whenever the negative sign is chosen for a frequency, it is necessary to take the complex conjugate of the corresponding field in Eqs. (9.4) and (9.5). One can, in principle of course, proceed to even higher-order terms in the expansion for the nonlinear polarization, but to date these terms have not been found useful.

The particular components of the nonlinear susceptibility tensors which are nonzero depend on the crystal symmetry of the material. For example, $\underline{x}^{(2)}$ is uniquely zero for media with centers of inversion symmetry. As a result, isotropic media cannot produce second harmonic generation with incident plane waves. All materials possess at least three nonzero $\underline{x}^{(3)}$ components.

The required nonlinear coefficients are known for a large range of materials. Second-order susceptibilities were tabulated by Singh [9] in the *Handbook of Lasers* and a sampling of more recent determinations can be found in, for example, Levine [10], Cassidy et al. [11], Kato [12], Halbout et al. [13], and Lalama et al. [14]. Typically, in MKS units, $\underline{x}^{(2)}$ is 10^{-11} to 10^{-13} m/V. There is no complete tabulation of $\underline{x}^{(3)}$ available, and one must go to review articles or to the original literature material by material to obtain numerical values. In fact, it is usually the coefficient for the intensity-dependent refractive index that has been reported.

9.2.2 Gradient Nonlinear Polarization Terms

In addition to the nonlinear polarization that occurs inside a material due to the product of electric fields, nonlinear source terms can also occur near surfaces where the field gradients are large [15]. These terms have symmetry properties similar to that of quadrapoles and hence are frequently called the quadrapolar contribution. The leading term in this case is

$$\underline{p}^{NL}(\underline{r},\omega) = \epsilon_0(\delta - \beta)[\underline{E}(\underline{r},\omega_1)\cdot\nabla]\underline{E}(\underline{r},\omega_2)$$
$$+ \beta\epsilon_0\underline{E}(\underline{r},\omega_1)[\underline{\nabla}\cdot\underline{E}(\underline{r},\omega_2)] + \alpha\underline{E}(\underline{r},\omega_1) \times \underline{H}(\underline{r},\omega_2) \qquad (9.6)$$

where α, β, and δ are scalars for isotropic media, and vectors for optically anisotropic media. It can easily be shown that this non-linear polarization term is zero for TE polarized waves. However, for TM modes, there are contributions in the bulk waveguiding media because there exists an electric field component parallel to the propagation wavevector. In addition, because the E_z TM field component is discontinuous across the boundary, there are also purely surface contributions due to a $\partial E_z/\partial z$ term. To date, these terms have not been included in any analysis of second-order nonlinear phenomena at surfaces.

The contribution of these gradient (quadrapole-like) terms is small compared to the usual $\underline{x}^{(2)}$ terms that enter into second-order nonlinear phenomena. The magnitude of the α and β coefficients is typically 10^{-21} to 10^{-22} m^2/V in MKS units. Since the gradient terms in the waveguiding media involve terms of magnitude k, which is typically 10^7 m^{-1}, the contribution to the nonlinear polarization is typically less than the usual $(\underline{x}^{(2)})$ terms by a factor of 10 to 10^4, depending on the material and mode. One would therefore expect the "surface" terms to be important only for cases in which $\underline{x}^{(2)}$ is uniquely zero because of symmetry reasons. Similarly for the purely surface terms, their contribution is typically equivalent to that of a tenth or less of a monolayer of material.

9.2.3 Guided-Wave Fields

The forms that the guided-wave fields can take are well known for a large number of guiding geometries. For simplicity in our analytical and numerical examples, we consider only waves guided by thin films.

A. TE Modes

For TE_m guided waves traveling along the x direction in the xy plane, the field distributions are given by

$$\underline{E}_m(\underline{r}, \omega_a) = \hat{j} C_{am} f_{am}(z) \exp(-i\beta_{am}x) \tag{9.7}$$

where C_{am} is chosen so that $|a_a(x)|^2$ is the guided-wave power in watts per meter of wavefront (along the y axis), that is,

$$1 = \frac{1}{2} \int_{-\infty}^{\infty} E_y(\underline{r}, \omega_a) H_z(\underline{r}, \omega_a)^* \, dz \tag{9.8}$$

Here β_{am} is the guided-wave wavevector, which is a solution to the appropriate dispersion relation, and $\underline{\triangledown} \times \underline{E} = -\partial B / \partial t$ is used to evaluate the magnetic fields.

The expressions for $f_{am}(z)$ and C_{am} depend on the particular guiding geometry. They are straightforward for a thin film of thickness h with boundaries at z = 0 and z = h. For example, for the cladding (c), film (f) and substrate (s),

$$n_c: \quad f_{am}(z) = \exp(s_{am}z) \tag{9.9a}$$

$$n_f: \quad f_{am}(z) = \cos(\kappa_{am}z) + \frac{s_{am}}{\kappa_{am}} \sin(\kappa_{am}z) \tag{9.9b}$$

$$n_s: \quad f_{am}(z) = \left[\cos(\kappa_{am}h) + \frac{s_{am}}{\kappa_{am}} \sin(\kappa_{am}h)\right]$$
$$\times \exp[-\rho_{am}(z - h)] \tag{9.9c}$$

with

$$s_{am}^2 = \beta_{am}^2 - n_c^2 \frac{\omega_a^2}{c^2} \qquad \kappa_{am}^2 = n_f^2 \frac{\omega_a^2}{c^2} - \beta_{am}^2$$

$$\rho_{am}^2 = \beta_{am}^2 - n_s^2 \frac{\omega_a^2}{c^2} \tag{9.10}$$

The normalization constant is

$$
C_{am} = \frac{2\kappa_{am}}{[(\beta_{am}/\omega_a\mu_0)(h + s_{am}^{-1} + \rho_{am}^{-1})(\kappa_{am}^2 + s_{am}^2)]^{1/2}}
\tag{9.11}
$$

and

$$
\tan(\kappa_{am}h - m\pi) = \frac{\kappa_{am}(s_{am} + \rho_{am})}{\kappa_{am}^2 - s_{am}\rho_{am}}
\tag{9.12}
$$

is the dispersion relation which must be satisfied by the allowed values of β_{am}. There are a discrete number of solutions for a given value of film thickness, which are labeled TE_0, TE_1, TE_2, and so on, where the subscript is the value of m in Eq. (9.12). We show typical field distributions in Fig. 9.1a since they will play a key role in understanding nonlinear cross sections involving mode conversion.

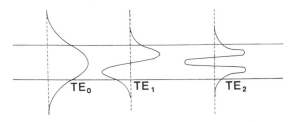

(a)

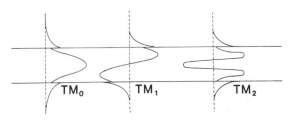

(b)

FIG. 9.1 Typical electric field distributions for (a) TE_m (E_y) and (b) TM_m (E_z) modes.

B. TM Guided Waves

For this polarization the magnetic field is polarized in the plane of the surfaces. The Fourier components are written as

$$\underline{H}_m(\underline{r}, \omega) = \hat{\underline{j}} C_{am} f_{am}(z) \exp(-i\beta_{am}x) \tag{9.13}$$

where

$$1 = \frac{1}{2} \int_{-\infty}^{\infty} E_z(\underline{r}, \omega_a) H_y(\underline{r}, \omega_a)^* \, dz \tag{9.14}$$

For the simple case of a thin film, the corresponding field distributions, dispersion relations, and so on, can be found in any standard reference on integrated optics (e.g., Sipe and Stegeman [16]). The solutions to the dispersion relations lead to a discrete set of modes labeled TM_0, TM_1, TM_2, and so on, for $m = 0, 1, 2, \ldots$ The electric field distributions important for the expansion of the nonlinear polarization are sketched in Fig. 9.1b and are obtained from the Maxwell equation $\underline{\nabla} \times \underline{H} = \varepsilon_0 n^2 (\partial E / \partial t)$.

C. Channel Waveguides

Light can also be guided by rectangular channels whose width W and height H both have dimensions of the order of optical wavelengths (see Fig. 9.2). Although approximate analytical solutions have been given by Marcitilli [17] and improved formulas by Kumar et al. [18], exact solutions for the dispersion relations and field distributions require numerical iterative techniques. The modes are usually classed as E^y with dominant E_y and H_z fields, and E^z with dominant E_z and H_y fields which degenerate into the TE and TM guided modes associated with thin films in the limit $W \to \infty$. Typical fields for E^y are also shown in Fig. 9.2. The fields have the same form as given by Eq. (9.1), that is

$$\underline{E}_{mn}(\underline{r}, \omega) = E_0 C_{mn} f_{mn}(y, z) \exp(-i\beta_{mn}x) \tag{9.15}$$

The guided-wave power in watts is given by $|a_{mn}(x)|^2$ with C_{mn} determined from the normalization condition,

$$1 = \frac{1}{2} \int_{-\infty}^{\infty} \int_{-\infty}^{\infty} [\underline{E}(\underline{r}, \omega) \times \underline{H}(\underline{r}, \omega)^*] \cdot \hat{i} \, dy \, dz \tag{9.16}$$

The discrete modes are now described by two resonances, one across each dimension, and hence two integers n and m are needed to

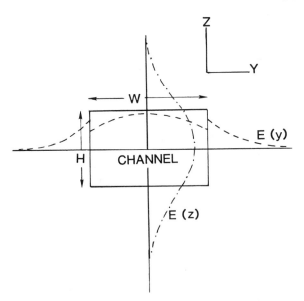

FIG. 9.2 Channel waveguide geometry and the variation in the E_y field distribution for an E_y mode with both cross-sectional coordinates.

describe the solutions to the dispersion relations. For details the reader is referred to the work of Goell [19], Marcitilli [17], and Kumar [18].

9.2.4 Conservation Laws

There are three conservation laws that must be taken into account when dealing with nonlinear guided-wave interactions. Two of these are identical to those encountered in plane-wave nonlinear interactions. The first is already clear from the preceding discussion, namely that the generated frequency is determined by the frequencies of the input waves. For example, for second-order processes, $\omega = \omega_1 \pm \omega_2$. The second is energy conservation; that is, the energy flux across any plane is a constant. An alternative form is that of the Manley-Rowe relations. This condition leads to coupled wave equations which must be solved in the case that pump depletion effects have to be taken into account.

The condition that is modified in the integrated optics case is wave vector conservation. Although the nonlinear interaction usually takes place over a surface area large with respect to an optical wavelength, the depth dimension is restricted. From standard

electromagnetic theory, the work done by a polarization field on an
electromagnetic wave (with amplitude a_3 leaving the interaction re-
gion) is given by the integral of $\underline{P} \cdot \underline{E}^*$ over the interaction volume.
For a three-wave interaction in which the polarization field is pro-
portional to exp $[-i(\underline{k}_1 + \underline{k}_2)x]$ and the generated field varies as
exp $(-i\underline{k}_3 x)$, and assuming for simplicity constant field amplitudes,

$$|a_3|^2 \propto \int \exp \; [i(\underline{k}_3 - \underline{k}_2 - \underline{k}_1) \cdot \underline{r}] \; dV$$

where V is the interaction volume and $\underline{k}_i$, i = 1, 2, 3, are generalized three-dimensional wave vectors. In the case of surface areas
many wavelengths in size, the integral over the surface area leads
to a δ function and to wave vector conservation in the plane of the
surface. The integral over the depth dimension leads to a distribution that is a function of $\Delta k_z h$. For example, if we model the waveguide interaction strength to be a constant over the film and zero
elsewhere,

$$|a_3|^2 \propto \delta(\underline{k}_{3p} - \underline{k}_{1p} - \underline{k}_{2p}) \; \frac{\sin \; (\Delta k_z h/2)}{\Delta k_z h/2} \qquad (9.17)$$

$$\Delta k_z = k_{3z} - k_{1z} - k_{2z} \qquad (9.18)$$

Wave vector is conserved only in the plane of the surface (i.e.,
$\underline{k}_{3p} = \underline{k}_{1p} + \underline{k}_{2p}^*$). Because the interaction volume is limited in the
depth dimension, wavevector does not have to be conserved normal
to the surface. Wave vector mismatch in this direction does, however, decrease the interaction efficiency, which is maximized when
$\Delta k_z = 0$. Therefore, interactions are allowed in which the wave
vectors for the interacting waves do not conserve wave vector in
three dimensions.

9.2.5 Catalog of Interactions

There are a large number of interactions that satisfy the conservation conditions. This includes the complete spectrum of nonlinear
plane-wave interactions, plus those unique to surfaces and film
geometries.

The second- and third-order nonlinear interactions discussed to
date are summarized schematically in Fig. 9.3. In terms of second-
order nonlinear phenomena, second harmonic generation of co-
directional and contra-directional guided waves, sum and difference
frequency generation, and parametric oscillation have been analyzed

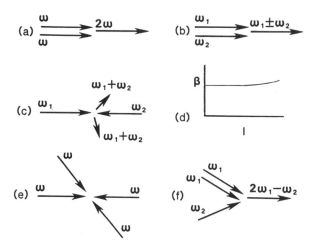

FIG. 9.3 The nonlinear guided wave interactions studied to date:
(a) Second harmonic generation; (b) Sum or difference frequency
generation; (c) Sum frequency plane wave generation; (d) Intensity
dependent wavevector; (e) Degenerate four wave mixing; (f) Coherent Anti-Stokes Raman Scattering.

and observed experimentally. It has also been predicted theoretically that parametric mixing by monolayers on thin-film waveguides
should lead to large signals and may be a useful tool for surface
spectroscopy.

Of the rich variety of third-order nonlinear interactions possible,
coherent anti-Stokes Raman scattering (CARS), degenerate four-
wave mixing, and intensity-dependent refractive index effects have
been observed. However, one can expect in the near future reports
on optical bistability, which has been already treated theoretically
[20].

9.2.6 Analysis of Growing Guided-Wave Fields

The most efficient interactions are those that start with input guided
waves and result in guided waves which are phase matched to the
nonlinear polarization field. This problem can be solved by explicitly solving the polarization-driven wave equation for the total fields
generated by the nonlinear polarization. One of the solution fields
evolves with propagation distance into a guided wave whose amplitude
grows linearly with propagation distance. Another and simpler approach is to use coupled-mode theory. In the limit of very small

depletion of the incident beam energies, the rate of growth of the mth mode at frequency ω_a is given by coupled-mode theory as [7,8,16]

$$\frac{d}{dx}a_{am}(x) + \beta_{amI}a_{am}(x) = \frac{k_a c}{4i} \int_{-\infty}^{\infty} \underline{P}^{NL}(\underline{r},\omega_a)$$
$$\cdot \underline{E}_{am}(\underline{r},\omega_a)^* \, dz \qquad (9.19)$$

where we have written $\beta_{am} = \beta_{amR} - i\beta_{amI}$ for the complex wave vector of the mth mode at the frequency ω_a. The uncoupled-mode propagation distance is $0.5 \, \beta_{amI}^{-1}$. There are limitations on the use of Eq. (9.19) in some cases and we will identify them as they are encountered. For most simple interactions of the type discussed here, Eq. (9.19) can be integrated with respect to both x and z and analytical results are possible.

9.2.7 Generation of Plane-Wave Radiation Fields

The interaction of guided-wave fields can also lead to the generation of plane waves which are radiated away from the surfaces. This case can also be analyzed by a coupled-mode-theory approach. Alternatively, one can explicitly solve the polarization-driven wave equation subject to the electromagnetic boundary conditions at the various surfaces that define the waveguide. In this particular case both approaches are of comparable difficulty for thin-film waveguides.

The calculations are too cumbersome to be reproduced in detail here. Instead we shall outline the sequence of steps followed and refer the interested reader to So et al. [21], Normandin et al. [22], and Liao et al. [23] for details. First the wave equation driven by the nonlinear polarization sources is solved analytically for the driven fields. These electric and magnetic fields do not a priori satisfy the usual tangential-boundary conditions. They must, however, be satisfied at the nonlinear frequency. It is therefore necessary to introduce normal-mode solutions to the wave equation whose amplitudes are adjusted so that the total fields, driven + normal modes, do satisfy the boundary conditions. Included in these normal modes are radiation fields which carry energy away from the waveguide region. It is these plane waves that are observed experimentally and hence are of interest in this chapter.

9.3 SECOND-ORDER PHENOMENA

In this section we discuss a variety of phenomena for which the nonlinear polarization is proportional to the product of two incident optical fields. This area of research is quite mature by now and there has been a great deal of progress, both experimental and theoretical. Certainly, the motivation has been to produce an efficient means of second harmonic generation. Sum and difference frequency generation has been demonstrated. Of more recent vintage are attempts to obtain parametric oscillation and the interaction of oppositely directed guided waves to perform convolution for the analysis of picosecond optical pulses. Very recently, it has been suggested that parametric mixing can be used to study monolayers on waveguide surfaces.

Traditionally, the nonlinear polarization is expressed in terms of $d_{i,jk}$ coefficients instead of $\underline{x}^{(2)}$. For two input waves of frequency ω_1 and ω_2,

$$p_i^{NL}(\underline{r}, \omega_1 \pm \omega_2) = 2\varepsilon_0 d_{i,jk} E_j(\underline{r},\omega_1) E_k(\underline{r},\pm\omega_2)$$
$$\times a_1(x) a_2(x) \qquad (9.20)$$

For the special case of harmonic generation there is only one input wave (i.e., $\omega_1 = \omega_2$), and the factor of 2 in Eq. (9.20) is removed. Since the materials with nonzero $d_{i,jk}$ must be non-centrosymmetric, they are crystalline and the orientation of crystal axes relative to the propagation direction and surface normal must be specified in each case. Here we shall assume that the nonlinear coefficients have been rotated via the Euler transformations into the geometry where z is normal to the surface and β lies along the x axis. Note that either the film or the substrate has to be a crystalline non-centrosymmetric material in order to obtain nonlinear waveguide effects.

The magnitude and variety of nonzero tensor elements depends on crystal symmetry and structure. Nonzero components of the $\underline{d}$ tensor vary typically from 10^{-14} to 10^{-10} m/V. Usually, these coefficients are written in the compressed Voigt notation in the form d_{iJ} where J is a contraction of the jk indices.

9.3.1 Mixing Between Co-Directional Waves

The mixing between two co-directional guided waves has been analyzed by a number of authors and by a variety of techniques (Suematsu [24], Boyd [25], Conwell [26], So et al. [27], and others). The results are identical to one another, as expected. Here we shall carry out the approach suggested in Sec. 9.2; namely, coupled-mode theory. Assuming that the two input waves are first overlapped at x = 0 and that there is no significant depletion of these waves due to the nonlinear generation of a third wave,

$$a_1(x) = a_1(0)e^{-\beta_{1I}x} \qquad a_2(x) = a_2(0)e^{-\beta_{2I}x} \qquad (9.21)$$

Evaluating Eq. (9.19) and assuming that $a_1(0)^2 = P_1/H$, $a_2(0)^2 = P_2/H$, and $a_3(L)^2 = P_3/H$,

$$P_3 = D^{NL} \frac{L^2}{H} |G|^2 P_1 P_2 \qquad (9.22)$$

$$|G|^2 = \frac{e^{-2\beta_{3I}L} + e^{-2(\beta_{1I}+\beta_{2I})L} - 2\cos(\Delta\beta L)e^{-(\beta_{1I}+\beta_{2I}+\beta_{3I})L}}{[\Delta\beta^2 + (\beta_{1I} + \beta_{2I})^2]L^2}$$

$$(9.23)$$

where the details of the nonlinear interaction are included in the cross-section term D^{NL} and G is an attenuation-dependent phase-mismatch term. Note that in the limit of small attenuation, $G \rightarrow \sin^2\Delta/\Delta^2$, where $\Delta\beta = \beta_{3R} - \beta_{2R} - \beta_{1R}$ is the usual phase-mismatch term, and $\Delta = \Delta\beta L/2$.

The details of the D^{NL} term depend on the details of the interacting waves. For sum and difference frequency generation,

$$D^{NL} = \frac{(\omega_1 \pm \omega_2)\varepsilon_0}{2i} \int_{-\infty}^{\infty} [d_{i,jk}E_j(\underline{r},\omega_1)E_k(\underline{r},\pm\omega_2)$$

$$\times E_i(\underline{r},\omega_3)^*] \, dz \qquad (9.24)$$

where for second harmonic generation the 2 in the denominator is replaced by 4. Note again that we have explicitly assumed that $\underline{E}(\underline{r},-\omega)^* = \underline{E}(\underline{r},\omega)$. For example, for all TE inputs, the term in the brackets is

$$d_{222}E_y(\underline{r},\omega_1)E_y(\underline{r},\pm\omega)E_y(\underline{r},\omega_1 \pm \omega_2)^* \qquad (9.25)$$

We note again that the mode numbers of the three waves need not be the same. It is straightforward to write down similar expressions for all TM and mixed TE-TM geometries.

The principal difficulty in achieving efficient sum or difference frequency generation is to obtain and maintain phase matching. Consider, for example, the case of second harmonic generation where $\beta_{am}(2\omega) > 2\beta_{am}(\omega)$, due to normal thin-film waveguide dispersion. Furthermore, the refractive index usually increases with decreasing λ. As a result, phase matching with all TE modes (or all TM) requires that $\beta_{m'}(2\omega) = 2\beta_m(\omega)$ with m' > m. However, because the mode distributions are different as shown in Fig. 9.1, when Eq. (9.24) is evaluated, cancellation effects that occur reduce the cross section considerably. Since $\beta_{mTE}(\omega) > \beta_{mTM}(\omega)$, phase matching can be obtained by using combinations of TE and TM modes. The possibility also exists that the same mode number can be used for all the waves and that the field overlap integral given by Eq. (9.24) can be optimized.

The second difficult problem is to maintain the phase-matching condition over long distances. The wave vector β depends on waveguide uniformity with respect to both refractive index and waveguide dimensions. Therefore, for a thin-film waveguide, both the film thickness and refractive index must be controlled very accurately.

A. Slab Waveguides

By slab waveguides we essentially mean waveguiding geometries that consist of discrete layers of different materials in the form of films and slabs. A number of nonlinear interactions have been reported in such structures, with both the film and the substrate serving as the nonlinear medium.

Second Harmonic Generation. Second harmonic generation was first observed in thin-film waveguides. The early work utilized thin amorphous films deposited on non-centrosymmetric substrates [27–32]. In this case the conversion efficiency η [= $P(2\omega)/P(\omega)$] was small because under typical phase-matching conditions only a small fraction of the incident power is carried in the substrate. Note that according to Eq. (9.22) the efficiency is a function of input power, and hence one must quote the incident power together with the efficiency when describing an experiment. Most of the published work is summarized in Table 9.1. As scanning this table shows, higher efficiencies are obtained by using the film instead of the substrate as the nonlinear medium [1,3,33–41], despite the fact that in many cases the film consisted of crystallites not all oriented in the same direction.

The second harmonic radiation has usually been observed as a guided wave which is coupled out by a prism, grating, or through the end of the film. However, in some cases in which phase matching to guided waves was not possible, but it was possible to radiate plane waves into the substrate, second harmonic radiation was observed in the form of Cerenkov radiation (see Fig. 9.4) into the substrate [27,30].

A variety of techniques have been used to facilitate phase matching of the harmonic to the fundamental wave. Assuming that the appropriate off-diagonal matrix elements are nonzero, the first and most obvious approach is to compute the dispersion relations for

both TE_m and TM_m modes at both frequencies and to search for
film thicknesses at which $\beta(2\omega) = 2\beta(\omega)$. In fact, for thick-enough
films one can usually find a number of such intersections of the dispersion relations. The real question is whether films can be manufactured precisely enough to match the required thickness. A typical variation in harmonic intensity with film thickness is shown in
Fig. 9.5. At issue here is not only film thickness, but also knowledge of the refractive indices of the film at both the harmonic and
fundamental frequencies. These values depend on preparation conditions. Typically, the refractive index is measured accurately at
the fundamental frequency and the harmonic refractive index is estimated from the Sellmeier relations using coefficients of the bulk
material from which the film was deposited. Some researchers have
gently tapered their films in a direction orthogonal to the propagation direction and then translated the beam in the film to find the
optimum phase-matching condition (Suematsu et al. [28], Shiosaki
et al. [38], and others who have not mentioned this technique speci-
fically in their papers). Phase-matching conditions can also be made
less critical by using multilayer films in which there are obviously
more degrees of freedom [33]. Another approach has been to use
a liquid as the cladding medium; then one can tune for phase matching by varying the refractive index of the liquid [29], or minimize
the phase-matching sensitivity by using an almost symmetrical waveguide [42]. Finally, we note that gratings can be used to supply
the extra wave vector component needed to obtain phase matching
between a doubled fundamental wave vector and the harmonic wave
vector [35,36].

The conversion efficiency depends on a large number of parameters. The critical ones are the phase-matching distance (listed
in Table 9.1), the guided-wave attenuations, the magnitude of the
nonlinear coefficients used, and the "overlap integral" between the
two modes. This last term is governed by Eq. (9.24). When the
modes used have different mode indices m, it can be seen from
FIG. 9.1 that destructive interference effects occur and the value
of the integral can be small. This can also be seen from Fig. 9.6,
which actually deals with a two-film waveguide. If in that case the
two films were one and the same material, the value of the integral
would be very small due to cancellation effects. This, and the
fact that for nonlinear substrates very little of the guided-wave
energy is carried in the substrate, are primarily responsible for
the small efficiencies listed in Table 9.1. Note that for the two
cases in which large values of η were found [3,33], the overlap
integral was optimized.

It has been known for a number of years that organic materials
can be engineered to have large nonlinearities [10-14]. Within the
last two years there have been two reports of nonlinear waveguides

TABLE 9.1 Summary of Second Harmonic Experiments with Slab Waveguides

References	Waveguide	λ (μm) Modes	h^a (μm)
27	ZnS-ZnO	1.06 $TE_0 \rightarrow C^d$	0.2
3	GaAs (slabs)	10.6 $TE_1 \rightarrow TE_2$	3.2
1	ZnO-glass	1.06 $TE_0 \rightarrow TM_1$	0.47
28,134	7059-α quartz	1.06 $TE_0 + TM_0 \rightarrow TM_2$	2.5
29	Liquid-TiO_2-α quartz	1.06 $TE_0 \rightarrow TM_0$	0.054
30	Al_2O_3-α quartz	0.6 $TE_0 \rightarrow TE_2$	
48	ZnS-BK7	1.05 $TE_1 \rightarrow TM_4$	1.3
34	AlGaAs- GaAs- AlGaAs	2.0 $TE_0 \rightarrow TM_2$	0.92
31	7059-α quartz	0.58	
36	MgF_2-nitro- benzene- MgF_2	1.06 $TM_3 \rightarrow TM_1$ $TM_2 \rightarrow TM_4$ $TM_4 \rightarrow TM_5$	
32	Al_2O_3-α quartz	1.06 $TE_0 \rightarrow TE_1$	0.48
35	GaAlAs- GaAs- GaAlAs	2.0 $TE_0 \rightarrow TM_0$	0.50
33	Air-TiO_2- ZnS-glass	0.93 $TE_0 \rightarrow TM_1$	0.12 μm (ZnS) 0.21 μm (TiO_2)
37	GaP ribbon	1.06 $TE \rightarrow TE$	<1

l_{eff}^b		Nonlinear medium	η^c (P_i)	Comments
1	cm	Substrate	$\sim 10^{-5}$ (2 W)	Harmonic appears as Cerenkov radiation
1	mm	Film	$\sim 10^{-2}$ (1 W)	
20	μm	Film	$\sim 10^{-9}$ (8 kW)	
12	mm	Substrate	$\sim 10^{-4}$ (300 W)	Taper used for phasic matching
3.5	mm	Substrate	10^{-4} (100 W)	Refractive index of liquid variable
~ 100	μm	Substrate		
~ 35	μm	Film	10^{-5} (300 W)	
		Film		Strongly focused beam
		Substrate	10^{-9} (2 W)	
2	cm	Film		Used grating for phase matching
~ 72	μm	Substrate	$\sim 10^{-6}$ (40 kW)	Used grating for phase matching
		Film		Used grating for phase matching
		Film (ZnS)	10^{-3} (100 mW)	Multilayer to facilitate phase matching
1	mm			
50	μm	Film		

TABLE 9.1   (Continued)

References	Waveguide	λ (μm) Modes	h^a (μm)
38	ZnO-SiO$_2$	1.06 TE $\rightarrow$ TE	0.97
41	Si-α quartz	1.06 TE$_0$ $\rightarrow$ C^d	0.29
39	*para*-Chloro-phenylurea-Corning 7059	0.84– TM$_0$ $\rightarrow$ TM$_2$ 0.92 TM$_0$ $\rightarrow$ TE$_2$	0.90
40	MNA-Corning 7059-fused silica	1.06 TE$_m$ $\rightarrow$ TM$_1$	1.68

[a]Film thickness.
[b]Effective coherence length (if stated); otherwise, sample length.
[c]Second harmonic generation efficiency and the input power level at which it was achieved.
[d]Second harmonic appears as Cerenkov radiation.
[e]$I_{2\omega}/I_\omega^2 = 0.3$.

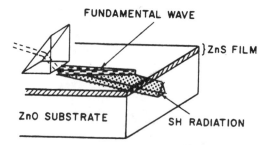

FIG. 9.4   Experimental arrangement for second harmonic generation via Cerenkov radiation from a thin film waveguide (From Ref. 27.)

l_{eff}^b	Nonlinear medium	η^c (P_i)	Comments
2.5 mm	Film	5×10^{-6} (150 mW)	Used taper for phase matching
	Substrate	10^{-9} (20 mW)	Cerenkov harmonic
	Film	0.3^a	Organic film
0.5 mm	Cladding	"Small"	MNA crystal on waveguide

containing organic materials [39,40]. Hewig and Jain [39] fabricated a thin polycrystalline film of *para*-chlorophenylurea and obtained second harmonic generation over the wavelength range 0.84 to 0.92 μm by phase matching the TE_0 and TM_0 waveguide modes. Their optimum conversion efficiency was 30%. Sasaki and colleagues [40] used a single crystal of MNA pressed onto a glass waveguide to obtain harmonic generation. These materials do hold considerable promise [43], for efficient second harmonic generation ($\eta \stackrel{\sim}{=} 10^{-2}$ with input powers of 10 mW).

The experiments discussed above have used prisms or gratings to couple the fundamental beam into the waveguide. As a result, the harmonic generation process is actually initiated inside the coupler. This case has been analyzed by Sarid [44], and subsequently extended by Malov et al. [45]. The principal result is that the harmonic signal generated inside the coupling region can be optimized by judicious choice of coupling parameters.

Parametric Mixing. There have been very few reports of sum or difference frequency generation in slab waveguides. A very

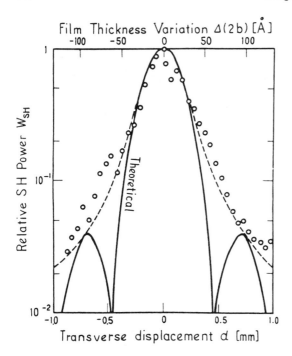

FIG. 9.5 Second harmonic power versus film thickness (or position in a tapered film). The phase matching film thickness is 2.5 μm so that 25 Å represents 0.1%. (From Ref. 28.)

intriguing case has been that of far-infrared generation by difference frequency mixing in planar GaAs waveguides [46]. Different emission lines from two CO_2 lasers were mixed in a series of GaAs slabs to obtain FIR wavelengths varying from 200 to 800 μm. The FIR power as a function of λ is shown in Fig. 9.7. For input powers of typically 1 kW, the efficiency was $\cong 10^{-6}$. The absolute power levels, however, are very competitive for that part of the spectrum.

B. Diffused LiNbO₃ Waveguides

It became evident in the mid and late 1970s that diffused $LiNbO_3$ waveguides were an excellent candidate for efficient nonlinear interactions. The $LiNbO_3$ nonlinearities are among the largest known for simple dielectric materials, the waveguides are very low loss and so deep that end-tire coupling techniques can be used efficiently, and the material is highly birefringent, so that phase matching between the lowest-order TE and TM modes is feasible.

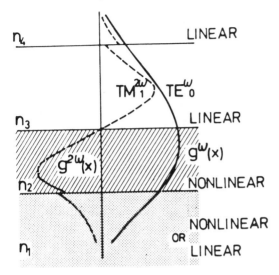

FIG. 9.6 Guided wave field overlap for a two film waveguide for harmonic generation for $TE_0 + TE_0 \rightarrow TM_1$. (From Ref. 33.)

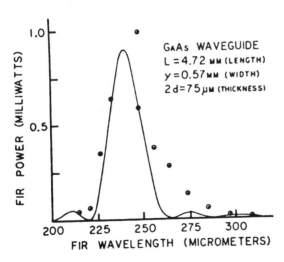

FIG. 9.7 Theoretical (x0.2) and measured FIR power versus FIR wavelength for difference frequency mixing in a GaAs slab. (From Ref. 46.)

The only major disadvantage is the cumulative optical damage, which is known to occur as a function of time in the visible. Despite this problem, major advances in nonlinear integrated optics have been made with this material.

Second Harmonic Generation. A representative survey of second harmonic generation in LiNbO$_3$ waveguides is given in Table 9.2. (Also included are results on a LiNbO$_3$ slab prepared by thinning and polishing [47] and LiNbO$_3$ as a substrate with a ZnS film on top [48,49]. The interest in this type of waveguide is obvious from the efficiencies listed in Table 9.2. To date the largest reported conversion is 25% by Sohler and Suche [50]. The highest efficiency with the smallest input power was also achieved by them in a waveguide resonator geometry; namely, $\eta \simeq 10^{-3}$ with a 1-mW input. The end faces of the waveguide were polished and coated to a reflectivity of 96% in order to build up the fundamental field inside the waveguide structure.

Typically, phase matching in these waveguides is obtained from the temperature or wavelength dependence of the material birefringence. This explains why most of the experiments are done at temperatures of 100 to 200°C. Because lithium niobate is an electro-optic material, an applied electric field can be used to fine tune the phase matching condition. This concept has been demonstrated by Uesugi et al. [51], Zolotov et al. [52], and Buritsky et al. [53]. Zolotov and co-workers [54] have also used angle tuning in the x-y plane to obtain phase matching in planar in-diffused waveguides. There has also been considerable effort to characterize the effect of fabrication procedures on second harmonic generation [55]. Very recently, De Michell [56,57] and colleagues have shown that the tuning range of diffused LiNbO$_3$ waveguides can be extended (1.08 to 1.24 μm in their case) by using a combination of the usual titanium in-diffusion technique with the proton exchange process. By properly weighting the two processes, they obtained waveguides that support both TE and TM waves with the change in the refractive index Δn_e larger than that obtained from the in-diffusion process alone.

Sohler and Suche [58] have predicted what appears to be the best efficiency that can be obtained for in-diffused LiNbO$_3$ waveguides. They have shown that by making a matched resonator the efficiency should increase to 10^{-2} for 1-mW input powers. The idea is to trap the fundamental wave completely in the resonator, where it is either absorbed or converted to a second harmonic.

A novel approach to obtaining second harmonic generation is to use two, identical parallel channel waveguides [59]. In this case there are two nondegenerate modes, one with a symmetric field distribution and the other with an antisymmetric field pattern. The difference in the propagation wave vectors can be used to implement

TABLE 9.2 Summary of Second Harmonic Generation in In-Diffused LiNbO$_3$ Waveguides

Reference	λ (μm)	Modes	l_{eff}	η (P_i)	Comments
47	1.06	TE $\to$ TE	8 mm	10^{-3} (10 W)	Used LiNbO$_3$ platelets
49	1.11	TE$_0$ $\to$ TE$_2$	$\sim$40 μm	10^{-7} (70 W)	Used ZnS films on LiNbO$_3$ substrate
135	1.06	TE$_{00}$ $\to$ TE$_{00}$	1 cm	1.5×10^{-4} (2 mW)	Channel waveguide
61	1.09	TE$_{00}$ $\to$ TM$_{00}$	2 cm	0.7×10^{-2} (60 mW)	Channel waveguide
50	1.08	TM$_0$ $\to$ TE$_1$	17 mm	0.25 (45 W)	
51	1.09	TE$_{00}$ $\to$ TM$_{00}$	17.3		Electric field tuning via electrooptic effect
55	1.06		4 mm		Examined effects of preparation conditions
136,137	1.08– 1.24	TM$_0$ $\to$ TE$_1$ TM$_7$ $\to$ TE$_8$			Studied combination of proton exchange and Ti in-diffusion on phase matching
58	1.06	$\to$ TE	24 mm	10^{-3} (1 mW)	Made resonator out of channel waveguide
54	1.06	TE$_1$ $\to$ TM$_2$	12 mm	10^{-4} (15 mW)	Room-temperature xz-plane phase match
52	1.06	TE$_{11}$ $\to$ TM$_{21}$ $\to$ TM$_{13}$ $\to$ TM$_{15}$		3×10^{-5} (2 mW)	Channel, electro-optic and temperature tuning
60	1.06	TM$_{11}$ $\to$ TE$_{13}$	5 mm		Two coupled channels, electro-optic tuning

phase matching. Maier's analysis shows that there are six possible
fundamental-harmonic combinations for this case. The problem, however, is that the overlap of the fundamental field with the harmonic
is reduced relative to the single-channel case and the effective power
density driving the harmonic is also diminished. Bozhevol'nyl et al.
[60] have verified this approach, with the addition of electro-optical
tuning to facilitate phase matching.

Parametric Mixing. Both sum and difference frequency generation has been reported by Uesugi et al. [61], Uesugi [62], and
Sohler and Suche [58,63]. They [61] used either a HeNe (1.19 μm)
or a Nd:YAG laser in conjunction with an optical parametric oscillator operating in the near infrared to produce tunable sum frequency
radiation from 0.532 to 0.545 μm. Their tuning characteristics are
shown in Fig. 9.8. In 1980, Uesugi reported the generation of a
difference frequency signal when an idler wave from a continuous
wave (CW) YAG ($\lambda \stackrel{\sim}{=} 1.32$ μm) laser and a dye laser ($\lambda \stackrel{\sim}{=} 0.58$ μm)
which served as the pump were mixed in a Ti-in-diffused LiNbO$_3$,

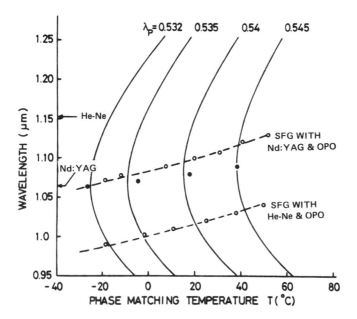

FIG. 9.8 Tuning characteristics for second harmonic generation (SHG)
and sum frequency generation (SFG) in Ti:in-diffused LiNbO$_3$ channel
waveguides. Closed and open circles are experimental data for SHG
and SFG respectively. The solid lines correspond to tuning curves.
The broken lines are theoretical curves for SFG. (From Ref. 51.)

waveguide. Using the mode combination $TE_0 + TM_0 \rightarrow TM_0$ with pump and idler powers of 70 and 1.7 mW, respectively, a generation efficiency of 0.014% was obtained at $\lambda \cong 1.035$ μm.

Sohler and coleagues have taken a somewhat different direction in studying similar phenomena. Their goal appears to be the fabrication of a parametric oscillator. Parametric gain can occur for the signal beam under appropriate conditions in a difference frequency experiment. They [63] demonstrated parametric amplification of a difference frequency signal beam at $\lambda = 1.15$ μm with a gain of up to 75% in a 32-mm-long channel waveguide. The pump laser in their experiment was a $\cong 200$-W dye laser and an idler of wavelength $\cong 1.5$ μm was generated. In more recent experiments, they obtained a maximum parametric gain of 16 dB with 150 W of pump power (pulsed to avoid damage) in a 20-μm-wide channel, 48 mm long.

Optical parametric oscillation has also been observed. It starts when the amplitudes of the signal and (or) idler waves reproduce themselves after one round-trip in a resonant structure. For large enough gains, oscillations starts from noise, and coherent radiation at the pump frequency is converted into coherent radiation at the signal and idler frequencies. This requires a very low loss waveguide with a net round-trip gain. The most recent results of Sohler and Suche [58] are shown in Fig. 9.9 for the output power and in Fig. 9.10 for the wavelengths generated.

9.3.2 Nonlinear Mixing of Contra-Directional Waves

The words "second harmonic generation" are almost synonymous with the mixing of co-directional waves. However, nonlinear polarizations are also produced by the mixing of two counterpropagating guided waves. For two waves traveling in opposite directions coupled into the waveguide at $x = 0$ and $x = L$, we write

$$a_1(x) = a_1(0)e^{-\beta_{1I}x} \qquad a_2(x) = a_2(L)e^{-\beta_{2I}(L-x)} \qquad (9.26)$$

Equation (9.20) gives the nonlinear polarization for this case with the factor of 2 retained even if the input waves are of the same frequency because they do represent different waves. When the two incident fields belong to the same mode and are multiplied together, the resulting polarization field is proportional to

$$[\exp(2i\omega t) + \exp(-2i\beta_R x)] \exp(-\beta_I L)a_1(0)a_2(L) \qquad (9.27)$$

This polarization field can only radiate normal to the waveguide surface because it has no wave vector component parallel to the surface. Plane waves are radiated into both the substrate and the

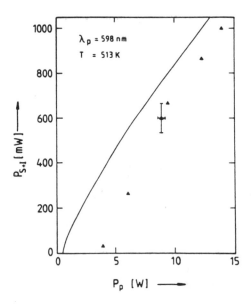

FIG. 9.9 Sum of the signal and idler powers versus pump wavelength for optical parametric oscillation in a Ti:in-diffused LiNbO₃ channel waveguides. The solid line is the theoretical curve. (From Ref. 58.)

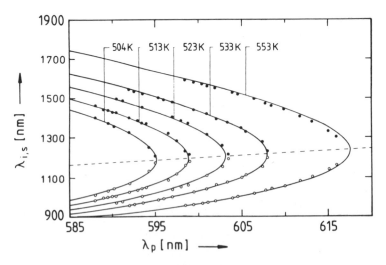

FIG. 9.10 The signal and idler wavelengths versus pump wavelength for optical parametric oscillation in a Ti:in-diffused LiNbO₃ channel waveguide at a series of temperatures. (From Ref. 58.)

cladding. The detailed analysis of this phenomenon is complicated and can be found in Normandin [64] for in-diffused LiNbO$_3$ waveguides and in Liao et al. [23] for slab waveguides. The results can be summarized in the form

$$P(2\omega) = D^{NL} \frac{L}{H} P_1 P_2 e^{-2\beta_I L} \qquad (9.28)$$

where L is the length of the overlap region and H is the width of the input beams so that the radiation comes from an area L × H. All the details of the interaction are contained in the D^{NL} term.

This process has been investigated experimentally by Normandin and Stegeman [64-68]. They used in-diffused LiNbO$_3$ waveguides and with a high-power Nd:YAG laser operating at λ = 1.06 μm they verified the existence of the radiated signal at 2ω. The measured values of D^{NL} were small, of the order of 5×10^{-14} W^{-1}, in good agreement with calculations. This nonlinear cross section was subsequently increased by a factor of 600 to $\cong 3 \times 10^{-11}$ by changing the field distributions of the guided waves. This was achieved by placing liquids with high refractive indices on the surface of LiNbO$_3$. The cross section can in principle be increased to $\cong 10^{-4}$ W^{-1} by using the new organic material MNA [43].

This phenomenon can be used to achieve the convolution of the two-input guided-wave pulses. If the temporal envelopes of the pulses are of the form $U_1(t - x/v)$ and $U_2(t + x/v)$, respectively, and if the total radiated signal is focused onto a fast detector [64,66],

$$U(2\omega,t) \propto \int |U_1(2t - \tau)U_2(\tau)|^2 \, d\tau \qquad (9.29)$$

which is the square convolution of the input pulses. This process is illustrated schematically in Fig. 9.11. If, instead, a series of detectors is placed parallel to and above the propagation surface and the signal is integrated on each over the duration of the interaction [64],

$$P(2\omega,x) \propto \int |U_1(x - x')U_2(x')|^2 \, dx' \qquad (9.30)$$

so that the convolution signal appears as a spatial distribution over the detector array. This concept was first demonstrated with a single movable detector [66] and then a CCD array was used to obtain the same signal [68]. The experimental geometry and resulting signal for 16-ps input pulses are shown in Figs. 9.12 and 9.13. The oscillation observed in the signal is due to multireflection interference effects in the multilayers which cover the sensitive areas of the CCD elements.

This interaction can also be used for other device purposes. With a δ-function input for one of the guided waves, the temporal

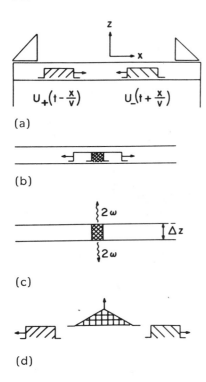

(a)

(b)

(c)

(d)

FIG. 9.11 Schematic diagram of signal convolution by the nonlinear mixing of two counter-propagating waves in an optical waveguide. (From Ref. 77.)

profile of the second pulse is displayed on the CCD array. [This can be seen easily from Eqs. (9.29) and (9.30).] Hence this device can act as an optical transient digitizer. If the two input waves are of different frequency, say ω_1 and ω_2, the sum frequency radiated into the air occurs at the angle θ from the surface normal given by

$$\sin \theta = \frac{(\beta_1 - \beta_2)c}{\omega_1 + \omega_2}$$

and hence a device based on this principle can act as a crude but very fast demultiplexer of high power radiation.

9.3.3 Parametric Mixing Spectroscopy

Nonlinear phenomena in and on optical waveguides are so efficient that it has been proposed [69] that they can be used to study

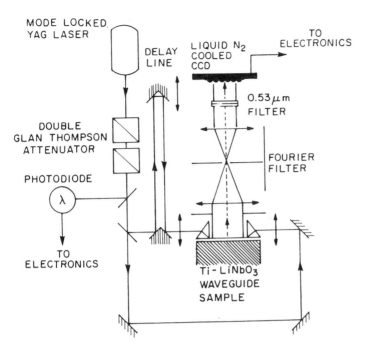

FIG. 9.12 Experimental apparatus used to demonstrate an optical transient digitizer. The laser source produced 16 picosecond pulses. (From Ref. 68.)

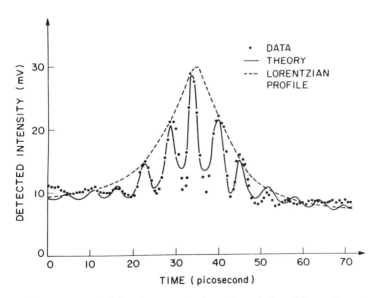

FIG. 9.13 Digitized convolution signal for 16 ps input pulses. Also shown is the computer model calculation for the response of the CCD array to a 16 ps pulse. (From Ref. 68.)

monolayers on waveguide surfaces. In the immediate vicinity of an
interface, translational symmetry is broken by the surface. For the
molecules right at this boundary, the response of the electrons to
an applied electromagnetic field is different when the field drives
the electrons toward the surface than when it drives them away from
the surface. Therefore, the molecules respond in a non-centrosymmetric fashion and nonlinear polarization fields can be created in
these molecules via an effective $x^{(2)}$. For the sum frequency case
with light incident in the near infrared and visible, the sum frequency lies in the blue and near-ultraviolet regions of the spectrum,
where typically, electronic transitions occur. Therefore, by tuning

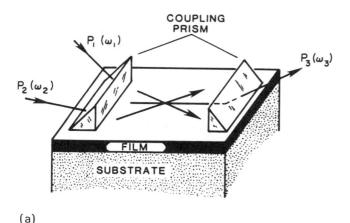

(a)

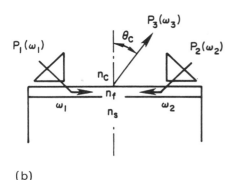

(b)

FIG. 9.14 Two guided wave geometries for parametric mixing of
light via monolayers on a film surface. (a) corresponds to the phase-
matched generation of a harmonic wave. (b) corresponds to the mixing of two guided waves to produce a sum frequency radiation field.

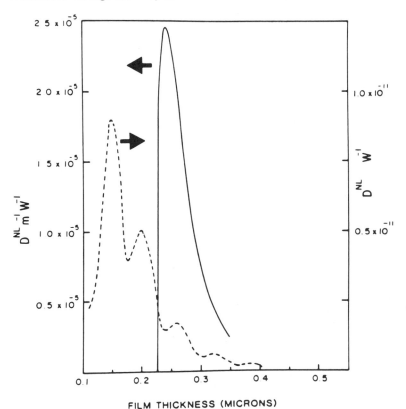

FIG. 9.15 The nonlinear cross-section D^{NL} for the sum frequency mixing of two guided waves via a monolayer on a Ta_2O_3 waveguide versus film thickness. _____ (lefthand scale) corresponds to the generation of a phase-matched harmonic wave. ----- (righthand scale) is for the generation of a sum frequency radiation field. (From Ref. 69.)

the sum frequency, one can study resonance effects via coupling to electronic transitions, that is,

$$\underline{x}^{(2)} \propto \underline{x}^{(2)}(\text{background}) + \sum_p \underline{x}_p^{(2)} \frac{\Gamma_p}{(\omega_1 + \omega_2 - \omega_p) + i\Gamma_p}$$

$$(9.31)$$

where the ω_p correspond to electronic transitions. The beauty of this technique is that only the molecules near the boundary experience the symmetry breaking, and hence in isotropic media they are the sole source of the sum frequency signal. (The gradient terms discussed in Sec. 9.2 will also contribute and will have to be taken into account.) This phenomenon has already been investigated by Chen and co-workers [70,71] on reflection from a surface.

The efficiency of this process has been calculated [69] for the two geometries shown in Fig. 9.14. The first corresponds to phase-matched sum frequency generation of a guided wave and the second to sum frequency generation of plane waves by oppositely directed guided waves. Typical nonlinear cross sections D^{NL} are given by Eqs. (9.24) and (9.28) and are shown in Fig. 9.15. As predicted by Sipe et al. [69], the signal from a monolayer can be as large as 10^{12} and 10^4 photons per second for the two geometries of Fig. 9.14, respectively, when studied with a CW dye laser. These are very large signal levels, and this approach should prove to be a very powerful tool for surface spectroscopy in the future.

9.4 THIRD-ORDER PHENOMENA

In this section we discuss phenomena that originate from nonlinear polarization terms which are third order in the incident optical fields. Although the experimental activity to date in this area has been limited to one-of-a-kind preliminary experiments, many numerical calculations have been reported predicting useful signal levels for a number of different phenomena. Since we are dealing here with fourth-rank tensors, the number of contributing tensor elements becomes large very quickly as the crystal symmetry of the medium is reduced. As a result, most of the calculations are based on isotropic media in which there are only three unique nonzero elements. Far away from any resonances, the situation is simplified further by the Kleinman symmetries, which interrelate the components in isotropic media.

The standard starting point is Eq. (9.5). For an isotropic material,

$$P_i^{NL}(\underline{r},\omega) = 2\varepsilon_0\chi_{1122}^{(3)}(-\omega;\pm\omega_3,\pm\omega_2,\pm\omega_1)E_i(\underline{r},\pm\omega_1)$$
$$\times E_j(\underline{r},\pm\omega_2)E_j(\underline{r},\pm\omega_1) + \varepsilon_0\chi_{1221}^{(3)}$$
$$\times (-\omega;\pm\omega_3,\pm\omega_2,\pm\omega_1)E_j(\underline{r},\pm\omega_1)E_j(\underline{r},\pm\omega_2)E_i(\underline{r},\pm\omega_3) \qquad (9.32)$$

Here it is understood that $\underline{E}(\underline{r},-\omega) = \underline{E}(\underline{r},\omega)^*$. This expression simplifies far from any resonant behavior in the $\chi^{(3)}$ terms. For $\omega = \omega_1 = \omega_2 = \omega_3$, $3\chi_{1122}^{(3)} = 3\chi_{1221}^{(3)} = \chi_{1111}^{(3)}$. For guided-wave fields, this expression for the nonlinear polarization is more complex than that for the plane-wave case. For example, with only one beam present at frequency ω (= ω_a) in an isotropic medium [20],

$$
P_{\gamma i}^{NL}(z) = c\varepsilon_0 n_\gamma^2 n_{2\gamma}^2 \left[\frac{2}{3} E_{\gamma i}(z) E_{\gamma j}(z) E_{\gamma j}^*(z) \right.
$$
$$
\left. + \frac{1}{3} E_{\gamma i}^*(z) E_{\gamma j}(z) E_{\gamma j}(z) \right] \tag{9.33}
$$

where the refractive index $n = n_\gamma + n_{2\gamma}S$ for the γth medium, S is the local intensity, and $n_{2\gamma}$ is the intensity-dependent refractive index. When dealing with a single-input beam we simplify the notation by dropping the ω dependence and retain only the z part of the $\underline{r}$ dependence. Note that we reserve the subscript γ = c, f, or s for identifying the medium as either the cladding, film (where appropriate), or substrate, respectively. For TE polarized waves

$$
P_{\gamma y}^{NL}(z) = c\varepsilon_0 n_\gamma^2 n_{2\gamma}^2 |E_{\gamma y}(z)|^2 E_{\gamma y}(z) \tag{9.34}
$$

just as in the plane-wave case. The situation for TM is much more complicated because there are two field components, E_x and E_z, which have a nonzero phase difference between them. The simplest case occurs for either the substrate or the cladding medium, where $E_x(z)$ and $E_z(z)$ are $\pi/2$ out of phase with one another. Evaluating Eq. (9.33) yields

$$
P_{\gamma x}^{NL} = c\varepsilon_0 n_\gamma^2 n_{2\gamma}^2 \left[|E_{\gamma x}(z)|^2 + \frac{1}{3} |E_{\gamma z}(z)|^2 \right] E_{\gamma x}(z) \tag{9.35a}
$$

$$
P_{\gamma z}^{NL} = c\varepsilon_0 n_\gamma^2 n_{2\gamma}^2 \left[|E_{\gamma z}(z)|^2 + \frac{1}{3} |E_{\gamma x}(z)|^2 \right] E_{\gamma z}(z) \tag{9.35b}
$$

Writing the dielectric displacement field as

$$
D_{\gamma i}(z) = \varepsilon_0 (n_\gamma^2 + \alpha_{ij} |E_{\gamma j}(z)|^2) E_{\gamma i}(z) \tag{9.36}
$$

and restricting the discussion to evanescent fields with either a TE or a TM wave present, we obtain

$$
\alpha_{xx} = \alpha_{zz} = \alpha_{yy} = c\varepsilon_0 n_\gamma^2 n_{2\gamma} \tag{9.37a}
$$

$$
\alpha_{xz} = \alpha_{zx} = \frac{1}{3} c\varepsilon_0 n_\gamma^2 n_{2\gamma} \tag{9.37b}
$$

Clearly, one of the consequences of the third-order nonlinearity is a field-dependent refractive index, and therefore an intensity-dependent guided-wave wave vector.

The values of $n_{2\gamma}$ vary widely both from material to material and with wavelength. Far from any resonance, $n_{2\gamma}$ varies from 10^{-16} to 10^{-21} m^2/W. Near the bandgap resonance in InSb ($\lambda = 5.5$ μm), $n_{2\gamma}$ can be as large as 10^{-7} m^2/W [72]. In the vicinity of resonance phenomena there is also additional attenuation which limits propagation distance and may render the material unsuitable for guided-wave applications, which usually require some coupling distance for waveguide excitation.

9.4.1 Degenerate Four-Wave Mixing

The classic four-wave mixing geometry is shown in Fig 9.16. There are three equal-frequency input waves and one output wave. For plane waves interacting in isotropic bulk media, phase matching of the interaction is ensured by making beams 1 and 2 exactly collinear. The key point in bulk media is that the magnitude of the wave vector is the same for all three waves. In the case of guided waves, different waveguide modes are possible, each characterized by a different value of β_m. To obtain phase matching in this case, beams 1 and 2 must be of the same polarization (TE or TM) and the same mode number m. Similarly, beams 3 and 4 must both be either $TE_{m'}$ or $TM_{m'}$ with m not necessarily equal to m'.

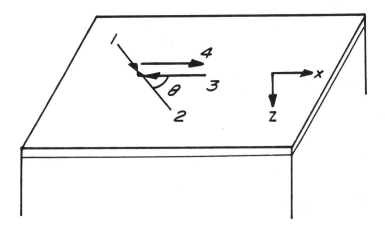

FIG. 9.16 The degenerate four wave mixing geometry. Beams 1, 2 and 3 are the input waves and 4 is the output signal. (From Ref. 74.)

A. Analysis

Assuming that beams 1 and 2 propagate along x', which is oriented at an angle θ to the x axis and that there is no pump beam depletion,

$$a_1(x') = a_1(0)e^{-\beta_{1I}x'} \tag{9.38a}$$

$$a_2(x') = a_2(L')e^{-\beta_{1I}(L'-x')} \tag{9.38b}$$

where the beams are coupled in at x' = 0 and x' = L', respectively. Furthermore, we assume that beam 3 has the form

$$a_3(x) = a_3(L)e^{-\beta_{3I}(L-x)} \tag{9.38c}$$

where beam 3 is injected at the point x = L.

The details of the solutions have been reported by Karaguleff et al. [73]. The result for the generated wave (4) leaving the interaction at x = L is

$$a_4(L) = \frac{A[e^{-\beta_{1I}L'} - e^{-(\beta_{1I}L'+2\beta_{3I}L)}]}{2\beta_{3I}} a_1(0)a_2(L')a_3(L) \tag{9.39}$$

where for all TE wave inputs,

$$A = \frac{kc^2\varepsilon_0^2}{6i} \int_{-\infty}^{\infty} n_\gamma^2(z)n_{2\gamma}(z)[(2 + \cos^2\theta)$$

$$\times |E_{m'y'}(\underline{r},\omega)|^2 |E_{my}(\underline{r},\omega)|^2] \, dz \tag{9.40}$$

Note that the nonlinearity can be in any of the media that constitute the waveguide. For all TM inputs, the terms in the brackets in Eq. (9.40) are replaced by

$$2[|E_{mx}(\underline{r},\omega_3)|^2 - |E_{mz}(\underline{r},\omega)|^2] \times [|E_{m'x'}(\underline{r},\omega)|^2$$

$$- |E_{m'z}(\underline{r},\omega)|^2] + |E_{mx}(\underline{r},\omega)E_{m'x'}(\underline{r},\omega) \cos\theta$$

$$- E_{m'z}(\underline{r},\omega)E_{mz}(\underline{r},\omega)|^2 \tag{9.41}$$

For mixed TE and TM wave geometries, the appropriate quantities
for the brackets are

$$| E_{m'y'}(\underline{r}, \omega)|^2 [\, | E_{mx}(\underline{r}, \omega)|^2 (2 + \sin^2 \theta)$$

$$- 2| E_{mz}(\underline{r}, \omega)|^2] \qquad (9.42a)$$

$$| E_{my}(\underline{r}, \omega)|^2 [\, | E_{m'x'}(\underline{r}, \omega)|^2 (2 + \sin^2 \theta)$$

$$- 2| E_{m'z}(\underline{r}, \omega)|^2] \qquad (9.42b)$$

respectively. Note that the plane-wave result is recovered in the
limit $E_x \to 0$. The minus signs appear with this term because $E_x^* = -E_x$ for a TM wave.

It is typically the beam 4 signal power that is of most interest.
Writing $a_1(0)^2 = P_1/H'$, $a_2(L')^2 = P_2/H'$, $a_3(L)^2 = P_3/H$, and $a_4(L)^2 = P_4/H$, where beams 1 and 2 are H' wide and beams 3 and 4 are H
wide,

$$P_4 = | A |^2 \left(\frac{L}{H'} \right)^2 | G |^2 P_1 P_2 P_3 \qquad (9.43)$$

where the effect of attenuation is included in

$$G = \frac{e^{-\beta_{1I}L'} - e^{-(\beta_{1I}L' + 2\beta_{3I}L)}}{2\beta_{3I}L} \qquad (9.44)$$

and $D^{NL} = | A |^2$. For $2\beta_{3I}L \ll 1$, that is, negligible attenuation
over the interaction region, $G \to 1$. For $L = L'$, $m = m'$, and all
waves either TE or TM polarized, which is the usual case, the optimum interaction distance is $L_{opt} = 1.099/2\beta_{3I}$. Under these conditions, the optimum signal power is

$$P_4 = D^{NL} \frac{1}{27H^2 \beta_I^2} P_1 P_2 P_3 \qquad (9.45)$$

One can estimate the input powers required for strong conversion of the input-to-output signals. From Eq. (9.43), this power is
given approximately by $P_1 = (D^{NL})^{-1/2}$. In this limit, however, it
is necessary to take incident beam depletion into account, which
leads to a series of coupled-mode equations between the amplitudes
of the various beams. This case has not been considered yet for
guided waves.

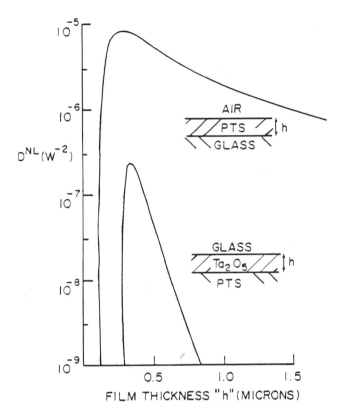

FIG. 9.17 The nonlinear cross-section D^{NL} versus film thickness for a PTS film deposited on a glass substrate, and a "thick" PTS overlayer on a Ta_2O_5 waveguide. (From Ref. 74.)

Sample calculations [74] for degenerate four-wave mixing using PTS ($\lambda \cong 1$ µm; [75]) are shown in Fig. 9.17. In every case there is a film thickness at which the interaction is optimized. For the diacetylene PTS, $\cong 100$ W of peak power is needed for strong signal conversion. For InSb [74], only $\cong 25$ µW of input power is needed.

The analysis can easily be extended to channel waveguides. In that case the parameters L, L', H, and H' are all defined by the channel waveguide dimensions and intersection angle. The principal difference will be in the term A, which will now consist of an integral over more than one dimension. One can easily construct the required form from Eq. (9.32).

Degenerate four-wave mixing is also possible with two waves guided and two waves incident from outside and through the waveguide boundaries [76]. Because two of the fields are not guided,

the power required for this process is larger than that for four
waves guided.

B. Applications

A number of applications of degenerate four-wave mixing to signal
processing have been suggested [77]. The case shown in Fig. 9.18a
corresponds to convolution. The two input beams being convolved
are 1 and 2, and beam 3 is a control beam whose presence is required to make the process possible via degenerate four-wave mixing.
For this case $\theta \cong 90°$ is preferable. If waves 1 and 2 have pulse
envelopes (in time) of the form $U_1(t)$ and $U_2(t')$, the total radiated
signal is of the form

$$U_4(t) \propto \int U_1(2t - \tau) U_2(\tau) \, d\tau \tag{9.46}$$

just as in the contra-directional SHG case.

Another example is time inversion. This case is illustrated in
Fig. 9.18b. The waveform to be inverted is coupled into a slow
mode (m') of the waveguide and a very short pulse (δ function in
time) in a fast mode (m < m') overtakes it essentially at a small

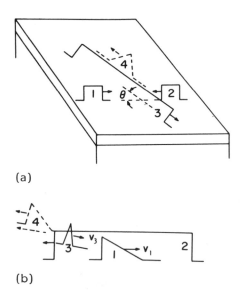

(a)

(b)

FIG. 9.18 The application of degenerate four wave mixing to (a)
convolution of pulses 1 and 2, and (b) to the time inversion of
beam 1. (From Ref. 77.)

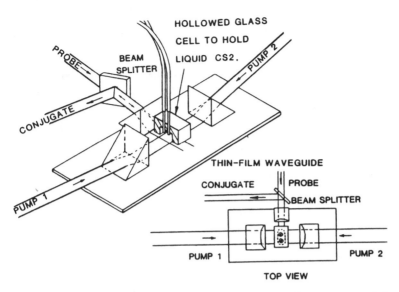

FIG. 9.19 Sample and coupling degenerate four wave mixing geometry showing the thin film waveguide (Corning 7059 sputtered onto a glass microscope slide), the arrangement of high index coupling prisms, and the glass cell for holding CS_2. The probe beam intersects the pump beams at 90° under the liquid cell. (From Ref. 73.)

angle from behind. In this case the control beam is 2 and the radiated signal is beam 4.

C. Experiment

The first experiments [73] on degenerate four-wave mixing in a planar waveguide geometry with all waves guided were carried out using liquid CS_2 as a nonlinear cladding material. Figure 9.19 shows the sample, the three coupling prisms, and the beam interaction geometry used. The waveguide consists of an $\cong 1$-μm-thick film of Corning 7059 glass deposited on a BK-7 substrate. A small quantity of CS_2 which has a large (relative to glass) third-order nonlinearity $x^{(3)} \cong 2 \times 10^{-20}$ m^2/V^2 was used as a nonlinear cover medium over the part of the waveguide where the beams intersect. Thus the evanescent tails of the guided beams in the CS_2 generate the nonlinear source polarization. The efficient coupling of three independent beams with prisms was found to be very difficult to achieve and resulted in reduced coupling efficiencies, of the order of 1%.

A passively Q-switched (pulse length $\cong 10$ ns), frequency-doubled ($\lambda = 0.53$ μm) Nd^+:YAG laser was used in the experiments.

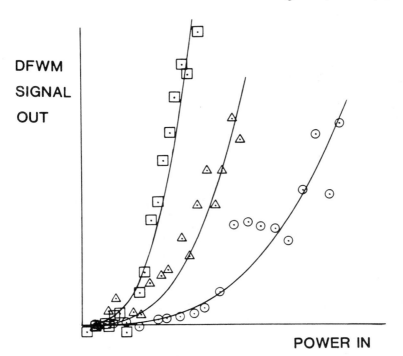

FIG. 9.20 Degenerate four wave mixing signal versus laser power for three different experiments with the background component subtracted out. The curves drawn represent the best cubic fit. (From Ref. 73.)

The results are shown in Fig. 9.20 and the cubic dependence of the DFWM signal on incident power is clear. The background signal due to stray light has been subtracted off. The signal beam reflectivity was only 10^{-9} because of the relatively low DFWM cross section obtained with CS_2 used in a cladding geometry. We expect that the use of more highly nonlinear materials such as PTS (74,75) for the film itself will lead to useful DFWM signals for all-optical signal processing.

9.4.2 Intensity-Dependent Refractive Index Phenomena

A high-power guided wave can effect its own propagation constant by virtue of its intensity. At low powers, the propagation wave vector and field distribution are calculated from the usual dispersion relations and so on. For high intensities, the local refractive index is power dependent, which affects both the dispersion relation and the field distribution. For most applications it is the change in propagation wave vector which is of interest [20,78,79] and a first-order perturbation theory calculation suffices. It is this approach that we describe first [20,78]. In Sec. F, exact solutions for the intensity-dependent refractive index problem, and their applications, are discussed.

A. Single-Waveguide Perturbation Analysis

The change in its own propagation wave vector produced by a high-power guided wave is easily calculated from Eqs. (9.33)–(9.35). For a planar waveguide it is given by

$$\frac{d}{dx}a(x) + \beta_I a(x) = -i\Delta\beta'(x)a(x) \tag{9.47}$$

$$\Delta\beta'(x) = \Delta\beta'_0 a^2(o)e^{-2\beta_I x} \tag{9.48}$$

$$\Delta\beta'_0 = \frac{\omega c \varepsilon_0^2}{4} \int_{-\infty}^{\infty} n_\gamma^2(z)n_{2\gamma}(z)\left[\frac{2}{3} |E_\gamma(z)|^2 |E_\gamma(z)|^2\right.$$
$$\left. + \frac{1}{3}(\underline{E}_\gamma(z)^* \cdot \underline{E}_\gamma(z)^*)(\underline{E}_\gamma(z) \cdot \underline{E}_\gamma(z))\right] dz \tag{9.49}$$

The net result is that the intensity-dependent wave vector is $\beta + \Delta\beta'(x)$, and hence the net phase after a propagation distance L depends on the guided-wave power as indicated by Eq. (9.48). The power-dependent phase change $\Delta\phi$ is given by

$$\Delta\phi = \int_0^L \Delta\beta'(x)\ dx \tag{9.50}$$

and the minimum required power $a_m(0)^2$ for a change $\Delta\phi$ as $L \to \infty$ is given by

$$a_m(0)^2 = \frac{2\Delta\phi\beta_I}{\Delta\beta'_0} \tag{9.51}$$

We note that for a TE mode,

$$\Delta\beta'_0 = \frac{\omega c \varepsilon_0^2}{4} \int_{-\infty}^{\infty} n_\gamma^2(z)n_{2\gamma}(z)|E(\underline{r},\omega)|^4\ dz \tag{9.52}$$

However, for TM modes,

$$\Delta\beta'_0 = \frac{\omega c \varepsilon_0^2}{4} \int_{-\infty}^{\infty} n_\gamma^2(z) n_{2\gamma}(z) \left[\frac{2}{3} |E(\underline{r},\omega)|^4 \right.$$

$$\left. + \frac{1}{3} |E_x(\underline{r},\omega)^2 + E_z(\underline{r},\omega)^2|^2 \right] dz \qquad (9.53)$$

which can be obtained correctly in the plane-wave approximation only in the limit $|E_z|^2 \gg |E_x|^2$.

For channel waveguides it is necessary to integrate Eq. (9.49) over both waveguide dimensions to obtain the appropriate value of $\Delta\beta'_0$. If the approximate formulas developed by Marcatili [17] are used, they are valid in the limit of strong field confinement in the channel, which corresponds to almost plane waves in the channel. Hence the plane-wave approximation should be valid when Marcatili's approximate formulas are used.

Sample calculations are shown in Fig. 9.21 for InSb waveguides for $\lambda \simeq 5.5$ μm, where $n_{2\gamma}$ is very large for this material. For phase changes of $\pi/2$, powers in the tens of microwatts should be sufficient for 1-mm-wide beams. Clearly, this effect can be very efficient in appropriate materials.

B. Distributed Feedback Gratings

One of the important ingredients required for producing optical bistability with guided waves is that the nonlinear medium should be placed inside a resonator. In terms of guided waves, this can easily be accomplished with distributed feedback gratings. These can be situated at both ends of the nonlinear medium to act solely as mirrors, or the grating can be fabricated in or on the nonlinear region. For the first case (i.e., reflectors at each end with the nonlinear medium between them), the analysis presented to date for the switching (minimum) power is appropriate.

The linear operational characteristics of gratings are well known (e.g., Kogelnik [8], Streifer et al. [80], and Stegeman et al. [81]). Because of its reflection properties, there are always two waves present inside a grating, which we label $a_+(x)$ and $a_-(x)$ for propagation along the +x and -x axes, respectively. For a sinusoidal surface grating centered on the plane $z = z'$ with a displacement given by $u = u_0 \sin \kappa x$,

$$i \frac{d}{dx} a_-(x) = \Gamma e^{-i\Delta\beta x} a_+(x) \qquad (9.54a)$$

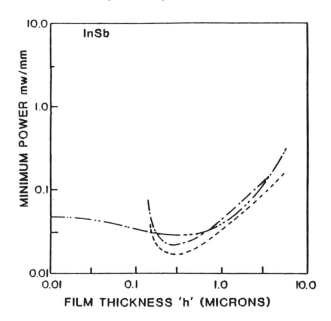

FIG. 9.21 The minimum power for a $\pi/2$ phase shift for a nonlinear InSb film at $\lambda \cong 5.5$ μm. Curves ------ and -·-··-· correspond to an InSb film on a ZnS substrate at two different wavelengths and -·-··-· corresponds to a 1000 Å InSb film on a ZnS-Al$_2$O$_3$ waveguide. (From Ref. 72.)

$$-i \frac{d}{dx} a_+(x) = \Gamma e^{i\Delta\beta x} a_-(x) \tag{9.54b}$$

where [81]

$$\Gamma = \frac{\omega\varepsilon_0}{8} u_0 [n^2(z > z') - n^2(z < z')] [\underline{E}_+(x = 0, \ z > z', \ \omega)$$

$$\times \ \underline{E}_-(x = 0, \ z < z', \ \omega)^*] \tag{9.55a}$$

for a planar waveguide, and

$$\Gamma = \frac{\omega\varepsilon_0}{8} u_0 \int_{-\infty}^{\infty} [n^2(z > z') - n^2(z < z')]$$

$$\times \ [\underline{E}_+(x = 0, \ y, \ z > z', \ \omega) \cdot \underline{E}_-(x = 0, \ y, \ z < z', \ \omega)^*] \ dy$$

$$\tag{9.55b}$$

for a channel guide [82] with

$$\Delta\beta = 2\beta - \kappa \tag{9.56}$$

Note that we have assigned different meanings to $\Delta\beta$, which in the present discussion is the grating wave vector mismatch. The parameter $\Delta\beta_0'$, first used in Eq. (9.47), is a constant describing an intensity-dependent wavevector, and $\Delta\beta'(x)$, introduced initially in Eq. (9.48), is the intensity- and position-dependent wave vector. For a linear grating L long,

$$R = \frac{a_-(0)^2}{a_+(0)^2} = \frac{(\Gamma L)^2}{\Delta\beta^2 L^2 + (\Gamma^2 - \Delta\beta^2)L^2 \coth^2[(\Gamma^2 L^2 - \Delta\beta^2 L^2)^{1/2}]} \tag{9.57a}$$

$$T = \frac{a_+(L)^2}{a_+(0)^2} = 1 - R \tag{9.57b}$$

The case of distributed feedback bistability is somewhat more complicated. It has been analyzed by Winful et al. [83] for the general case and Stegeman et al. [82] for the guided-wave case. Including both the forward and backward traveling waves, the appropriate coupled-mode equations are

$$i\frac{d}{dx} a_+(x) = \Gamma e^{-i\Delta\beta x} a_-(x) + \Delta\beta_0'[a_+(x)^2 + 2a_-(x)^2]a_+(x) \tag{9.58a}$$

$$-i\frac{d}{dx} a_-(x) = \Gamma e^{i\Delta\beta x} a_+(x) + \Delta\beta_0'[2a_+(x)^2$$

$$+ a_-(x)^2]a_-(x) \tag{9.58b}$$

These equations have been solved analytically by Winful et al. [83] and Stegeman et al. [82] in terms of an incident switching power $a_{sw}(0)^2$ given by

$$a_{sw}(0)^2 = \frac{2\beta c}{3\Delta\beta_0'\omega L} \tag{9.59}$$

This is typically the power required to obtain switching. Writing the incident and transmitted power as $I = a_+(0)^2/a_{sw}(0)^2$ and $J = a_+(L)^2/a_{sw}(0)^2$, then for $\Delta\beta = 0$ the grating transmission is given by

$$T = \frac{J}{I} = \frac{2}{1 + nd\left[2 \sqrt{\Gamma^2 L^2 + J^2}; \; (1 + J^2/\Gamma^2 L^2)^{-1}\right]} \tag{9.60}$$

Here $nd(u;m)$ is one of the tabulated Jacobian elliptic functions. As shown in Fig. 9.22, switching occurs for $a_+(0)^2 \cong a_{sw}(0)^2$, provided that the feedback parameter ΓL is of the order of unity or larger. The variation in the switching power with both ΓL and $\Delta \beta L$ was obtained by Winful et al. [83] and is shown in Fig. 9.23.

A host of other devices based on nonlinear gratings are possible [84,85]. They include optical switching, logic, amplifiers, scanners, and so on.

Sample calculations have been reported [82] for InSb ridge and buried channel waveguides in which the surface corrugation was assumed to be at the channel-air boundary. Typical results are shown in Fig. 9.24. Clearly, bistability should be possible with power levels of a few tens of nanowatts. Polydiacetylene (PTS) ridge waveguides have also been treated [82] and power levels of hundreds of milliwatts appear feasible for switching.

Optically induced gratings in waveguides have been demonstrated in semiconductors but not in the context discussed here. For example, Bykovskii et al. [86] created a dynamic grating in $GaAs_{1-x}P_x$

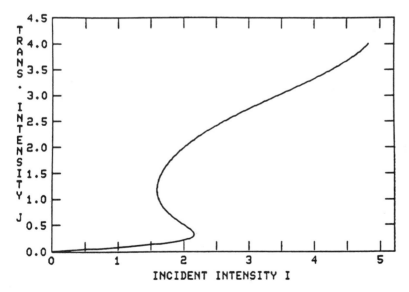

FIG. 9.22 The transmitted versus incident power for distributed feedback bistability with $\Gamma L = 2$. (From Ref. 82.)

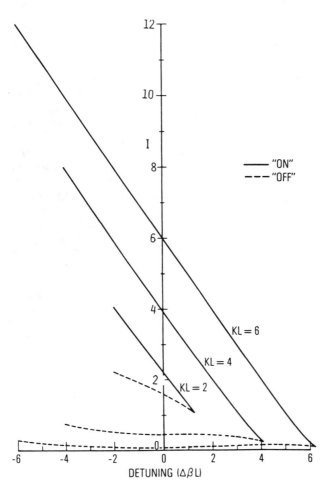

FIG. 9.23 The "on" and "off" switching powers for distributed feedback bistability as a function of initial offset $\Delta\beta L$ for a variety of feedback parameters ΓL. (From Ref. 83.)

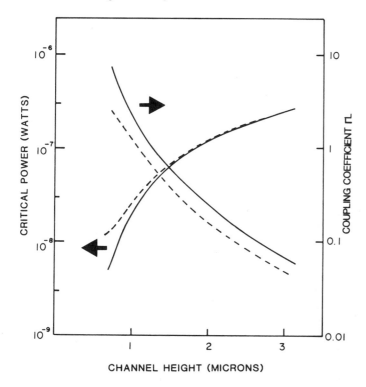

FIG. 9.24 The critical (minimum) power $a_m^2(0)$ and the feedback parameter ΓL versus the channel width for H/W = 0.4 for an InSb channel with air above and Al_2O_3 as the substrate. The solid line is for a buried waveguide. (From Ref. 116.)

via an interference pattern written into the material by a Nd:YAG laser which illuminated the surface from above. Two-photon absorption was the mechanism used and a guided wave was deflected through a small angle in analogy to the acousto-optic effect. A diffraction efficiency of 45% was obtained with incident power levels of only 0.9 MW/cm^2.

C. Nonlinear Coherent Coupler

An intensity-dependent refractive index can be used to alter the phase-matching condition between two channel waveguides [87,88]. For example, consider the channel waveguide coupler shown in Fig. 9.25. If the two waveguides are identical, the two waveguides are synchronously coupled. When light is injected into one channel,

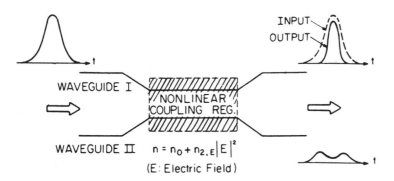

FIG. 9.25 Schematic of the two coupled channel waveguides including a nonlinear material in the coupling region. (From Ref. 90.)

the overlap of that guided-wave field with the second waveguide
results in power transfer into the second waveguide. By placing
a nonlinear material within reach of the guiding fields, the synchronous coupling condition becomes power dependent and interesting characteristics with device potential result [87–90].

This case has been treated analytically by Jensen [87,88].
Assuming identical waveguides with guided field amplitudes described
by $a_1(x)$ and $a_2(x)$, respectively, and neglecting attenuation, the
interaction can be described by the coupled wave equations

$$-i \frac{d}{dx} a_1(x) = \gamma a_2(x) + [\Delta\sigma_0 a_1(x)^2 + 2\Delta\sigma_0' a_2(x)^2] a_1(x) \quad (9.61a)$$

$$-i \frac{d}{dx} a_2(x) = \gamma a_1(x) + [\Delta\sigma_0 a_2(x)^2 + 2\Delta\sigma_0' a_1(x)^2] a_2(x) \quad (9.61b)$$

where in the plane-wave approximation

$$\Delta\sigma_0 = \frac{\omega c \varepsilon_0^2}{4} \int_{-\infty}^{\infty} \int_{-\infty}^{\infty} n_\gamma^2(y,z) n_{2\gamma}(y,z) |\underline{E}_1(\underline{r})|^4 \, dy \, dz \quad (9.62a)$$

$$\Delta\sigma_0' = \frac{\omega c \varepsilon_0^2}{4} \int_{-\infty}^{\infty} \int_{-\infty}^{\infty} n_\gamma^2(y,z) n_{2\gamma}(y,z) |\underline{E}_1(\underline{r})|^2$$

$$\times |\underline{E}_2(\underline{r})|^2 \, dy \, dz \quad (9.62b)$$

$$\gamma = \frac{\omega \varepsilon_0}{4} \int_{-\infty}^{\infty} \int_{-\infty}^{\infty} n_\gamma^2(y,z) \underline{E}_1(\underline{r}) \cdot \underline{E}_2(\underline{r})^* \, dy \, dz \qquad (9.62c)$$

Here γ is the usual linear coupling between the two channels and $\Delta\sigma_0$ and $\Delta\sigma_0'$ describe the nonlinear effects due to each guided-wave field by itself, and the overlap of the two fields, respectively. For the special case that light is incident in one channel only [i.e., $a_1(0)^2 > 0$ and $a_2(0)^2 = 0$], the analytical solution is [87,88]

$$a_1(x)^2 = a_1(0)^2 \frac{1 + cn(2x;m)}{2} \qquad (9.63)$$

where $m = a_1(0)^2/a_c(0)^2$, the critical power to which everything scales, is $a_c(0)^2 = 4\gamma/(\Delta\sigma_0 - 2\Delta\sigma_0')$ and $cn(2x;m)$ is a Jacobi elliptic function. Typical solutions are shown in Fig. 9.26 for a series of input powers and variable propagation distance. Note the variety of conditions that can be achieved, depending on the input power level. It can be used for AND and XOR gates, for pulse compression [90], for discriminators and limiters, and so on. For materials such as GaAs, the critical powers should be of the order of milliwatts.

This phenomenon has recently been demonstrated using indiffused channel waveguides in LiNbO$_3$ [91]. Although $n_{2\gamma}$ is quite small for this material, a measurable transfer, as shown in Fig. 9.27, has been observed by Haus and colleagues. Note the picosecond response of the device, which makes it very attractive for optical logic and signal processing. Lattes et al. [91] have proposed an all-optical logic gate based on this device. Its operation and the associated truth table are shown in Fig. 9.28. Although the required power levels are high for the LiNbO$_3$ case, the phenomenon has clearly been demonstrated, and future experiments based on more nonlinear materials such as quantum well structures in GaAs can be anticipated in the future.

D. Bistable Ring

It is possible to make a resonator out of a channel waveguide by making it into a ring [92,93]. As shown in Fig. 9.29, light can be coupled into the ring via evanescent fields from an adjacent input channel waveguide. The output signal is also sampled via evanescent field coupling to an output channel guide. If the material in the vicinity of the ring is characterized by a large value of $n_{2\gamma}$, the ring resonances, which appear every time the round-trip ring phase changes by 2π, can be tuned by varying the input power. Such a device can be used for tapping information from a fiber, as an analog-to-digital converter, as a logic element, and so on.

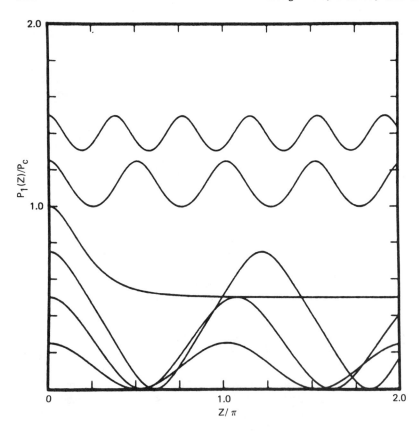

FIG. 9.26 The amount of power remaining in waveguide 1 as it propagates along a nonlinear coherent coupler. Distinctly different solutions are obtained for $a_c^2(0) > a_1^2(0)$ and for $a_1^2(0) > a_c^2(0)$. The normalized propagation distance is z/π, $P_1(z) \to a_1^2(x)$ and $P_c \to a_c^2(0)$. (From Ref. 88.)

This device was first discussed by Garmire [92]. The relation between the input (a_i^2) and output a_o^2) channel power is given by

$$a_i^2 = a_o^2 \xi [1 + \Lambda \sin^2(\theta_0 + \phi)] \tag{9.64}$$

$$\xi = (1 - R)^2 \gamma^{-4} \exp(\beta_I l) \tag{9.65a}$$

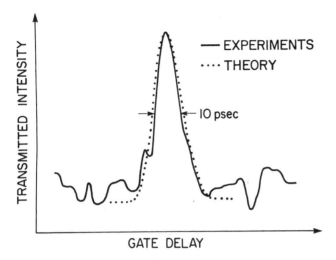

FIG. 9.27 Picosecond response of the all-optical logic gate. The experimental trace is a smoothed version of multiple experiments. The agreement with the theoretical calculation is good. (From Ref. 91.)

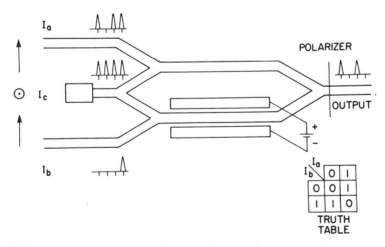

FIG. 9.28 Schematic of an optical logic gate. A continuous stream of pulses is modulated by the information carrying pulses incident in waveguides a and b. (From Ref. 91.)

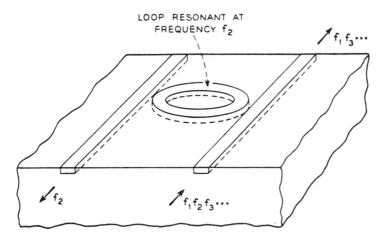

FIG. 9.29 A bistable ring waveguide with input and output coupling channel waveguides. (From Ref. 17.)

$$\Lambda = \frac{4R}{(1 - R)^2} \qquad\qquad (9.65b)$$

$$\phi = \frac{\Delta\sigma_0 l\gamma^{-2}}{2} a_o^2 \exp(\beta_I l) \qquad\qquad (9.65c)$$

where γ and $\Delta\sigma_0$ are given by Eqs. (9.62a) and (9.62c) respectively, and θ_0 is an initial offset phase term. The circumference of the ring is given by l and $R = (1 - \gamma^2) \exp(-\beta_I l)$. This device exhibits switching properties characteristic of optical bistability. Detailed calculations based on InSb have predicted switching powers of tens of microwatts, with outputs of tens of nanowatts [93]. To date no experimental verification of this device concept has been reported.

E. Power-Dependent Coupling

The coupling of an external radiation field into a guided wave is usually achieved by distributed couplers such as prisms or gratings. Phase matching is required for these devices to be efficient. In the prism case, the projection of the plane-wave wave vector in the prism onto the base of the prism must equal the guided-wave wave vector. However, it was shown above that the guided-wave wave vector changes with guided-wave power if one of the guiding media has an intensity dependent refractive index. Therefore, the phase-matching condition changes as the guided wave power grows under the prism, and for a fixed angle of incidence for the input beam, the coupling efficiency becomes power dependent and decreases [94–100]. Furthermore, the incidence angle for optimum coupling will change.

This nonlinear coupling phenomenon was first demonstrated for grating coupling to surface plasmons guided by the interface between a metal and a nonlinear medium [94,96]. The power-dependent shift in the optimum coupling angle, an example of which is shown in Fig. 9.30, coupled with a detailed analysis of the phenomenon [95], was used to measure both the sign and the magnitude of the third-order nonlinearity in semiconductors at wavelengths away from their bandgap resonances. A similar approach was used to measure the nonlinear properties of organic materials [97]. These experiments were performed by measuring the uncoupled reflected light, that is, via a standard attenuated total reflection geometry (ATR).

The variation in the guided-wave power emanating from under a coupling prism with power incident onto a nonlinear waveguide through the prism at a fixed angle has also been measured [100]. The nonlinearity was in the form of a liquid crystal filling the gap between the prism and the thin film. The result is shown in Fig. 9.31, and the decrease in coupling efficiency is clearly shown. It was also verified that the coupling efficiency could be reoptimized at a different coupling angle, and that when the nonlinear coupler was used to couple light out of a waveguide, this efficiency was independent of power.

This phenomenon has been analyzed in detail [98,99], the salient features of which are reproduced here. The coupling between the guided-wave and incident fields is described by [101]

$$\frac{\partial a(x)}{\partial x} = \hat{t} a_{inc}(x) e^{i\Delta\beta(|a[x]|^2)x} - (l^{-1} + \alpha)a(x) \tag{9.66}$$

with

$$\Delta\beta = \frac{\omega}{c} n_p \sin\theta - \beta(|a[x]|^2) \tag{9.67}$$

where $\hat{t}$ is a transfer coefficient, α the waveguide attenuation coefficient, and l the characteristic distance for reradiation of the guided wave field into the prism. Assuming that the guided-wave power remains small enough so that Eq. (9.48) is valid,

$$\beta(|a[x]|^2) = \beta_0 + \Delta\beta_0'|a(x)|^2 \tag{9.68}$$

The field incident onto the prism/waveguide interface at an angle θ to the normal has a field distribution at the base of the prism

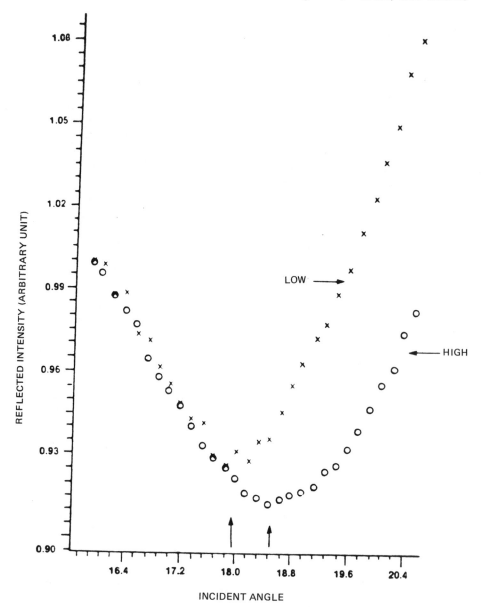

FIG. 9.30 Angular scans of the radiation reflected from a corrugated silver metal surface overcoated with Si. The curves marked high and low correspond to input energies of 70 μJ and 10 μJ per pulse at a wavelength of 1.06 μm. The arrows mark the optimum excitation angles which correspond to the excitation of surface plasmons at the silver-silicon interface. (From Ref. 94.)

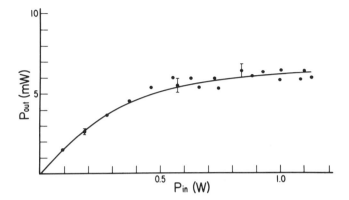

FIG. 9.31 The power prism-coupled into the nonlinear waveguide versus incident power for the TE_1 wave. The liquid crystal MBBA (n_c = 1.55, n_{2c} = 10^{-9} m^2/W) was the nonlinear cladding, the film was 1.7 μm Corning 7059 glass (n_f = 1.57) on a pyrex substrate. (From Ref. 100.)

(refractive index n_p) typically given by $a_{inc}(x) \propto \exp(-x^2/w_0^2)$. assuming that $\Delta\beta$ can be set to zero at low powers by adjusting the coupling conditions, $\Delta\beta = \Delta\beta_0' |a(x)|^2$. Therefore, as the guided-wave power increases, the synchronous coupling condition $\Delta\beta = 0$ is no longer valid, phase mismatch occurs between the incident and guided-wave fields, and the coupling efficiency is reduced. The power levels at which nonlinear coupling should occur can be estimated from the phase-mismatch term in Eq. (9.68). The maximum phase shift is

$$\Delta\phi \cong \Delta\beta_0' \int_{-\infty}^{\infty} |a(x)|^2 \, dx \cong 2w_0 \eta \, \Delta\beta_0' P_{inc} \qquad (9.69)$$

where P_{inc} is the incident power per unit length (along the y axis) in the prism and η is the optimized low-power coupling efficiency. For phase shifts of the order of $\pi/2$ or more, the coupling efficiency becomes power dependent. Readjusting the incidence angle can improve the coupling efficiency at a given power level. Because this effect is cumulative with distance, the efficiency is always less than that obtained at low powers. An example of how the optimized coupling efficiency varies with incident power is shown in Fig. 9.32. Note that the beam profile in the plane of the surface is distorted [99].

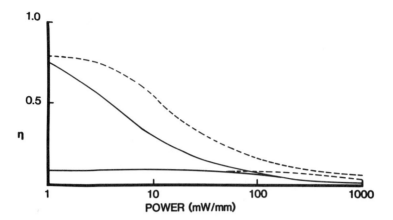

FIG. 9.32 Coupling efficiency η versus incident power for a variety of angle tuning conditions and material parameters. The solid lines are for zero detuning with no waveguide losses (upper curve) and with $\alpha = 100$ cm^{-1} (lower curve). The dashed lines are obtained when the angle tuning is adjusted to give optimum efficiency at each power with the lossless waveguide case given by the upper curve and the $\alpha = 100$ cm^{-1} case corresponding to the lower curve. (From Ref. 95.)

Some calculations [102,103] which predict bistability in the coupling process have been reported on surface plasmon waves excited in a prism geometry. Such a phenomenon has not yet been observed. It was assumed in these calculations that the beam cross section is infinite, and in view of the growing nature of the guided wave and subsequent mismatch with distance discussed above, it is not clear whether bistability can be observed with finite cross-section beams in the coupling process.

F.　Power–Dependent Waves

The effect of an intensity-dependent refractive index on the propagation properties of a guided wave can either be calculated approximately using coupled-mode theory as discussed in Sec. A, which assumes that the field distribution is not affected by the nonlinearity, or exactly by solving the nonlinear wave equation and boundary conditions. A coupled-mode-theory approach can always be used when the maximum change in index that can be produced optically, $\Delta n_{sat} \geqslant n_{2\gamma}S$, is much less than the index differences $n_f - n_c$ or $n_f - n_s$ which exist at low powers between the film and the bounding media. Otherwise, the field distributions are also affected by the

high optical intensities and the more exact theory outlined here must be used.

A great deal of progress has been made recently in solving the nonlinear wave equation [104,105] as applied to thin-film waveguides [106-119]. The solutions for the TE case are exact; the appropriate formulation for the TM case is still under development [118, 120]. Assuming a TE wave, the nonlinear wave equation for $E_{\gamma y}(z)a(0)$ in the absence of attenuation so that $a(x)$ is a constant (γth medium) is

$$\nabla^2 E_{\gamma y}(z) + \frac{\omega^2}{c^2}[n_\gamma^2 + \alpha_{\gamma yy} E_{\gamma y}(z)^2 a(0)^2]E_{\gamma y}(z) = 0 \qquad (9.70)$$

The solutions to this equation are by now well known [104,105]. For example, in a nonlinear cladding ($z < 0$) with $\alpha_{\gamma yy} = \alpha_\gamma$ and $k_0 = \omega/c$,

$$E_{cy}(z)a(0) = \sqrt{\frac{2}{\alpha_c}} \frac{q}{\cosh[k_0 q(z_c - z)]} \qquad (9.71a)$$

$$E_{cy}(z)a(0) = \sqrt{\frac{2}{|\alpha_c|}} \frac{q}{\sinh[k_0 q(z_c - z)]} \qquad (9.71b)$$

for $n_{2c} > 0$ (self-focusing medium) and $n_{2c} < 0$ (self-defocusing medium), respectively, with $\alpha_\gamma \cong n_\gamma^2 c \varepsilon_0 n_{2\gamma}$. Here $q^2 = \beta^2 - n_c^2$ and z_c is a constant that depends on the total guided-wave power unit distance along the wavefront. (Note that we define β the effective index for this section so that the guided-wave wave is βk_0.) Because of this power dependence of z_γ, the field tributions also vary with guided-wave power. In the gener both the film and substrate can also be nonlinear. For b ing media nonlinear, the dispersion relation is

$$\tan(k_0 \kappa h) = \frac{\kappa[q \tanh^{\pm 1}(k_0 q z_c) + s \tanh^{\pm 1}(k_0 s z_s)]}{\kappa^2 - q \tanh^{\pm 1}(k_0 q z_c) s \tanh^{\pm 1}(k_0 s)}$$

for $\beta < n_f$, where $\kappa^2 = |n_f^2 - \beta^2|$, $s^2 = \beta^2 - n_s^2$ correspond to $n_{2\gamma} > 0$ and $n_{2\gamma} < 0$, respectively features of nonlinear guided waves is that a limi tions also exist for $n_f > \beta$, in which case $\tan$ and $\kappa^2 \to -\kappa^2$ in Eq. (9.72). Note that the li ponds to $z_\gamma \to \infty$, in which limit one recovers

relationships of integrated optics. The guided-wave power per unit length along the y axis is obtained in the usual way by integrating the Poynting vector over the depth dimension, that is,

$$P = \int_{-\infty}^{\infty} \underline{E} \times \underline{H} \; dz = P_c + P_f + P_s$$

The solutions for nonlinear films that involve Jacobi elliptic integrals have been obtained [107,108,121], and have been evaluated numerically in just one case to date [107]. No device possibilities have been predicted to date for the nonlinear thin-film case.

These dispersion relations predict very interesting behavior for waves guided by thin dielectric films. In this case there are always low-power solutions for some of the waves, the usual integrated optics modes. If one of the bounding media has a self-defocusing nonlinearity (e.g., $n_{2c} < 0$), there is an upper limit to the power that can be transmitted by the waveguide for each mode (if $n_c > n_s$)

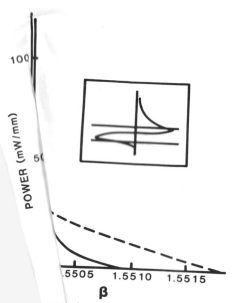

FIG. 31 guided wave power versus effective index β. Here $n_c = 1.57$ and $n_{2c} = -10^{-9}$ m^2/W. For the dashed line h = 1.6 = 1.45. For the solid line $n_s = 1.55$ and h = 1.2 µm. n b he film thickness is just above the cut-off thickness. From

[117,119]. An example is shown in Fig. 9.33. Note that at cutoff, the field distribution does not degenerate into a plane wave which extends to infinity as it does in the power-independent case. If $n_s > n_c$ (and $n_{2c} < 0$), no limiting action occurs. If both bounding media have $n_{2\gamma} < 0$, there is always an upper limit to the power that can be transmitted. This phenomenon is clearly applicable to optical limiters.

This situation is even more interesting for self-focusing bounding media. If the cladding has $n_{2c} > 0$ and $n_{2s} = 0$, there is an upper limit to the power that can be transmitted for TE_m waves, $m \geqslant 1$ [119], and at high powers the TE_0 mode degenerates into a nonlinear surface wave guided by the film-cladding interface [106,116,119]. An example is shown in Fig. 9.34. Of potential device interest are waveguides whose thickness is below the low-power cutoff for the TE_0 mode [85]. The power at which these waveguides begin to transmit can be controlled by varying the film thickness; that is, they can be used as controllable lower-threshold optical devices (see Fig. 9.35). For two nonlinear, self-focusing bounding media, numerous new solutions are obtained, all with power thresholds. Examples are shown in Fig. 9.36. There are two waves with TE_0 properties and three for TE_m for $m \geqslant 0$. This multiplicity of solutions arises because self-focusing can occur in either of the nonlinear media or in both [106]. Of most interest here are the TE_0 branches, which, depending on appropriate choices for the material parameters, can take the form shown in Fig. 9.37. In principle it may be possible to obtain optical switching between the two branches. Note also the existence of higher-order waves (TE_2 in Fig. 9.36) which exist only over a specific power range if the film thickness is below their low-power cutoff value.

The existence of such nonlinear guided waves has been verified experimentally using a nonlinear liquid-crystal medium [122] on top of a glass waveguide. In separate experiments the TE_0 and TE_1 modes were coupled into the linear waveguide region with strontium titanate prisms, were guided into, through, and out off the region with the liquid crystal on top, and then were coupled out with a second strontium titanate prism, as shown in Fig. 9.38. The results are shown in Fig. 9.39. For the TE_0 case, the transmitted power is linear in the incident power, with some saturation effects evident at the highest power levels investigated. On the other hand, in the TE_1 case there is a pronounced saturation effect, as well as hysteresis with respect to increasing versus decreasing incident power. Increasing the power initially moves the transmission along the upper branch. When the maximum power transmission point is reached, the transmitted power saturates, as indicated in Fig. 9.39. Now as the guided-wave power is decreased, waves on both the upper and lower branches are excited and hence the net transmission drops. Because the field becomes progressively more localized in

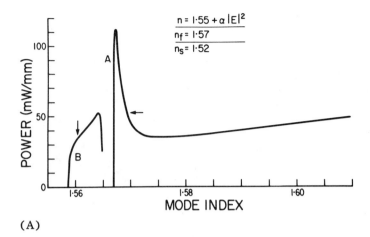

(A)

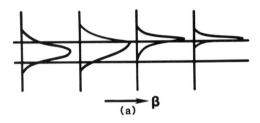

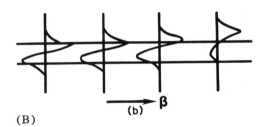

(B)

FIG. 9.34   (A) The guided wave power versus the effective mode
index for TE_0 (a) and TE_1 (b) waves guided by a 2.0 μm film of
Corning 7059 glass on a soda lime glass substrate (Ref. 116).   The
nonlinear refractive index for the cladding is $n_{2c} = 10^{-9}$ m^2/W.
(B) Nonlinear guides wave field distributions with increasing effective index β for TE_0 (a) and TE_1 (b).   (From Ref. 119.)

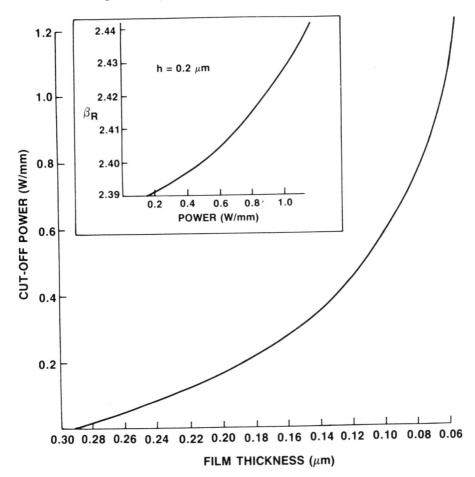

FIG. 9.35 The cut-off power versus film thickness for the TE_0 wave for a nonlinear self-focussing ZnS cladding medium ($n_{2c} = 3 \times 10^{-11}$ m²/W, $n_c = 2.39$) with some composite film characterized by $n_f = 2.40$ (for example containing $SrTiO_3$), and substrate $n_s = 2.38$. The inset shows the variation in effective index β with guided wave power for a film thickness of 0.2 µm. At low powers, the TE_0 mode for this structure is cut-off at a film thickness of $\cong 0.30$ µm. (From Ref. 85.)

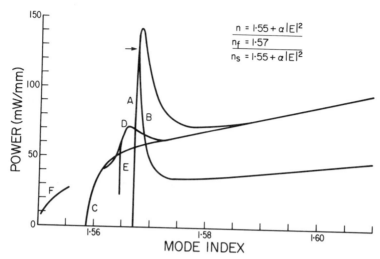

FIG. 9.36 The mode index β versus guided wave power for TE_0-like waves (A,B), TE_1-like waves (C,D,E) and TE_2-like waves (F) for a glass film with $n_f = 1.57$ with identical bounding nonlinear media ($n_c = n_s = 1.55$ and $n_{2c} = n_{2S} = 10^{-9}$ m^2/W). (From Ref. 116.)

the liquid crystal as power is decreased along the lower branch, progressively less of this mode is excited at the liquid crystal/air boundary as the prism launched TE_1 wave encounters the liquid-crystal bead. This results in the transmission curve approaching the linear transmission curve.

These first experiments verified the basic structure of the non-linear guided waves. Note, however, that it has been explicitly assumed that no self-focusing (or defocusing) occurs in the plane of the film; that is, the width of the beam has been assumed large enough to be considered a plane wave over the propagation distance of the waveguide. Self-focusing has not been analyzed for thin-film waveguides, although some results are available for surface plasmons [123,124]. Note also that the prospects of soliton propagation have also been considered [125,126]. We expect further progress in the near future in the area of upper- and lower-power-limiting devices, and perhaps even applications to optical switching.

9.4.3 Coherent Anti-stokes Raman Scattering: CARS

One of the most unusual applications of nonlinear integrated optics is to vibrational Raman spectroscopy. It is made possible by virtue

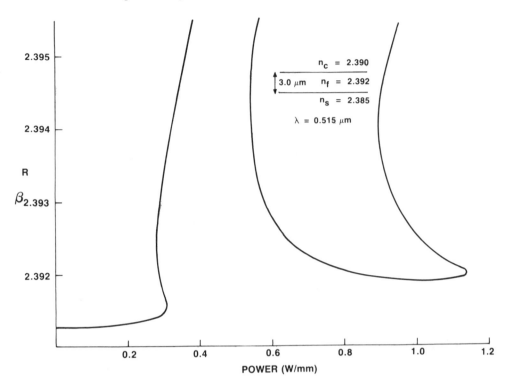

FIG. 9.37 The effective index β versus guided wave power for TE_0 waves assuming both nonlinear ZnS cladding and substrate media, but with different parameters. Here $n_f = 2.392$, $n_c = 2.390$, $n_{2c} = n_{2s} = 3 \times 10^{-11}$ m^2/W and $n_s = 2.385$ (different evaporation or sputtering conditions). (From Ref. 85.)

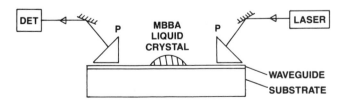

FIG. 9.38 The experimental geometry for demonstrating nonlinear guided waves with a nonlinear cladding medium.

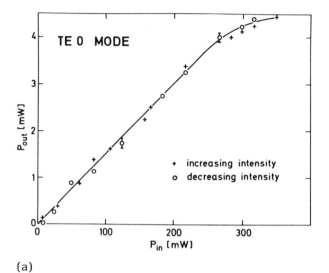

(a)

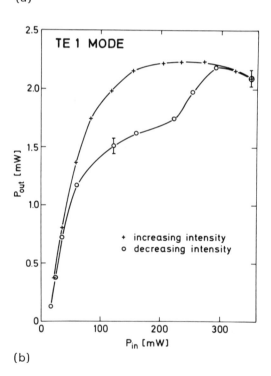

(b)

FIG. 9.39 The guided wave power for (a) TE_0 and (b) TE_1 waves transmitted through a waveguide with a nonlinear cladding. The waveguide consisted of the liquid crystal MBBA as the cladding ($n_c = 1.55$, $n_{2c} \cong 10^{-9}$ m^2/W), a 1.0 μm film of borosilicate glass ($n_f = 1.61$) and a soda lime glass substrate ($n_s = 1.52$). (From Ref. 122.)

of the structure of the third-order susceptibility $\chi^{(3)}$. It contains resonance terms which include vibrational resonances. As a result, when a difference frequency generated by two incident light fields is approximately equal to a vibrational frequency, the nonlinear susceptibility is resonantly enhanced. Thus by tuning this frequency difference, the Raman spectrum of the species sampled by the waveguide fields can be obtained.

The CARS process works approximately as follows. Two optical fields at the frequencies ω_1 and ω_2 ($\omega_1 \geqslant \omega_2$) are incident onto a molecule and excite vibrational motions at the difference frequency $\omega_1 - \omega_2$. If $\omega_1 - \omega_2 \cong \omega_p$, the vibration is resonantly excited. This vibration then interacts with another photon from the ω_1 field via a normal Raman scattering process and the scattered field appears at the frequency $2\omega_1 - \omega_2$. In terms of the guided-wave wave vectors, the nonlinear polarization field occurs at $2\underline{\beta}_1 - \underline{\beta}_2$. If this wave vector is equal to that of a guided wave of frequency $2\omega_1 - \omega_2$ (i.e., $\underline{\beta}_3 = 2\underline{\beta}_1 - \underline{\beta}_2$), the process is phase matched and the signal beam amplitude grows linearly with interaction distance.

The detailed analysis of this phenomenon has already been reported [127]. The appropriate third-order susceptibility has the form

$$\underline{\chi}^{(3)}(-2\omega_1 + \omega_2; -\omega_2, \omega_1, \omega_1) = \underline{\chi}_b^{(3)} + \sum_p \underline{\chi}_p^{(3)}$$

$$\times \frac{\Gamma_p}{(\omega_1 - \omega_2 - \omega_p) + i\Gamma_p} \qquad (9.73)$$

where $\underline{\chi}_b^{(3)}$ is a background electronic term and the vibrational transitions are characterized by $\chi_p^{(3)}$, a frequency ω_p, and lifetime Γ_p^{-1}. Typically, the individual resonance terms can be up to 40 times that of the background terms. For example, for the 992-cm^{-1} transition in benzene, $\chi_p^{(3)} \cong 1.7 \times 10^{-20}$ m^2/V^2 and $\chi_b^{(3)} \cong 8 \times 10^{-22}$ m^2/V^2. Assuming that the ω_1 and ω_2 beams are injected at $x = 0$ and that the beam powers are $P_1 = a_1(0)^2 H$, $P_2 = a_2(0)^2 H$, and $P_3 = a_3(0)^2 H$, it has been shown for the geometry in Fig. 9.40 that

$$P_3 = D^{NL} \left[\frac{L}{H}\right]^2 |G|^2 P_1^2 P_2 \qquad (9.74)$$

$$D^{NL} = \frac{\omega \varepsilon_0}{4} |F|^2 \qquad (9.75a)$$

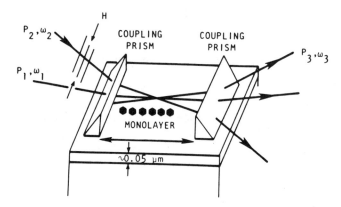

FIG. 9.40 CARS geometry, complete with coupling prisms (after Stegeman 1983a).

$$F = \int_{-\infty}^{\infty} \chi^{(3)}{}_{ijkl}(-2\omega_1 + \omega_2, -\omega_2, \omega_1, \omega_1)E_i(z, 2\omega_1 - \omega_2)^* \quad (9.75b)$$

$$\times \; E_j(z, \omega_1)E_k(z, \omega_1)E_l(z, \omega_2)^* \; dz$$

$$|G|^2 = \frac{e^{-2\beta_{3I}L} + e^{-2[2\beta_{1I} + \beta_{2I}]L} - 2\cos(\Delta\beta'L)e^{-[2\beta_{1I} + \beta_{2I} + \beta_{3I}]L}}{[\Delta\beta'^2 + (2\beta_{1I} + \beta_{2I})^2]L^2}$$

$$(9.75c)$$

$$\Delta\beta' = (\underline{\beta}_{3R} - 2\underline{\beta}_{2R} - \underline{\beta}_{1R}) \cdot \underline{L} \qquad (9.75d)$$

where $|\underline{L}|$ is the interaction distance.

Calculations have been made for two sample cases [127]. Shown in Fig. 9.41 is the nonlinear cross section D^{NL} and the fraction of incident energy converted into the CARS signal for a polystyrene waveguide on glass. For incident power levels in the waveguide of $\cong 100$ MW/cm^2, which requires sub-millijoule energies in 15-ns pulses, conversion efficiencies of fractions of a percent are predicted for $L \cong 1$ cm on the peak of the 992-cm^{-1} benzene ring resonance. If, instead, one considers a monolayer of benzene molecules placed on the top of a waveguide, very large signal levels are predicted, as shown in Fig. 9.42. The signal levels predicted are enormous by surface spectroscopy standards. Also note that the background term

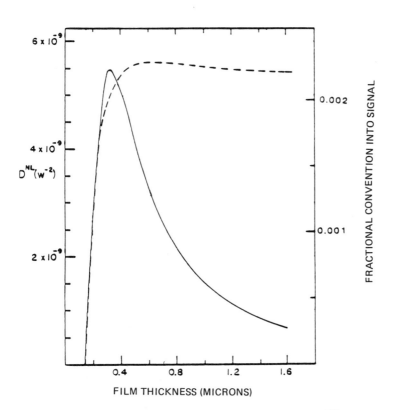

FIG. 9.41 The nonlinear CARS cross-section D^{NL} (———) and the fraction of beam ω_2 converted into a CARS signal (------) as a function of polystyrene film thickness. The assumed power density is 100 MW/cm^2 and the coherence length $\cong$ 1 cm. (From Ref. 129.)

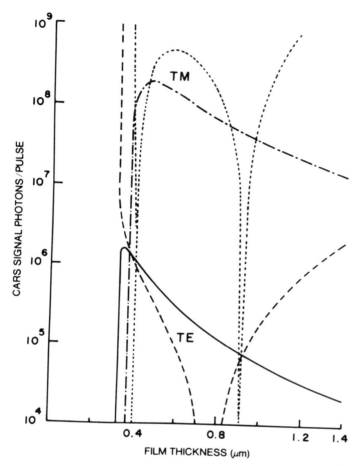

FIG. 9.42 The resonant and background CARS signals in photons/ pulse from a benzene monolayer on a Nb_2O_3-SiO_2 waveguide versus Nb_2O_3 film thickness. The power density assumed is 100 MW/cm^2. For TE waves: (———) resonant CARS signal; (------) background CARS signal. For TM waves: (-·--·-·) resonant CARS signal; (-- -- --) background CARS signal. The mode combinations used were $TE_2(\omega_1)$, $TE_2(\omega_2)$ and $TE_2(\omega_3)$; and $TM_2(\omega_1)$, $TM_1(\omega_2)$ and $TM_2(\omega_3)$. (From Ref. 129.)

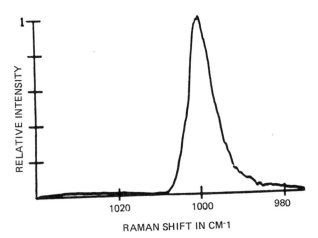

FIG. 9.43 Experimental CARS signal from a polystyrene waveguide. The peak corresponds to 0.2% conversion efficiency. (From Ref. 129.)

due to the film can be reduced substantially, if not eliminated, by an appropriate choice of modes.

The large-signal levels associated with this phenomenon have recently been verified experimentally [128,129]. The signal obtained from a 2-μm-thick polystyrene film is shown in Fig. 9.43 using pulses from two tunable dye lasers with pulse widths of $\simeq$ 100 ps. The conversion efficiency of 0.2% agrees very well with theoretical predictions. This process has the potential to turn into a very powerful diagnostic technique in surface science.

9.5 SUMMARY

In this chapter we have reviewed the field of nonlinear integrated optics. Second harmonic generation utilizing guided waves is now a very mature field. The current thrust into Ti-in-diffused $LiNbO_3$ channel waveguides will soon culminate in practical devices in which 10% conversion efficiencies with 10-mW input powers should be possible in matched resonator geometries. Another device will most surely be a parametric oscillator along the lines currently being developed by Sohler and colleagues. There also exists the possibility that highly nonlinear organic materials can be incorporated into Langmuir-Blodgett films [77,130]. This would lead to very efficient harmonic generation in nonlinear films which can be fabricated one monolayer at a time. If this type of molecular engineering becomes a reality, it would bring a large impetus to the field. For example, with such materials 10% conversion efficiencies might be feasible with sub-milliwatt input powers. Another application would be to an optical transient digitizer with 1-mW output powers produced by 10-W input pulses.

Nonlinear devices based on third-order nonlinearities should be an area of explosive growth in the next few years. One can expect it to be very important in the area of optical bistability because it offers the promise of the lowest-possible powers for achieving devices. Along the same lines, devices based on nonlinear coherent couplers and nonlinear gratings can be used for a large variety of operations in signal processing and logic. The recent observation of the large n_2 in quantum well structures [131,132] promises very efficient devices with materials compatible with guided-wave application. Another large impetus in the materials area is the development of polydiacetylene films [75,133] which are characterized by large, sub-picosecond nonlinearities. In addition, one can expect in the near future the application of degenerate four-wave mixing to picosecond serial signal processing.

Another exciting new thrust into surface spectroscopy—surface coherent Raman scattering and surface parametric mixing—can be expected. Its goal is not devices but rather, basic science. The signal levels for studying both the vibrational and electronic properties of molecules in and on thin films are expected to be enormous by surface spectroscopy standards. We expect the next decade to be a very exciting one in the field of nonlinear integrated optics.

ACKNOWLEDGMENTS

This research was supported by the Army Research Office and the Air Force Office of Scientific Research under the JSOP program and by the National Science Foundation (ECS-8117483, ECS-8304749, and DMR-8300599).

REFERENCES

1. S. Zemon, R. R. Alfano, S. L. Shapiro, and E. Conwell, *Appl. Phys. Lett.*, *21*:327 (1972).

2. R. Normandin, V. C.-Y. So, G. A. Teh, and G. I. Stegeman, *Appl. Phys. Lett.*, *34*:200 (1979).

3. D. B. Anderson and J. T. Boyd, *Appl. Phys. Lett.*, *19*:266 (1971).

4. R. Stolen, *Proc. IEEE, 68*:1232 (1980).

5. R. Stolen, *J. Fiber Integ. Opt., 3*:21 (1980).

6. R. Stolen, Fiber raman lasers, *Fiber and Integrated Optics* (D. B. Ostrowsky, ed.), Penum Press, New York, p. 157 (1980).

7. D. Marcuse, *Theory of Dielectric Optical Waveguides*, Academic Press, New York (1974).

8. H. Kogelnik, *Topics in Applied Physics, Vol. 7, Integrated Optics* (T. Tamir, ed.), Springer-Verlag, Berlin, p. 66 (1975).

9. S. Singh, *Handbook of Lasers with Selected Data on Optical Technology* (R. J. Pressley, ed.), Chemical Rubber Company, Cleveland, Ohio, p. 489 (1971).

10. B. F. Levine, C. G. Bethea, and R. A. Logan, *Appl. Phys. Lett., 26*: 375 (1979).

11. C. Cassidy, J. M. Halbout, W. Donaldson, and C. L. Tang, *Opt. Commun., 29*:1176 (1979).

12. K. Kato, *IEEE J. Quantum Electron., QE-16*:1288 (1980).

13. J. M. Halbout, S. Blit, and C. L. Tang, *IEEE J. Quantum Electron., QE-17*:513 (1981).

14. S. J. Lalama, K. D. Singer, A. F. Garito, and K. N. Desai, *Appl. Phys. Lett., 39*: 940 (1981).

15. N. Bloembergen, R. K. Chang, S. S. Jha, and C. H. Lee, *Phys. Rev., 174*: 813 (1968).

16. J. E. Sipe and G. I. Stegeman, *J. Opt. Soc. Am., 69*:1676 (1979).

17. E. A. J. Marcatili, *Bell Syst. Tech. J., 48*:2071 (1969).

18. A. Kumar, K. Thyagarajan, and A. K. Ghatak, *Opt. Lett., 8*: 63 (1983).

19. J. E. Goell, *Bell Syst. Tech. J., 48*:2133 (1969).

20. G. I. Stegeman, *IEEE J. Quantum Electron., QE-18*:1610 (1982).

21. V. C.-Y. So, R. Normandin, and G. I. Stegeman, *J. Opt. Soc. Am., 69*:1166 (1979).

22. R. Normandin, V. C.-Y. So, N. Rowell, and G. I. Stegeman, *J. Opt. Soc. Am., 69*:1153 (1979).

23. C. Liao, P. Bundman, and G. I. Stegeman, *J. Appl. Phys., 54*: 6213 (1983).

24. Y. Suematsu, *Jpn. J. Appl. Phys.*, *9*:798 (1970).

25. J. T. Boyd, *IEEE J. Quantum Electron.*, *QE-8*:788 (1972).

26. E. M. Conwell, *IEEE J. Quantum Electron.*, *QE-9*:867 (1973).

27. P. K. Tien, R. Ulrich, and R. J. Martin, *Appl. Phys. Lett.*, *17*:447 (1970).

28. Y. Suematsu, Y. Sasaki, and K. Shibata, *Appl. Phys. Lett.*, *23*:137 (1973).

29. W. K. Burns and A. B. Lee, *Appl. Phys. Lett.*, *24*:222 (1974).

30. B-U. Chen, C. L. Tang, and J. M. Telle, *Appl. Phys. Lett.*, *25*:495 (1974).

31. B-U. Chen and C. L. Tang, *IEEE J. Quantum Electron.*, *QE-11*:177 (1975).

32. B-U. Chen, C. C. Ghizoni, and C. L. Tang, *Appl. Phys. Lett.*, *28*:651 (1976).

33. H. Ito and H. Inaba, *Opt. Lett.*, *2*:139 (1978).

34. J. P. van der Ziel, R. C. Miller, R. A. Logan, W. A. Nordland, Jr., and R. M. Mikulyak, *Appl. Phys. Lett.*, *25*:238 (1974).

35. J. P. van der Ziel, M. Ilegems, P. W. Foy, and R. M. Mikulyak, *Appl. Phys. Lett.*, *29*:775 (1976).

36. B. F. Levine, C. G. Bethea, and R. A. Logan, *Appl. Phys. Lett.*, *26*:375 (1979).

37. J. Stone, C. A. Burrus, and R. D. Standley, *J. Appl. Phys.*, *50*:7906 (1979).

38. T. Shiosaki, S. Fukuda, K. Sakai, H. Kuroda, and A. Kawabata, *Jpn. J. Appl. Phys.*, *19*:2391 (1980).

39. G. H. Hewig and K. Jain, *Opt. Commun.*, *47*:347 (1983).

40. K. Sasaki, T. Kinoshita, and N. Karasawa, *Appl. Phys. Lett.*, *45*:333 (1984).

41. J. C. Peuzin, M. Olivier, R. Cuchet, and A. Chenevas-Paule, *Proc. SPIE*, *400*:136 (1983).

42. W. K. Burns and R. A. Andrews, *Appl. Phys. Lett.*, *22*:143 (1973).

43. G. I. Stegeman and C. Liao, *Appl. Opt.*, *22*:2518 (1983).

44. D. Sarid, *Appl. Phys. Lett.*, *37*:117 (1980).

45. V. V. Malov, A. V. Turovtsev, and L. V. Iogansen, *Sov. Phys. Tech. Phys.*, *28*:174 (1983).

46. D. E. Thompson and P. D. Coleman, *IEEE Trans. Microwave Theory Tech. MTT-22*:995 (1974).

47. A. T. Reutov and P. P. Tarashchenko, *Opt. Spectrosc. 37*:447 (1974).

48. H. Ito, N. Uesugi, and H. Inaba, *Appl. Phys. Lett.*, *25*:385 (1974).

49. H. Ito and H. Inaba, *Opt. Commun.*, *15*:104 (1975).

50. W. Sohler and H. Suche, *Appl. Phys. Lett.*, *33*:518 (1978).

51. N. Uesugi, K. Kaikoku, and K. Kubota, *Appl. Phys. Lett.*, *34*:60 (1979).

52. E. M. Zolotov, A. M. Prokhorov, and V. A. Chernykh, *Sov. Tech. Phys. Lett.*, *7*:129 (1981).

53. K. S. Buritsky, E. M. Zolotov, A. M. Prokhorov, and V. A. Chernykh, 1982, "Study of Second Harmonic Generation in a Channel Ti:LiNbO$_3$ Optical Waveguide," *Digest of Integrated and Guided Wave Optics*, Asilomar, Calif., p. ThA3-1 (1982).

54. E. M. Zolotov, V. G. Mikhalevich, V. M. Pelekhatyi, A. M. Prokhorov, V. A. Chernykh, and E. A. Shcherbakov, *Sov. Tech. Phys. Lett.*, *4*:89 (1978).

55. J. Noda, M. Fukuma, and Y. Ito, *J. Appl. Phys.*, *51*:1379 (1980).

56. M. De Mitchell, J. Botineau, S. Neveu, P. Sibillot, and D. B. Ostrowsky, *Opt. Lett.*, *8*:116 (1983).

57. M. De Mitchell, *J. Opt. Commun.*, *4*:25 (1983).

58. W. Sohler and H. Suche, *Integrated Optics III* (L. D. Hutcheson and D. G. Hall, eds.), *Proc. SPIE, 408*:163 (1983).

59. A. A. Maler, *Sov. J. Quantum Electron.*, *10*:925 (1980).

60. S. I. Bozhevol'nyi, K. S. Buritskii, E. M. Zolotov, and V. A. Chernykh, *Sov. Tech. Phys. Lett.*, *7*:278 (1981).

61. N. Uesugi, K. Kaikoku, and M. Fukuma, *J. Appl. Phys.*, *49*:4945 (1978).

62. N. Uesugi, *Appl. Phys. Lett.*, *36*:178 (1980).

63. W. Sohler and H. Suche, *Appl. Phys. Lett.*, *37*:255 (1980).

64. R. Normandin, Ph.D. thesis, University of Toronto, Toronto, Ontario (1980).

65. R. Normandin and G. I. Stegeman, *Opt. Lett.*, *4*:58 (1979).

66. R. Normandin and G. I. Stegeman, *Appl. Phys. Lett.*, *36*:253 (1980).

67. R. Normandin, P. J. Vella, and G. I. Stegeman, *Appl. Phys. Lett.*, *38*:759 (1981).

68. R. Normandin and G. I. Stegeman, *Appl. Phys. Lett.*, *40*:759 (1982).

69. J. E. Sipe, G. I. Stegeman, C. Karaguleff, R. Fortenberry, R. Moshrefzadeh, W. M. Hetherington III, and N. E. Van Wyck, *Opt. Lett.*, *8*:461 (1983).

70. C. K. Chen, T. F. Heinz, D. Ricard, and Y. R. Shen, *Phys. Rev. Lett.*, *46*:1010 (1981).

71. T. F. Heinz, C. K. Chen, D. Ricard, and Y. R. Shen, *Phys. Rev. Lett.*, *48*:478 (1982).

72. D. A. B. Miller, C. T. Seaton, M. E. Prise, and S. D. Smith, *Phys. Rev. Lett.*, *47*:197 (1981).

73. C. Karaguleff, G. I. Stegeman, R. Zanoni, and C. T. Seaton, *Appl. Phys. Lett.* *46*:621 (1985).

74. C. Karaguleff and G. I. Stegeman, *IEEE J. Quantum Electron.*, *QE-20*:716 (1984).

75. C. Sauteret, J.-P. Hermann, R. Frey, F. Pradere, and J. Cucuing, *Phys. Rev. Lett.*, *36*:956 (1976).

76. E. Weinert-Racyka, *Opt. Commun.*, *49*:245 (1984).

77. G. I. Stegeman, *J. Opt. Commun.*, *4*:20 (1988).

78. G. I. Stegeman, *Appl. Phys. Lett.*, *41*:214 (1982).

79. S. A. Shakir, *Phys. Lett.*, *93A*:510 (1983).

80. W. Streifer, D. R. Scitres, and R. D. Burnham, *IEEE J. Quantum Electron.*, *QE-11*:867 (1975).

81. G. I. Stegeman, D. Sarid, J. J. Burke, and D. G. Hall, *J. Opt. Soc. Am.*, *71*:1497 (1981).

82. G. I. Stegeman, C. Liao, and H. G. Winful, Distributed feedback bistability in channel waveguides, *Optical Bistability*, Vol. 3, (C. M. Bowden, H. M. Gibbs, and S. L. McCall, eds.) Plenum Press, New York, p. 389 (1984).

83. H. G. Winful, J. H. Marburger, and E. Garmire, *Appl. Phys. Lett.*, *35*:379 (1979).

84. H. G. Winful, and G. I. Stegeman, "Applications of Nonlinear Periodic Structures in Guided Wave Optics," Proceedings of the First International Conference on Integrated Optical Engineering, *Proc. SPIE, 517*:214 (1984).

85. C. T. Seaton, Xu Mai, G. I. Stegeman, and H. G. Winful, *Opt. Eng., 24*: 593 (1985).

86. Y. A. Bykovskii, Y. Vaitkus, E. P. Gaubas, Y. N. Kul'chin, V. L. Smirnov, and K. Y. Yarashyunas, *Sov. J. Quantum Electron., 12*:418 (1982).

87. S. M. Jensen, "The Nonlinear Coherent Coupler, a New Optical Logic Device," Digest of the Topical Meeting on Integrated and Guided-Wave Optics, Incline Village, Nev., p. MB4-1 (1980).

88. S. M. Jensen, *IEEE J. Quantum Electron., QE-18*:1580 (1982).

89. D. Sarid and M. Sargent III, *J. Opt. Soc. Am., 72*: 835 (1982).

90. K. Kitayama and S. Wang, *Appl. Phys. Lett., 43*:17 (1983).

91. A. Lattes, H. A. Haus, F. J. Leonberger, and E. P. Ippen, *IEEE J. Quantum Electron., QE-19*:1718 (1983).

92. E. Garmire, "Bistable Guided Wave Ring," Digest of the Sergio Porto Memorial Symposium, Brazil (1980).

93. D. Sarid, *Opt. Lett., 6*:552 (1981).

94. Y. J. Chen and G. M. Carter, *Appl. Phys. Lett., 41*:307 (1982).

95. Y. J. Chen and G. M. Carter, *Solid State Commun., 45*:277 (1983).

96. Y. J. Chen and G. M. Carter, *J. Phys. (Paris) Colloq., C5*: 164 (1984).

97. G. M. Carter, Y. J. Chen, and S. K. Tripathy, *Appl. Phys. Lett., 43*:891 (1983).

98. C. Liao and G. I. Stegeman, *Appl. Phys. Lett., 44*:164 (1984).

99. C. Liao, G. I. Stegeman, C. T. Seaton, R. L. Shoemaker, J. D. Valera, and H. G. Winful, *J. Opt. Soc. Am. B, 2*:590 (1985).

100. J. D. Valera, C. T. Seaton, G. I. Stegeman, R. L. Shoemaker, Xu Mai, and C. Liao, *Appl. Phys. Lett., 45*:1013 (1984).

101. R. Ulrich, *J. Opt. Soc. Am., 60*:1337 (1970).

102. G. M. Wysin, H. J. Simon, and R. T. Deck, *Opt. Lett., 6*:30 (1981).

103. P. Martinot, S. Laval, and A. Koster, *J. Phys. (Paris)*, 45:597 (1984).

104. A. E. Kaplan, *JETP Lett.*, 24:114 (1976).

105. A. E. Kaplan, *Sov. Phys. JETP*, 45:896 (1977).

106. N. N. Akhmediev, *Sov. Phys. JETP*, 56:299 (1982).

107. N. N. Akhmediev, K. O. Bolter, and V. M. Eleonskii, *Opt. Spektrosk.*, 53:906, 1097 (1982).

108. A. D. Boardman and P. Egan, *J. Phys. (Paris) Colloq.*, C5: 291 (1984).

109. F. Lederer, U. Langbein, and H.-E. Ponath, *Appl. Phys. b*, 31:187 (1983).

110. A. A. Maradudin, *Optical and Acoustic Waves in Solids-Modern Topics* (M. Borissov, ed.), World Scientific Publishers, Singapore, p. 72 (1983).

111. U. Langbein, F. Lederer, and H.-E. Ponath, *Opt. Commun.*, 46:167 (1983).

112. U. Langbein, F. Lederer, H.-E. Ponath, and U. Trutschel, *J. Mol. Struct.*, 115:493 (1984).

113. D. J. Robbins, *Opt. Commun.*, 47:309 (1983).

114. D. Mihalache and H. Totla, *Rev. Roum. Phys.*, 29:365 (1984).

115. D. Mihalache, R. G. Nazmitdinov, and V. K. Fedyanin, *Phys. Scr.*, 29:269 (1984).

116. G. I. Stegeman, C. T. Seaton, J. Chilwell, and S. D. Smith, *Appl. Phys. Lett.*, 44:830 (1984).

117. C. T. Seaton, J. D. Valera, R. L. Shoemaker, G. I. Stegeman, J. Chilwell, and S. D. Smith, *Appl. Phys. Lett.*, 45: 1162 (1984).

118. C. T. Seaton, J. D. Valera, B. Svenson, and G. I. Stegeman, *Opt. Lett.*, 10:149 (1985).

119. C. T. S aton, J. D. Valera, R. L. Shoemaker, G. I. Stegeman, J. Chilwell, and S. D. Smith, *IEEE J. Quantum Electron.*, QE21:774 (1985).

120. K. M. Leung, *Phys. Rev. B*, 32:5093 (1985). .

121. F. Fedyanin and D. Mihalache, *Z. Phys. B*, 47:167 (1982).

122. H. Vach, C. T. Seaton, G. I. Stegeman, and I. C. Khoo, *Opt. Lett.*, 9:238 (1984).

123. V. M. Agranovich, Proceedings of the 7th International Conference on Raman Spectroscopy, North-Holland, Amsterdam, p. 372 (1980).

124. V. Ya. Chernyak, *Sov. Phys. Solid State*, *25*:351 (1983).

125. H.-E. Ponath and M. Schubert, *Opt. Acta*, *30*:1139 (1983).

126. A. D. Boardman, G. S. Cooper, and P. Egan, *J. Phys. (Paris) Colloq.*, *C5*:197 (1984).

127. G. I. Stegeman, R. Fortenberry, C. Karaguleff, R. Moshrefzadeh, W. M. Hetherington III, N. E. Van Wyck, and J. E. Sipe, *Opt. Lett.*, *8*:295 (1983).

128. W. M. Hetherington, N. E. Van Wyck, E. W. Koenig, G. I. Stegeman, and R. Fortenberry, *Opt. Lett.*, *9*:88 (1984).

129. G. I. Stegeman, R. Fortenberry, R. Moshrefzadeh, W. M. Hetherington III, N. E. Van Wyck, and E. W. Koenig, Proceedings of the 1983 Los Alomos Conference on Optics, *SPIE Proc.*, *380*:212 (1983).

130. F. Kajzar, J. Messier, J. Zyss, and I. Ledoux, *Opt. Commun.*, *45*:133 (1983).

131. H. M. Gibbs, S. S. Tarng, J. L. Jewell, D. A. Weinberger, K. Tai, A. C. Gossard, S. L. McCall, A. Passner and W. Wiegmann, *Appl. Phys. Lett.*, *41*:221 (1982).

132. D. A. B. Miller, D. S. Chemia, D. J. Ellenberger, P. W. Smith, A. C. Gossard, and W. Wiegmann, *Appl. Phys. Lett.*, *42*:925 (1983).

133. G. M. Carter, Y. J. Chen, and S. K. Tripathy, in D. J. Williams (ed.), Nonlinear Optical Properties of Organic and Polymeric Materials, *ACS Symposia Series*, American Chemical Society, Washington, D.C., pp. 213 (1983).

134. Y. Suematsu, Y. Sasaki, K. Furuya, K. Shibata, and S. Ibukuro, *IEEE J. Quantum Electron.*, *10*:222 (1974).

135. N. Uesugi and T. Kumura, *Appl. Phys. Lett.*, *29*:572 (1976).

136. M. De Mitchelli, J. Botineau, S. Neveu, P. Sibillot, and D. B. Ostrowsky, *Opt. Lett.*, *8*:116 (1983).

137. M. DeMitchelli, *Journ. Opt. Commun.*, *4*:25 (1983).

138. P. J. Vella, R. Normandin, and G. I. Stegeman, *Appl. Phys. Lett.*, *38*:759 (1981).

139. H. G. Winful and G. D. Cooperman, *Appl. Phys. Lett.*, *40*: 298 (1982).

140. A. F. Garito, K. D. Singer, K. Hayes, G. F. Lipscomb, S. J. Lalama, and K. N. Desai, *J. Opt. Soc. Am.*, *70*: 1399 (1980).

Index